T0281225

Graduate Texts in Mathematics 79

Editorial Board
S. Axler   F.W. Gehring   K.A. Ribet

Springer
New York
Berlin
Heidelberg
Barcelona
Hong Kong
London
Milan
Paris
Singapore
Tokyo

Graduate Texts in Mathematics

(continued after index)

Peter Walters

An Introduction to Ergodic Theory

With 8 Illustrations

 Springer

Peter Walters
Mathematics Institute
University of Warwick
Coventry CV4 7AL
England

Editorial Board

S. Axler
Mathematics Department
San Francisco State University
San Francisco, CA 94132
USA

F.W. Gehring
Mathematics Department
East Hall
University of Michigan
Ann Arbor, MI 48109-1109
USA

K.A. Ribet
Mathematics Department
University of California at Berkeley
Berkeley, CA 94720-3840
USA

AMS Subject Classification: 28-01, 28DXX, 47A35, 54H20

Library of Congress Cataloging-in-Publication Data
Walters, Peter, 1943–
        An introduction to ergodic theory.

        (Graduate texts in mathematics; 79)
        Previously published as: Ergodic theory. 1975.
        Bibliography: p.
        Includes index.
        1. Ergodic theory.   I. Title.   II. Series.
QA313.W34   1981   515.4'2   81-9319
ISBN 0-387-95152-0   AACR2

Printed on acid-free paper.

First softcover printing, 2000.
© 1982 Springer-Verlag New York, Inc.
All rights reserved. This work may not be translated or copied in whole or in part without
the written permission of the publisher (Springer-Verlag New York, Inc., 175 Fifth Avenue,
New York, NY 10010, USA), except for brief excerpts in connection with reviews or schol-
arly analysis. Use in connection with any form of information storage and retrieval, elec-
tronic adaptation, computer software, or by similar or dissimilar methodology now known
or hereafter developed is forbidden.
The use of general descriptive names, trade names, trademarks, etc., in this publication,
even if the former are not especially identified, is not to be taken as a sign that such names,
as understood by the Trade Marks and Merchandise Marks Act, may accordingly be used
freely by anyone.

Printed and bound by Edwards Brothers, Inc., Ann Arbor, MI.
Printed in the United States of America.

9 8 7 6 5 4 3 2 1

ISBN 0-387-95152-0   SPIN 10783040

Springer-Verlag   New York Berlin Heidelberg
A member of BertelsmannSpringer Science+Business Media GmbH

Preface

In 1970 I gave a graduate course in ergodic theory at the University of Maryland in College Park, and these lectures were the basis of the Springer Lecture Notes in Mathematics Volume 458 called "Ergodic Theory— Introductory Lectures" which was published in 1975. This volume is now out of print, so I decided to revise and add to the contents of these notes. I have updated the earlier chapters and have added some new chapters on the ergodic theory of continuous transformations of compact metric spaces. In particular, I have included some material on topological pressure and equilibrium states. In recent years there have been some fascinating inter- actions of ergodic theory with differentiable dynamics, differential geometry, number theory, von Neumann algebras, probability theory, statistical mechanics, and other topics. In Chapter 10 I have briefly described some of these and given references to some of the others. I hope that this book will give the reader enough foundation to tackle the research papers on ergodic theory and its applications.

I would like to dedicate this volume to the memory of Rufus Bowen who died on July 30, 1978 at the age of 31. He made outstanding contributions to ergodic theory and his friendship enhanced the lives of all who knew him.

April, 1981 PETER WALTERS

Contents

CHAPTER 0
Preliminaries

§0.1 Introduction

In its broadest interpretation ergodic theory is the study of the qualitative properties of actions of groups on spaces. The space has some structure (e.g. the space is a measure space, or a topological space, or a smooth manifold) and each element of the group acts as a transformation on the space and preserves the given structure (e.g. each element acts as a measure-preserving transformation, or a continuous transformation, or a smooth transformation).

To see how this type of study arises consider a system of k particles moving in R^3 under known forces. Suppose that the state of the system at a given time is determined by knowing the positions and the momenta of each of the k particles. Thus at a given time the system is determined by a point in R^{6k}. As time continues the system alters according to the differential equations governing the motion, e.g., Hamilton's equations

$$\frac{dq_i}{dt} = \frac{\partial H}{\partial p_i}, \qquad \frac{dp_i}{dt} = -\frac{\partial H}{\partial q_i}.$$

If we are given an initial condition and if the equations can be uniquely solved then the corresponding solution gives us the entire history of the system, which is determined by a curve in R^{6k}.

If x is a point in the state space representing the system at a time t_0, let $T_t(x)$ denote the point of the state space representing the system at time $t + t_0$. From this we see that T_t is a transformation of the state space and, moreover, $T_0 = id$ and $T_{t+s} = T_t \circ T_s$. Thus $t \to T_t$ is an action of the group R on the state space. Because the Hamiltonian H is constant along solution curves, each energy surface $H^{-1}(e)$ is invariant for the transformation T_t

1

so that we get an action of R on each energy surface. One is interested in the asymptotic properties of the action i.e. in T_t for large t. The transformations $T_t | H^{-1}(e)$ are continuous and are smooth if $H^{-1}(e)$ is smooth. Measure theory enters this picture via a theorem of Liouville which tells us that if the forces are of a certain type one can choose coordinates in the state space so that the usual $6k$-dimensional measure in these coordinates is preserved by each transformation T_t.

The word "ergodic" was introduced by Boltzmann to describe a hypothesis about the action of $\{T_t | t \in R\}$ on an energy surface $H^{-1}(e)$ when the Hamiltonian H is of the type that arises in statistical mechanics. Boltzmann had hoped that each orbit $\{T_t(x) | t \in R\}$ would equal the whole energy surface $H^{-1}(e)$ and he called this statement the ergodic hypothesis. The word "ergodic" is an amalgamation of the Greek words ergon (work) and odos (path). Boltzmann made the hypothesis in order to deduce the equality of time means and phase means which is a fundamental algorithm in statistical mechanics. The ergodic hypothesis, as stated above, is false. The property the flow needs to satisfy in order to equate time means and phase means of real-valued functions is what is now called ergodicity.

It is common to use the name ergodic theory to describe only the qualitative study of actions of groups on measure spaces. The actions on topological spaces and smooth manifolds are often called topological dynamics and differentiable dynamics. This measure theoretic study began in the early 1930's and the ergodic theorems of Birkhoff and von Neumann were proved then. The next major advance was the introduction of entropy by Kolmogorov in 1958. The proof, by Ornstein in 1969, that entropy was complete for Bernoulli shifts revitalised the work on the isomorphism problem. During recent years ergodic theory had been used to give important results in other branches of mathematics.

We shall study actions of the group Z of integers on a space X i.e. we study a transformation $T: X \to X$ and its iterates T^n, $n \in Z$. This is simpler than studying the actions of R. Of course, if $\{T_t | t \in R\}$ is an action of R on X, then by choosing $t_0 \neq 0$ and observing the system at the times $\ldots, -t_0, 0, t_0, 2t_0, 3t_0, \ldots$, we are considering $(T_{t_0})^n$, $n \in Z$.

In the following sections we summarise some of the background ideas and notation we shall be using.

We shall use Z to denote the set of integers, Z^+ to denote the non-negative integers, R to denote the real numbers, R^+ to denote the non-negative reals, and $\mathbb{C}$ to denote the complex numbers. The empty set will be denoted by $\varnothing$.

If A,B are subsets of a set X, then $B \backslash A$ denotes the difference set $\{x \in X \,|\, x \in B, x \notin A\}$, and $A \triangle B$ denotes the symmetric difference $(A \backslash B) \cup (B \backslash A)$. We use 2^X to denote the collection of all subsets of X.

We use "iff" to denote "if and only if." We number lemmas and theorems in a single sequence (Theorem 5.6 is the sixth theorem in Chapter 5) but give a corollary the same number as the corresponding theorem (Corollary 5.6.2 is the second corollary of Theorem 5.6).

§0.2 Measure Spaces

We shall generally refer to Kingman and Taylor [1] and Parthasarathy [2].
Let X be a set. A *σ-algebra* of subsets of X is a collection $\mathscr{B}$ of subsets of X
satisfying the following three conditions: (i) $X \in \mathscr{B}$; (ii) if $B \in \mathscr{B}$ then $X \backslash B \in \mathscr{B}$;
(iii) if $B_n \in \mathscr{B}$ for $n \geq 1$ then $\bigcup_{n=1}^{\infty} B_n \in \mathscr{B}$.

We then call the pair $(X, \mathscr{B})$ a *measurable space*. A *finite measure* on $(X, \mathscr{B})$
is a function $m: \mathscr{B} \to R^+$ satisfying $m(\varnothing) = 0$ and $m(\bigcup_{n=1}^{\infty} B_n) = \sum_{n=1}^{\infty} m(B_n)$
whenever $\{B_n\}_1^{\infty}$ is a sequence of members of $\mathscr{B}$ which are pairwise disjoint
subsets of X. (Actually the latter condition implies $m(\varnothing) = 0$ since m is
finite-valued.) A finite measure space is a triple $(X, \mathscr{B}, m)$ where $(X, \mathscr{B})$ is a
measurable space and m is a finite measure on $(X, \mathscr{B})$. We say $(X, \mathscr{B}, m)$ is a
probability space, or a *normalised measure space*, if $m(X) = 1$. We then say
m is a probability measure on $(X, \mathscr{B})$. We shall usually consider only probability spaces.

A *finite signed measure* on a measurable space $(X, \mathscr{B})$ is a function $m: \mathscr{B} \to R$
satisfying $m(\varnothing) = 0$ and $m(\bigcup_{n=1}^{\infty} B_n) = \sum_{n=1}^{\infty} m(B_n)$ whenever $\{B_n\}_1^{\infty}$ is a
sequence of members of $\mathscr{B}$ which are pairwise disjoint subsets of X. The
Jordan decomposition says that a finite signed measure m on $(X, \mathscr{B})$ can be
written as the difference $m = m_1 - m_2$ of two finite measures on $(X, \mathscr{B})$ which
are uniquely determined by m (see Kingman and Taylor [1], pp. 62 and 64).

Measurable spaces are usually constructed by having a collection $\mathscr{S}$ of
interesting subsets of a set X (such as the collection of all subintervals of
$[0, 1]$) and then considering the smallest $σ$-algebra $\mathscr{B}$ containing all these
subsets. This makes sense because 2^X is a $σ$-algebra and any intersection of
$σ$-algebras of subsets of X is also a $σ$-algebra of subsets of X. It is then usually
difficult to decide which subsets of X are in $\mathscr{B}$. When constructing a measure
on a measurable space $(X, \mathscr{B})$ obtained in this way, one usually knows what
values the measure should take on members of $\mathscr{S}$ and then one needs to
extend it to be defined on $\mathscr{B}$. We now describe the basic extension theorem of
this type. This involves discussion of the properties the collection $\mathscr{S}$ should
have.

A collection $\mathscr{S}$ of subsets of X is called a *semi-algebra* if the following three
conditions hold: (i) $\varnothing \in \mathscr{S}$; (ii) if $A, B \in \mathscr{S}$, then $A \cap B \in \mathscr{S}$; (iii) if $A \in \mathscr{S}$,
then $X \backslash A = \bigcup_{i=1}^{n} E_i$ where each $E_i \in \mathscr{S}$ and $E_1, \ldots, E_n$ are pairwise disjoint
subsets of X. For example, the collection of all subintervals of $[0, 1]$ is a semi-
algebra. Also, the collection of all subintervals of $[0, 1]$ of the forms $[0, b]$
and $(a, b]$, with $0 \leq a < b \leq 1$, forms a semi-algebra.

A collection $\mathscr{A}$ of subsets of X is called an *algebra* if the following three
conditions hold: (i) $\varnothing \in \mathscr{A}$; (ii) if $A, B \in \mathscr{A}$, then $A \cap B \in \mathscr{A}$; (iii) if $A \in \mathscr{A}$,
then $X \backslash A \in \mathscr{A}$. Clearly every algebra is a semi-algebra and every $σ$-algebra
is an algebra. In the definition of an algebra we can replace (ii) by the condition that whenever $A_1, \ldots, A_n \in \mathscr{A}$ then $\bigcup_{i=1}^{n} A_i \in \mathscr{A}$.

Since the intersection of any family of algebras of a set X is again an
algebra of subsets of X it makes sense to speak of the algebra generated by

any given collection of subsets of X. There is the following simple theorem (Parthasarathy [2], p. 19)

Theorem 0.1. *Let $\mathscr{S}$ be a semi-algebra of subsets of X. The algebra, $\mathscr{A}(\mathscr{S})$, generated by $\mathscr{S}$ consists precisely of those subsets of X that can be written in the form $E = \bigcup_{i=1}^{n} A_i$ where each $A_i \in \mathscr{S}$ and $A_1, \ldots, A_n$ are disjoint subsets of X.*

Suppose $\mathscr{S}$ is a semi-algebra of subsets of X. A function $\tau : \mathscr{S} \to R^+$ is called *finitely additive* if $\tau(\varnothing) = 0$ and $\tau(\bigcup_{i=1}^{n} E_i) = \sum_{i=1}^{n} \tau(E_i)$ whenever $E_1, \ldots, E_n$ are members of $\mathscr{S}$ which are pairwise disjoint subsets of X and $\bigcup_{i=1}^{n} E_i \in \mathscr{S}$. Such a map τ is called *countably additive* if the second condition is replaced by the requirement that $\tau(\bigcup_{i=1}^{\infty} E_i) = \sum_{i=1}^{\infty} \tau(E_i)$ whenever $\{E_i\}_1^{\infty}$ are members of $\mathscr{S}$ which are pairwise disjoint subsets of X and $\bigcup_{i=1}^{\infty} E_i \in \mathscr{S}$. If $\mathscr{S}$ is the semi-algebra of all subintervals of $[0, 1]$ of the form $[0, b]$ and $(a, b]$ then the length function is countably additive. A simple application of Theorem 0.1 gives the following (Parthasarathy [2], pp. 20 and 59).

Theorem 0.2. *If $\mathscr{S}$ is a semi-algebra of subsets of X and $\tau : \mathscr{S} \to R^+$ is finitely additive then there is a unique finitely additive function $\tau_1 : \mathscr{A}(\mathscr{S}) \to R^+$ which is an extension of τ (i.e. $\tau_1 = \tau$ on $\mathscr{S}$). If τ is countably additive then so is τ_1.*

It could be that $X \notin \mathscr{S}$ but we always have that X is a disjoint union of a finite number of members $E_1, \ldots, E_n$ of $\mathscr{S}$ so $\tau_1(X) = 1$ if $\sum_{i=1}^{n} \tau(E_i) = 1$.

There is the following theorem on extension from an algebra $\mathscr{A}$ to the σ-algebra $\mathscr{B}(A)$ generated by $\mathscr{A}$. ($\mathscr{B}(A)$ is the intersection of all σ-algebras that contain $\mathscr{A}$.) (See Parthasarathy [2], pp. 70 and 71).

Theorem 0.3. *Let $\mathscr{A}$ be an algebra of subsets of X and let $\tau_1 : \mathscr{A} \to R^+$ be countably additive and $\tau_1(X) = 1$. Then there is a unique probability measure τ_2 on $(X, \mathscr{B}(\mathscr{A}))$ which extends τ_1.*

By combining Theorems 0.2 and 0.3 we see that a countably additive function τ on a semi-algebra $\mathscr{S}$ can be uniquely extended to a probability measure on $(X, \mathscr{B}(\mathscr{S}))$ if $\sum_{i=1}^{n} \tau(E_i) = 1$ when $X = \bigcup_{i=1}^{n} E_i$ is a disjoint union of members of $\mathscr{S}$. As an example the length function defined on the semi-algebra of all subintervals of $[0, 1]$ of the form $[0, b]$ and $(a, b]$ can be uniquely extended to a probability measure, called the Lebesgue measure, defined on the Borel subsets of $[0, 1]$.

In checking that the extension works for particular examples the most difficult part is usually showing countable additivity. This can sometimes be done for τ on the semi-algebra $\mathscr{S}$ (as for Lebesgue measure) but it is sometimes more convenient to prove that τ is finitely additive and that the finitely additive extension $\tau_1 : \mathscr{A}(\mathscr{S}) \to R^+$ is countably additive. The main tool for this is the following theorem (Kingman and Taylor [1], p. 56).

Theorem 0.4. *Let $\mathscr{A}$ be an algebra of subsets of X and let $\tau_1 : \mathscr{A} \to R^+$ be finitely additive and let $\tau_1(X) = 1$. Then τ_1 will be countably additive if for every decreasing sequence $E_1 \supset E_2 \supset E_3 \supset \cdots$ of members of $\mathscr{A}$ with $\bigcap_{n=1}^\infty E_n = \varnothing$ we have $\tau_1(E_n) \to 0$.*

One situation where this theorem is used is in defining product measures on a countable product of probability spaces. For $i \in Z$ let $(X_i, \mathscr{B}_i, m_i)$ be a probability space. Let $X = \prod_{i=-\infty}^\infty X_i$. So a point of X is a bisequence $\{x_i\}_{-\infty}^\infty$ with $x_i \in X_i$ for each i. We now define a σ-algebra $\mathscr{B}$ of subsets of X called the product of the σ-algebras $\mathscr{B}_i$. Let $n \geq 0$, let $A_j \in \mathscr{B}_j$ for $|j| \leq n$, and consider the set

$$\prod_{i=-\infty}^{-(n+1)} X_i \times \prod_{j=-n}^{n} A_j \times \prod_{i=n+1}^{\infty} X_i = \{(x_i)_{-\infty}^\infty \in X \mid x_j \in A_j \text{ for } |j| \leq n\}.$$

Such a set is called a *measurable rectangle* and the collection of all such subsets of X forms a semi-algebra $\mathscr{S}$. The σ-algebra $\mathscr{B}$ is the σ-algebra generated by $\mathscr{S}$. We write $(X, \mathscr{B}) = \prod_{i=-\infty}^\infty (X_i, \mathscr{B}_i)$. If we define $\tau : \mathscr{S} \to R^+$ by giving the above rectangle the value $\prod_{j=-n}^n m_j(A_j)$, then one can use Theorems 0.2 and 0.4 (see Kingman and Tayler [1], p. 140) to extend τ to a probability measure m on $(X, \mathscr{B})$. The probability space $(X, \mathscr{B}, m)$ is called the *direct product* of the spaces $(X_i, \mathscr{B}_i, m_i)$ and is sometimes denoted $\prod_{i=-\infty}^\infty (X_i, \mathscr{B}_i, m_i)$. The corresponding construction holds for a product $\prod_{i=0}^\infty (X_i, \mathscr{B}_i, m_i)$.

A special type of product space will be important for us. Here each space $(X_i, \mathscr{B}_i, m_i)$ is the same space $(Y, \mathscr{C}, \mu)$ and Y is the finite set $\{0, 1, \ldots, k-1\}$, $\mathscr{C} = 2^Y$, and μ is given by a probability vector $(p_0, p_1, \ldots, p_{k-1})$ where $p_i = \mu(\{i\})$. We can take elementary rectangles where each A_j (in the description above) is taken to be one point of Y. So if $n \geq 0$ and $a_j \in Y$, $|j| \leq n$, such an elementary rectangle has the form $\{(x_i)_{-\infty}^\infty \mid x_j = a_j \text{ for } |j| \leq n\}$. We shall denote this set by $_{-n}[a_{-n}, a_{-(n+1)}, \ldots, a_{n-1}, a_n]_n$ and call it a *block* with end points $-n$ and n. The collection of all these sets form a semi-algebra which generates the product σ-algebra $\mathscr{B}$. We have $m(_{-n}[a_n, \ldots, a_n]_n) = \prod_{j=n}^n p_j$. The measure m is called the $(p_0, \ldots, p_{k-1})$-product measure. Sometimes we consider *blocks* with end points h and l where $h \leq l$. Such a set is one of the form $_h[a_h, \ldots, a_l]_l = \{(x_i)_{-\infty}^\infty \mid x_i = a_i \text{ for } h \leq i \leq l\}$. It has measure $\prod_{i=h}^l p_i$.

Theorem 0.4 can also be used to obtain further measures on the space $(X, \mathscr{B})$ where $X = \prod_{-\infty}^\infty Y$, $Y = \{0, 1, \ldots, k-1\}$, and $\mathscr{B}$ is the product σ-algebra described above. The following is a special case of the Daniell–Kolmogorov consistency theorem (Parthasarathy [2], p. 119).

Theorem 0.5. *Fix $k \geq 1$ and let $Y = \{0, 1, \ldots, k-1\}$ and $(X, \mathscr{B}) = \prod_{-\infty}^\infty (Y, 2^Y)$. For each natural number n and $a_0, \ldots, a_n \in Y$ suppose a non-negative real number $p_n(a_0, \ldots, a_n)$ is given so that*

(a)
$$\sum_{a_0 \in Y} p_0(a_0) = 1$$

and

(b) $p_n(a_0, \ldots, a_n) = \sum_{a_{n+1} \in Y} p_{n+1}(a_0, \ldots, a_n, a_{n+1}).$

Then there is a unique probability measure m on $(X, \mathcal{B})$ with $m(_h[a_h, \ldots, a_l]_l) = p_{l-h}(a_h, \ldots, a_l)$ for all $h \leq l$ and all $a_i \in Y, h \leq i \leq l$.

The proof boils down to showing that the function naturally defined on the algebra of all finite unions of elementary rectangles is countable additive by using Theorem 0.4.

There is another way, which is useful in some proofs, of describing the σ-algebra $\mathcal{B}(\mathcal{A})$ generated by an algebra $\mathcal{A}$. A collection M of subsets of X is called a *monotone class* if whenever $E_1 \subset E_2 \subset E_3 \subset \cdots$ all belong to M then so does $\bigcup_{n=1}^{\infty} E_n$ and whenever $F_1 \supset F_2 \supset F_3 \supset \cdots$ all belong to M then so does $\bigcap_{n=1}^{\infty} F_n$. Since the intersection of any family of monotone classes is a monotone class, we can speak of the monotone class generated by any given collection of subsets of X.

Theorem 0.6. *Let $\mathcal{A}$ be an algebra of subsets of X. Then $\mathcal{B}(\mathcal{A})$ equals the monotone class generated by $\mathcal{A}$.*

As we have seen we usually know the elements of an algebra $\mathcal{A}$ but we do not know which subsets of X belong to $\mathcal{B}(\mathcal{A})$. This problem can sometimes be overcome by using the following approximation theorem (Kingman and Taylor [1], p. 84).

Theorem 0.7. *Let $(X, \mathcal{B}, m)$ be a probability space and let $\mathcal{A}$ be an algebra of subsets of X with $\mathcal{B}(\mathcal{A}) = \mathcal{B}$. Then for each $\varepsilon > 0$ and each $B \in \mathcal{B}$ there is some $A \in \mathcal{A}$ with $m(A \triangle B) < \varepsilon$.*

Note that when $m(A \triangle B) < \varepsilon$ then $|m(A) - m(B)| < \varepsilon$ because $m(A) = m(A \backslash B) + m(A \cap B)$ and $m(B) = m(B \backslash A) + m(A \cap B)$, so that $|m(A) - m(B)| \leq m(A \triangle B)$.

§0.3 Integration

Let $\mathcal{B}(R)$ denote the σ-algebra of Borel subsets of R. This is the σ-algebra generated by all open subsets of R and is also generated by the collection of all intervals, or by the collection of all intervals of the form (c, ∞).

Let $(X, \mathcal{B}, m)$ be a measure space. A function $f: X \to R$ is *measurable* if $f^{-1}(D) \in \mathcal{B}$ whenever $D \in \mathcal{B}(R)$ or equivalently if $f^{-1}(c, \infty) \in \mathcal{B}$ for all $c \in R$. A function $f: X \to \mathbb{C}$ is measurable if both its real and imaginary parts are measurable. If X is a topological space and $\mathscr{B}$ the σ-algebra generated by the open subsets of X, then any continuous function $f: X \to \mathbb{C}$ is measurable. We say $f = g$ a.e. if $m(\{x: f(x) \neq g(x)\}) = 0$. Suppose X is a topological space, $\mathscr{B}(X)$ its σ-algebra of Borel sets and m a measure on $(X, \mathscr{B}(X))$ with the property that each non-empty open set has non-zero measure. Then for two continuous functions $f, g: X \to R$, $f = g$ a.e. implies $f = g$ because $\{x: f(x) - g(x) \neq 0\}$ is an open set of zero measure.

Let $(X, \mathscr{B}, m)$ be a probability space. A function $f: X \to R$ is a *simple function* if it can be written in the form $\sum_{i=1}^{n} a_i \chi_{A_i}$, where $a_i \in R$, $A_i \in \mathscr{B}$, the sets A_i are disjoint subsets of X, and χ_{A_i} denotes the characteristic function of A_i. Simple functions are measurable. We define the integral for simple functions by:

$$\int f \, dm = \sum_{i=1}^{n} a_i m(A_i).$$

This value is independent of the representation $\sum_i a_i \chi_{A_i}$.

Suppose $f: X \to R$ is measurable and $f \geq 0$. Then there exists an increasing sequence of simple functions $f_n \nearrow f$. For example, we could take

$$f_n(x) = \begin{cases} \dfrac{i-1}{2^n}, & \text{if } \dfrac{i-1}{2^n} \leq f(x) < \dfrac{i}{2^n} \quad i = 1, \ldots, n2^n \\ n, & \text{if } f(x) \geq n. \end{cases}$$

We define $\int f \, dm = \lim_{n \to \infty} \int f_n \, dm$ and note that this definition is independent of the chosen sequence $\{f_n\}$. We say f is integrable if $\int f \, dm < \infty$.

Suppose $f: X \to R$ is measurable. Then $f = f^+ - f^-$ where $f^+(x) = \max\{f(x), 0\} \geq 0$ and $f^-(x) = \max\{-f(x), 0\} \geq 0$. We say that f is *integrable* if $\int f^+ \, dm, \int f^- \, dm < \infty$ and we then define

$$\int f \, dm = \int f^+ \, dm - \int f^- \, dm.$$

We say $f: X \to \mathbb{C}$ is integrable ($f = f_1 + i f_2$) if f_1 and f_2 are integrable and we define

$$\int f \, dm = \int f_1 \, dm + i \int f_2 \, dm.$$

Observe that f is integrable if and only if $|f|$ is integrable. If $f = g$ a.e. then one is integrable if the other is and $\int f \, dm = \int g \, dm$.

The two basic theorems on integrating sequences as functions are the following.

Theorem 0.8 (Monotone Convergence Theorem). *Suppose $f_1 \leq f_2 \leq f_3 \leq \cdots$ is an increasing sequence of integrable real-valued functions on $(X, \mathscr{B}, m)$. If $\{\int f_n \, dm\}$ is a bounded sequence of real numbers then $\lim_{n \to \infty} f_n$ exists a.e. and is integrable and $\int (\lim f_n) \, dm = \lim \int f_n \, dm$. If $\{\int f_n \, dm\}$ is an unbounded sequence then either $\lim_{n \to \infty} f_n$ is infinite on a set of positive measure or $\lim_{n \to \infty} f_n$ is not integrable.*

Theorem 0.9 (Fatou's Lemma). *Let $\{f_n\}$ be a sequence of measurable real-valued functions on $(X, \mathscr{B}, m)$ which is bounded below by an integrable function. If $\liminf_{n \to \infty} \int f_n \, dm < \infty$ then $\liminf_{n \to \infty} f_n$ is integrable and $\int \liminf f_n \, dm \le \liminf \int f_n \, dm$.*

Corollary 0.9.1 (Dominated Convergence Theorem). *If $g : X \to R$ is integrable and $\{f_n\}$ is a sequence of measurable real-valued functions with $|f_n| \le g$ a.e. $(n \ge 1)$ and $\lim_{n \to \infty} f_n = f$ a.e. then f is integrable and $\lim \int f_n \, dm = \int f \, dm$.*

We denote by $L^1(X, \mathscr{B}, m)$ (or $L^1(m)$) the space of all integrable functions $f : X \to \mathbb{C}$ where two such functions are identified if they are equal a.e. However we write $f \in L^1(X, \mathscr{B}, m)$ to denote that $f : X \to \mathbb{C}$ is integrable. The space $L^1(X, \mathscr{B}, m)$ is a Banach space with norm $\|f\|_1 = \int |f| \, dm$.

If $f \in L^1(X, \mathscr{B}, m)$, then $\int_A f \, dm$ denotes $\int f \cdot \chi_A \, dm$.

If m is a finite signed measure on $(X, \mathscr{B})$ and $m = m_1 - m_2$ is its unique Jordan decomposition into the difference of two finite measures, then we can define $\int f \, dm = \int f \, dm_1 - \int f \, dm_2$ for $f \in L^1(m_1) \cap L^1(m_2)$.

§0.4 Absolutely Continuous Measures and Conditional Expectations

Let $(X, \mathscr{B})$ be a measurable space and suppose μ, m are two probability measures on $(X, \mathscr{B})$. We say μ is *absolutely continuous* with respect to $m (\mu \ll m)$ if $\mu(B) = 0$ whenever $m(B) = 0$. The measures are *equivalent* if $\mu \ll m$ and $m \ll \mu$. The following theorem characterises absolute continuity.

Theorem 0.10 (Radon–Nikodym Theorem). *Let μ, m be two probability measure on the measurable space $(X, \mathscr{B})$. Then $\mu \ll m$ iff there exists $f \in L^1(m)$, with $f \ge 0$ and $\int f \, dm = 1$, such that $\mu(B) = \int_B f \, dm \; \forall B \in \mathscr{B}$. The function f is unique a.e. (in the sense that any other function with these properties is equal to f a.e.).*

The function f is called the Radon–Nikodym derivative of μ with respect to m and denoted by $d\mu/dm$.

The "opposite" notion to absolute continuity is as follows. Two probability measures μ, m on $(X, \mathscr{B})$ are said to be *mutually singular* $(\mu \perp m)$ if there is some $B \in \mathscr{B}$ with $\mu(B) = 0$ and $m(X \backslash B) = 0$. There is the following decomposition theorem.

Theorem 0.11 (Lebesgue Decomposition Theorem). *Let μ, m be two probability measures on $(X, \mathscr{B})$. There exists $p \in [0, 1]$ and probability measures μ_1, μ_2 on $(X, \mathscr{B})$ such that $\mu = p\mu_1 + (1 - p)\mu_2$ and $\mu_1 \ll m$, $\mu_2 \perp m$. $(\mu = p\mu_1 + (1 - p)\mu_2$ means $\mu(B) = p\mu_1(B) + (1 - p)\mu_2(B) \; \forall B \in \mathscr{B}$). The number p and probabilities μ_1, μ_2 are uniquely determined.*

The Radon–Nikodym theorem allows us to define conditional expectations. Let $(X, \mathscr{B}, m)$ be a measure space and let $\mathscr{C}$ be a sub-σ-algebra of $\mathscr{B}$. We now define the conditional expectation operator $E(\cdot / \mathscr{C}): L^1(X, \mathscr{B}, m) \to L^1(X, \mathscr{C}, m)$. If $f \in L^1(X, \mathscr{B}, m)$ takes non-negative real values then $\mu_f(C) = a^{-1} \int_C f \, dm$ (where $a = \int_X f \, dm$) defines a probability measue, μ_f, on $(X, \mathscr{C}, m)$ and $\mu_f \ll m$. By Theorem 0.10 there is a function $E(f / \mathscr{C}) \in L^1(X, \mathscr{C}, m)$ such that $E(f / \mathscr{C}) \geq 0$ and $\int_C E(f / \mathscr{C}) \, dm = \int_C f \, dm \; \forall C \in \mathscr{C}$. Moreover $E(f / \mathscr{C})$ is unique a.e. If f is real-valued we can consider the positive and negative parts of f and define $E(f / \mathscr{C})$ linearly. Similarly when f is complex-valued we can use the real and imaginary parts to define $E(f / \mathscr{C})$ linearly. Therefore if $f \in L^1(X, \mathscr{B}, m)$ then $E(f / \mathscr{C})$ is the only $\mathscr{C}$-measurable function with $\int_C E(f / \mathscr{C}) \, dm = \int_C f \, dm \; \forall C \in \mathscr{C}$. The following properties of the map $E(\cdot / \mathscr{C}): L^1(X, \mathscr{B}, m) \to L^1(X, \mathscr{C}, m)$ hold (Parthasarathy [2], p. 225):

(i) $E(\cdot / \mathscr{C})$ is linear.
(ii) If $f \geq 0$, then $E(f / \mathscr{C}) \geq 0$.
(iii) If $f \in L^1(X, \mathscr{B}, m)$ and g is $\mathscr{C}$-measurable and bounded,

$$E(fg / \mathscr{C}) = gE(f / \mathscr{C}).$$

(iv) $|E(f / \mathscr{C})| \leq E(|f| / \mathscr{C}), \quad f \in L^1(X, \mathscr{B}, m)$
(v) If $\mathscr{C}_2 \subset \mathscr{C}_1$, then $E(E(f / \mathscr{C}_1) / \mathscr{C}_2) = E(f / \mathscr{C}_2), f \in L^1(X, \mathscr{B}, m)$.

§0.5 Function Spaces

One way to deal with some problems on a measure space is to use certain natural Banach spaces of functions associated with the measure space.

Let $(X, \mathscr{B}, m)$ be a measure space and let $p \in R$ with $p \geq 1$. Consider the set of all measurable functions $f: X \to \mathbb{C}$ with $|f|^p$ integrable. This space is a vector space under the usual addition and scalar multiplication of functions. If we define an equivalence relation on this set by $f \sim g$ iff $f = g$ a.e. then the space of equivalence classes is also a vector space. Let $L^p(X, \mathscr{B}, m)$ denote the space of equivalence classes, although we write $f \in L^p(X, \mathscr{B}, m)$ to denote that the function $f: X \to \mathbb{C}$ has $|f|^p$ integrable. The formula $\|f\|_p = [\int |f|^p \, dm]^{1/p}$ defines a norm on $L^p(X, \mathscr{B}, m)$ and this norm is complete. Therefore $L^p(X, \mathscr{B}, m)$ is a Banach space. If $L^p_R(X, \mathscr{B}, m)$ denotes those equivalence classes containing real-valued functions then $L^p_R(X, \mathscr{B}, m)$ is a real Banach space. The bounded measurable functions are dense in $L^p(X, \mathscr{B}, m)$. If $m(X) < \infty$ and $1 \leq p < q$ then $L^q(X, \mathscr{B}, m) \subset L^p(X, \mathscr{B}, m)$. We sometimes write $L^p(m)$ or $L^p(\mathscr{B})$ instead of $L^p(X, \mathscr{B}, m)$ when no confusion can arise.

A Hilbert space $\mathscr{H}$ is a Banach space in which the norm is given by an inner product, i.e., $\mathscr{H}$ is a Banach space and there is a map $(\cdot, \cdot): \mathscr{H} \times \mathscr{H} \to \mathbb{C}$ such that $(\cdot, \cdot)$ is bilinear, $(f, g) = \overline{(g, f)} \; \forall g, f \in \mathscr{H}, (f, f) \geq 0 \; \forall f \in \mathscr{H}$, and $f = (f, f)^{1/2}$ is the norm on $\mathscr{H}$.

The Banach space $L^p(X, \mathscr{B}, m)$ is a Hilbert space iff $p = 2$. The inner product in $L^2(X, \mathscr{B}, m)$ is given by $(f, g) = \int f \bar{g} \, dm$.

In every Hilbert space $\mathscr{H}$ we have the Schwarz inequality:

$$|(f, g)| \leq \|f\| \cdot \|g\| \qquad \forall f, g \in \mathscr{H}.$$

Separable Hilbert spaces (i.e. those having a countable dense set) are the simplest. The space $L^2(X, \mathscr{B}, m)$ is separable iff $(X, \mathscr{B}, m)$ has a *countable basis*, in the sense that there is a sequence of elements $\{E_n\}_1^\infty$ of $\mathscr{B}$ such that for every $\varepsilon > 0$ and every $B \in \mathscr{B}$ with $m(B) < \infty$ there is some n with $m(B \triangle E_n) < \varepsilon$. If X is a metric space and $\mathscr{B}$ is the σ-algebra of Borel subsets of X (the σ-algebra generated by the open sets) and m is any probability measure on $(X, \mathscr{B})$ then $(X, \mathscr{B}, m)$ has a countable basis. (This follows from Theorem 6.1.) Therefore most of the spaces we shall deal with have $L^2(X, \mathscr{B}, m)$ separable.

Any separate Hilbert space $\mathscr{H}$ contains a basis $\{e_n\}_1^\infty$, i.e. $(e_n, e_k) = 0$ if $n \neq k$ and only the zero element is orthogonal to all the e_n. If $\{e_n\}_1^\infty$ is a basis then each $v \in \mathscr{H}$ is uniquely expressible as $v = \sum_{n=1}^\infty a_n e_n$ where $a_n \in \mathbb{C}$. We have

$$\|v\|^2 = \sum_{n=1}^\infty |a_n|^2 \quad \text{so that} \quad \sum_{n=1}^\infty |a_n|^2 < \infty.$$

An isomorphism between two Hilbert spaces $\mathscr{H}_1$, $\mathscr{H}_2$ is a linear bijection $W: \mathscr{H}_1 \to \mathscr{H}_2$ that preserves norms ($\|Wv\| = \|v\| \, \forall v \in \mathscr{H}_1$). The norm-preserving condition can also be written as $(Wu, Wv) = (u, v) \, \forall u, v \in \mathscr{H}_1$. Any two separable Hilbert spaces are isomorphic if they both have a basis with an infinite number of elements. A Hilbert space with a basis of k elements is isomorphic to $\mathbb{C}^k$. An isomorphism of a Hilbert space $\mathscr{H}$ to itself is called a unitary operator.

If V is a closed subspace of a Hilbert space $\mathscr{H}$ then $V^\perp = \{h \in \mathscr{H} \,|\, (v, h) = 0 \; \forall v \in V\}$ is a closed subspace of $\mathscr{H}$ and $V \oplus V^\perp = \mathscr{H}$ (i.e. each $f \in \mathscr{H}$ has a unique representation $f = f_1 + f_2$ where $f_1 \in V$ and $f_2 \in V^\perp$.) The linear operator $P: \mathscr{H} \to V$ given by $P(f) = f_1$ is called the orthogonal projection of $\mathscr{H}$ onto V. In fact $P(f)$ is the unique member of V that satisfies $\|f - P(f)\| = \inf \{\|f - v\| \,|\, v \in V\}$. We have $P|V = id$ and $(Pf, g) = (f, Pg) \; \forall f, g \in \mathscr{H}$.

Let $(X, \mathscr{B}, m)$ be a probability space and recall from §0.4 that if $\mathscr{C}$ is a sub-σ-algebra of $\mathscr{B}$ then the conditional expectation operator

$$E(\cdot / \mathscr{C}): L^1(X, \mathscr{B}, m) \to L^1(X, \mathscr{C}, m)$$

is defined. Since $L^2(X, \mathscr{B}, m) \subset L^1(X, \mathscr{B}, m)$ the conditional expectation operator acts on $L^2(X, \mathscr{B}, m)$ and the following result describes what this restriction is.

Theorem 0.12. *Let $(X, \mathscr{B}, m)$ be a probability space and let $\mathscr{C}$ be a sub-σ-algebra of $\mathscr{B}$. The restriction of the conditional expectation operator $E(\cdot / \mathscr{C})$ to $L^2(X, \mathscr{B}, m)$ is the orthogonal projection of $L^2(X, \mathscr{B}, m)$ onto $L^2(X, \mathscr{C}, m)$.*

PROOF. For $f \in L^1(X, \mathcal{B}, m)$ we know $E(f/\mathcal{C})$ is the only $\mathcal{C}$-measurable function h such that $\int_C h \, dm = \int_C f \, dm \; \forall C \in \mathcal{C}$. Let P denote the orthogonal projection of $L^2(X, \mathcal{B}, m)$ onto the closed subspace $L^2(X, \mathcal{C}, m)$. If $f \in L^2(X, \mathcal{B}, m)$ then Pf is $\mathcal{C}$-measurable and if $C \in \mathcal{C}$

$$\int_C f \, dm = (f, \chi_c) = (f, P\chi_c) = (Pf, \chi_c) = \int_C Pf \, dm.$$

Therefore $Pf = E(f/\mathcal{C})$.   □

§0.6 Haar Measure

There is a probability measure on a compact group G which ties in with the group structure on G. This measure is defined on the σ-algebra $\mathcal{B}(G)$ of all Borel subsets of G. It will also have the property of regularity. Recall that a measure m on the Borel σ-algebra $\mathcal{B}(X)$ of a compact topological space X is *regular* if for every $\varepsilon > 0$ and every $E \in \mathcal{B}(X)$ there is a compact set M and an open set U with $M \subset E \subset U$ and $m(U \backslash M) < \varepsilon$. It suffices to require that for each $\varepsilon > 0$ and $E \in \mathcal{B}(X)$ there is a compact set M with $M \subset E$ and $m(E \backslash M) < \varepsilon$ (since we can also apply this to $X \backslash E$ to get an open set U with $E \subset U$ and $m(U \backslash E) < \varepsilon$). If X is metrisable then any probability measure on $(X, \mathcal{B}(X))$ is regular (see Theorem 6.1).

Theorem 0.13. *Let G be a compact topological group. There exists a probability measure m defined on the σ-algebra $\mathcal{B}(G)$ of Borel subsets of G such that $m(xE) = m(E) \; \forall x \in G \; \forall E \in \mathcal{B}(G)$ and m is regular. There is only one regular rotation invariant probability measure on $(G, \mathcal{B}(G))$.*

This unique measure is called Haar measure. Notice we have required Haar measure to be a probability measure. The Haar measure m also satisfies $m(Ex) = m(E) \; \forall x \in G, \; \forall E \in \mathcal{B}(G)$, because for each fixed $x \in G$ the measure m_x defined by $m_x(E) = m(Ex)$ is rotation invariant and regular and hence equals m. If G is metrisable then, as mentioned above, any probability measure on $(G, \mathcal{B}(G))$ is regular so we can omit the regularity assumption from the statement about uniqueness of Haar measure.

The rotation invariance of m can also be expressed by requiring $\int_G f(xy) \, dm(y) = \int_G f(y) \, dm(y) \; \forall f \in L^1(m), \; \forall x \in G$.

If U is a non-empty open subset of G then it has non-zero Haar measure, because $G = \bigcup_{g \in G} gU = g_1 U \cup g_2 U \cup \cdots \cup g_k U$ by compactness.

For the circle group $K = \{z \in \mathbb{C} \mid |z| = 1\}$ the Haar measure is the normalised circular Lebesgue measure. For the n-torus K^n the Haar measure is the direct product of the Haar measure on K.

If m_i is the Haar measure on G_i, $i \in Z$, then the direct product of the measures m_i is the Haar measure on the direct product group $\prod_{i=-\infty}^{\infty} G_i$.

For the two point group $\{0, 1\}$ the Haar measure gives each point measure $\frac{1}{2}$ so the Haar measure on the direct product group $\prod_{i=-\infty}^{\infty} \{0, 1\}$ is the direct product of this measure.

On any compact metrisable group G there is a metric ρ which is rotation invariant in the sense that $\rho(gx, gy) = \rho(x, y) = \rho(xg, yg)\ \forall g, x, y \in G$. If d is any metric on G and m is Haar measure we could take $\rho(x, y) = \int(\int d(gxh, gyh)\ dm(g))\ dm(h)$.

§0.7 Character Theory

Many of our examples will be rotations, endomorphisms, or affine transformations of compact groups. (We mean endomorphism in the sense of topological groups, i.e., an abstract group endomorphism which is continuous.) In some proofs we will use the character theory of compact abelian groups which we summarise in this section. For those not familiar with character theory, proofs in the later sections involving characters will usually be preceded by the proof in a special case where the group used is the unit circle and then classical Fourier analysis will be used. The proofs of all results quoted in this section can be found in Hewitt and Ross [1].

Let G be a locally compact abelian group. Let $\hat{G}$ denote the collection of all continuous homomorphisms of G into the unit circle K. The members of $\hat{G}$ are the *characters* of G. Under the operation of pointwise multiplication of functions $\hat{G}$ is an abelian group. With the compact open topology $\hat{G}$ becomes a locally compact abelian group.

In §0.8 we shall show that when $G = K = \{z \in \mathbb{C} \mid |z| = 1\}$ each element of $\hat{K}$ is of the form $z \to z^n$ for some $n \in Z$. Hence $\hat{K} \simeq Z$. We also show that the character group of the n-torus K^n is isomorphic to Z^n and each $\gamma \in \hat{K}^n$ is of the form

$$\gamma(z_1, \ldots, z_n) = z_1^{p_1}, z_2^{p_2}, \ldots, z_n^{p_n} \quad \text{for some } (p_1, p_2, \ldots, p_n) \in Z^n.$$

We have the following results.

(1) G has a countable topological basis iff $\hat{G}$ has a countable topological basis

(2) G is compact iff $\hat{G}$ is discrete.

Combining (2) with (1) we have G is compact and metrisable iff $\hat{G}$ is a discrete countable group. This allows us to transform some problems about compact abelian metrisable groups to problems about discrete countable abelian groups.

(3) (Duality Theorem). $(\hat{\hat{G}})$ is naturally isomorphic (as a topological group) to G, the isomorphism being given by the map $\alpha \to a$ where $\alpha(\gamma) = \gamma(a)$ for all $\gamma \in \hat{G}$.

(4) If G is compact then G is connected iff $\hat{G}$ is torsion free (i.e. has no elements of finite order apart from the identity element.)

(5) If G_1, G_2 are locally compact abelian groups then $\widehat{G_1 \times G_2} = \hat{G}_1 \times \hat{G}_2$. (Here "$\times$" denotes direct product.) Hence all characters of $G_1 \times G_2$ are of the form $(x, y) \to \gamma(x)\delta(y)$ where $\gamma \in \hat{G}_1$, $\delta \in \hat{G}_2$.

(6) If Γ is a subgroup of $\hat{G}$ then $H = \{g \in G \,|\, \gamma(g) = 1 \; \forall \gamma \in \Gamma\}$ is a closed subgroup of G and $(\widehat{G/H}) = \Gamma$. Notice it makes sense for elements of Γ to act on G/H and this result says that these are the only continuous homomorphisms of G/H into K.

(7) If H is a closed subgroup of G and $H \neq G$ there exists a $\gamma \in \hat{G}$, $\gamma \neq 1$ such that $\gamma(h) = 1 \; \forall h \in H$. (We shall write this $\gamma(H) = 1$.)

(8) Let G be compact. The members of $\hat{G}$ are mutually orthogonal members of $L^2(m)$, where m is Haar measure.

PROOF. It suffices to show

$$\int_G \gamma(x)\, dm(x) = 0 \quad \text{if } \gamma \neq 1.$$

If $a \in G$, then, since m is a Haar measure,

$$\int \gamma(x)\, dm(x) = \int \gamma(ax)\, dm(x) = \gamma(a) \int \gamma(x)\, dm(x).$$

Choosing a so that $\gamma(a) \neq 1$ we have $\int \gamma(x)\, dm(x) = 0$. $\quad\square$

(9) If G is compact, the members of $\hat{G}$ form an orthonormal basis for $L^2(m)$ where m is Haar measure.

This is part of the Peter–Weyl theorem and can be easily deduced from the Stone–Weierstrass theorem, which implies that finite linear combinations of characters are dense in $C(G)$, the space of complex-valued continuous functions of G.

Therefore for each $f \in L^2(m)$ there are uniquely determined complex numbers a_γ such that $f = \sum_{\gamma \in \hat{G}} a_\gamma \gamma$ in $L^2(m)$. This means $\forall \varepsilon > 0$ there is a finite subset J_0 of $\hat{G}$ such that if J is finite and $J_0 \subset J$ then $\left\| f - \sum_{\gamma \in J} a_\gamma \gamma \right\|_2 < \varepsilon$. Only a countable number of a_γ are non-zero. We have $\sum_{\gamma \in \hat{G}} |a_\gamma|^2 = \|f\|_2^2 < \infty$. This representation of f is called the Fourier series of f. When $G = K = \{z \in \mathbb{C} \,|\, |z| = 1\}$ the Fourier series of f is the classical Fourier series $f(z) = \sum_{n=-\infty}^{\infty} a_n z^n$ since $\hat{G}$ consists of the maps $\gamma(z) = z^n$, $n \in Z$.

(10) If $A : G \to G$ is an endomorphism we can define the dual endomorphism $\hat{A} : \hat{G} \to \hat{G}$ by $\hat{A}\gamma = \gamma \circ A$, $\gamma \in \hat{G}$. It is easy to see that A is one-to-one if and only if $\hat{A}$ is onto, and A is onto if and only if $\hat{A}$ is one-to-one. Therefore A is an automorphism if and only if $\hat{A}$ is an automorphism.

Recall that for compact groups G, G is metrisable iff G has a countable topological base.

§0.8 Endomorphisms of Tori

We shall view the n-torus in two ways: multiplicatively as K^n and additively as R^n/Z^n where R^n is the additive group of n-dimensional Euclidean space and Z^n is the subgroup of R^n consisting of the points with integer coordinates. A topological group isomorphism from K^n to R^n/Z^n is given by

$$(e^{2\pi i x_1}, \ldots, e^{2\pi i x_n}) \longmapsto (x_1, \ldots, x_n) + Z^n.$$

Theorem 0.14

(i) *Every closed subgroup of K is either K or is a finite cyclic group consisting of all p-th roots of unity for some integer $p > 0$.*

(ii) *The only automorphisms of K are the identity and the map $z \mapsto z^{-1}$.*

(iii) *The only homomorphisms of K are the maps $\phi_n(z) = z^n$, $n \in Z$.*

(iv) *The only homomorphisms of K^n to K are maps of the form*

$$(z_1, \ldots, z_n) \mapsto z_1^{m_1} \cdot \cdots \cdot z_n^{m_n} \quad \text{where } m_1, \ldots, m_n \in Z.$$

PROOF. Let d denote the usual Euclidean metric on K which is a rotation invariant metric on K.

(i) Let H be a closed subgroup of K. If H is infinite it has a limit point so $\forall \varepsilon > 0 \ \exists a, b \in H$ with $d(a, b) < \varepsilon$ and $a \neq b$. Then $d(b^{-1}a, 1) < \varepsilon$, and therefore the powers of $b^{-1}a$ are ε-dense in K. Therefore H is ε-dense in K and $H = K$.

If H is finite and has p elements then $a^p = 1 \ \forall a \in H$. So each element of H is a p-th root of unity, and since there are p elements in H, H must consist of all the p-th roots of unity.

(ii) Let $\theta \colon K \to K$ be an automorphism. We have $\theta(1) = 1$. Since -1 is the only element of K of order 2 we have $\theta(-1) = -1$. Since i, $-i$ are the only elements of order 4 either $\theta(i) = i$ and $\theta(-i) = -i$ or $\theta(i) = -i$ and $\theta(-i) = i$. Consider the first case. Since θ maps intervals to intervals, the interval $\overrightarrow{[1, i]}$ from 1 to i is either mapped to itself or to $\overrightarrow{[i, 1]}$ (all intervals go anticlockwise). But since $\overrightarrow{[1, i]}$ does not contain -1 it cannot be mapped to $\overrightarrow{[i, 1]}$ so $\theta \overrightarrow{[1, i]} = \overrightarrow{[1, i]}$. The only element of order 8 in $\overrightarrow{[1, i]}$ is $e^{\pi i/4}$ and so this must be fixed by θ. Therefore $\theta \overrightarrow{[1, e^{\pi i/4}]} = \overrightarrow{[1, e^{\pi i/4}]}$. By induction one shows that $\theta(e^{2\pi i/2^k}) = e^{2\pi i/2^k}$ for each $k > 0$. It follows that θ fixes all the 2^k-th roots of unity $\forall k > 0$ and hence is the identity. In the second case one shows that $\theta(e^{2\pi i/2^k}) = e^{-2\pi i/2^k} \ \forall k > 0$ and hence $\theta(z) = z^{-1}$, $z \in K$.

(iii) Let $\theta \colon K \to K$ be an endomorphism. If θ is non-trivial, its image, $\theta(K)$, is a closed connected subgroup of K and so $\theta(K) = K$ by (i). The kernel Ker θ is a closed subgroup of K so either Ker $\theta = K$ or Ker $\theta = H_p$, the group of all p-th roots of unity, for some p. The first case corresponds to trivial θ. If Ker $\theta = H_p$ let $\alpha_p \colon K/H_p \to K$ be the isomorphism given by $\alpha_p(zH_p) = z^p$, and let $\theta_1 \colon K/H_p \to K$ be the isomorphism induced by θ ($\theta_1(zH_p) = \theta(z)$). Then $\theta_1 \alpha_p^{-1}$ is an automorphism of K and by (ii) either $\theta_1 \alpha_p^{-1}(z) = z \ \forall z \in K$ or

$\theta_1 \alpha_p^{-1}(z) = z^{-1} \ \forall z \in K$. Hence either $\theta(z) = \theta_1(zH_p) = \theta_1 \alpha_p^{-1}(z^p) = z^p \ \forall z \in K$ or $\theta(z) = z^{-p} \ \forall z \in K$.

(iv) Let $\gamma_i : K \to K^n$ be the homomorphism that imbeds K in the i-th component of K^n, i.e. $\gamma_i(z) = (1, 1, \ldots, 1, z, 1, \ldots, 1)$, where z appears in the i-th component. If $\theta : K^n \to K$ is a homomorphism then $\theta \circ \gamma_i : K \to K$ is an endomorphism and so $\theta \circ \gamma_i(z) = z^{m_i}$ for some $m_i \in Z$ by (iii). Hence

$$\begin{aligned} \theta(z_1, \ldots, z_n) &= \theta(\gamma_1(z_1) \cdot \gamma_2(z_2) \cdot \cdots \cdot \gamma_n(z_n)) \\ &= \theta\gamma_1(z_1) \cdot \theta\gamma_2(z_2) \cdot \cdots \cdot \theta\gamma_n(z_n) \\ &= z_1^{m_1} \cdot z_2^{m_2} \cdot \cdots \cdot z_n^{m_n}. \end{aligned} \qquad \square$$

The following theorem says that each endomorphism of K^n is determined by an $n \times n$ matrix with integer entries.

Theorem 0.15

(i) *Every endomorphism $A : K^n \to K^n$ is of the form*:

$$A(z_1, \ldots, z_n) = (z_1^{a_{11}} \cdot \cdots \cdot z_n^{a_{1n}}, \ldots, z_1^{a_{n1}} \cdot \cdots \cdot z_n^{a_{nn}})$$

where each $a_{ij} \in Z$. In additive notation,

$$A\left(\begin{pmatrix} x_1 \\ \vdots \\ x_n \end{pmatrix} + Z^n \right) = [a_{ij}] \begin{pmatrix} x_1 \\ \vdots \\ x_n \end{pmatrix} + Z^n,$$

where $[a_{ij}]$ denotes the $n \times n$ matrix with (i, j)-th element a_{ij}.

(ii) *A maps K^n onto K^n iff $\det[a_{ij}] \neq 0$.*

(iii) *A is an automorphism of K^n iff $\det[a_{ij}] = \pm 1$.*

PROOF

(i) Let $\pi_i : K^n \to K$ be the projection to the i-th coordinate. Then $\pi_i \circ A : K^n \to K$ is a homomorphism, so by (iv) of the previous theorem

$$\pi_i \circ A(z_1, \ldots, z_n) = z_1^{a_{i1}} \cdot z_2^{a_{i2}} \cdot \cdots \cdot z_n^{a_{in}}$$

where $a_{ij} \in Z$.

(ii) Assume $\det[a_{ij}] = 0$. Since the rows of $[a_{ij}]$ are linearly dependent over the rationals there exist integers $m_1, \ldots, m_n$ not all zero such that $m_1 A_1 + \cdots + m_n A_n = 0$ where A_i is the i-th row of $[a_{ij}]$. Then each point $(\omega_1, \ldots, \omega_n)$ of K^n in the image of A satisfies $\omega_1^{m_1} \cdots \omega_n^{m_n} = 1$. Thus $A(K^n) \neq K^n$. If $\det[a_{ij}] \neq 0$ then the linear map of R^n determined by the matrix $[a_{ij}]$ maps R^n onto R^n and hence A maps K^n onto K^n.

(iii) If A is an automorphism represented by a matrix $[A]$ then A^{-1} is also an automorphism represented by a matrix $[B]$, and since $AA^{-1} = I = A^{-1}A$ we have that $[B] = [A]^{-1}$. Since $[B]$ is an integer matrix, $\det[A] = \pm 1$. Conversely, if $\det[A] = \pm 1$, $[A]^{-1}$ has integer entries and hence defines an endomorphism B of K^n satisfying $AB = BA = I$. $\qquad \square$

Therefore the surjective endomorphisms of K^n are in one-to-one correspondence with the $n \times n$ integral matrices with non-zero determinant.

If A is an endomorphism of the n-torus, $[A]$ will always denote the associated matrix and $\tilde{A}$ will denote the linear transformation of R^n determined by $[A]$. So if $\pi: R^n \to R^n/Z^n$ is the natural projection ($\pi(x) = x + Z^n$) we have $\pi\tilde{A} = \tilde{A}\pi$.

Let $A: K^n \to K^n$ be an endomorphism. We now consider how the map $\hat{A}: \hat{K}^n \to \hat{K}^n$ (introduced in §0.7) acts as a map of Z^n when $\hat{K}^n$ is identified with Z^n by the isomorphism:

$$\gamma \to \begin{pmatrix} m_1 \\ m_2 \\ \vdots \\ m_n \end{pmatrix} \quad \text{when } \gamma(z_1, z_2, \ldots, z_n) = z_1^{m_1} \cdot z_2^{m_2} \cdot \cdots \cdot z_n^{m_n}.$$

One readily checks that the endomorphism $\hat{A}: Z^n \to Z^n$ is given by

$$\hat{A}\begin{pmatrix} m_1 \\ \vdots \\ m_n \end{pmatrix} = [A]_t \begin{pmatrix} m_1 \\ \vdots \\ m_n \end{pmatrix}$$

where $[A]_t$ denotes the transpose of the matrix $[A]$.

§0.9 Perron–Frobenius Theory

Let $A = [a_{ij}]$ be a $k \times k$ matrix. We say A is non-negative if $a_{ij} \geq 0$ for all i, j. Such a matrix is called *irreducible* if for any pair i, j there is some $n > 0$ such that $a_{ij}^{(n)} > 0$ where $a_{ij}^{(n)}$ is the (i,j)-th element of A^n. The matrix A is *irreducible* and *aperiodic* if there exists $n > 0$ such that $a_{ij}^{(n)} > 0$ for all i, j. We shall use the following result (see Gantmacher [1])

Theorem 0.16 (Perron–Frobenius Theorem). *Let $A = [a_{ij}]$ be a non-negative $k \times k$ matrix.*

(i) *There is a non-negative eigenvalue λ such that no eigenvalue of A has absolute value greater than λ.*

(ii) *We have $\min_i (\sum_{j=1}^k a_{ij}) \leq \lambda \leq \max_i (\sum_{j=1}^k a_{ij})$.*

(iii) *Corresponding to the eigenvalue λ there is a non-negative left (row) eigenvector $u = (u_1, \ldots, u_k)$ and a non-negative right (column) eigenvector*

$$v = \begin{pmatrix} v_1 \\ \vdots \\ v_k \end{pmatrix}.$$

(iv) *If A is irreducible then λ is a simple eigenvalue and the corresponding eigenvectors are strictly positive (i.e. $u_i > 0$, $v_i > 0$ all i).*

(v) *If A is irreducible then λ is the only eigenvalue of A with a non-negative eigenvector.*

Let A be an irreducible non-negative matrix and let λ, u, v be as in Theorem 0.16. Define the $k \times k$ matrix $P = [p_{ij}]$ by $p_{ij} = a_{ij}v_j/\lambda v_i$. Then $0 \le p_{ij} \le 1$ and $\sum_{j=1}^k p_{ij} = 1$ for all i, so that P is a stochastic matrix. Normalise u and v so that if $p_i = u_i v_i$ then $\sum_{i=1}^k p_i = 1$. Then $p = (p_1, \dots, p_k)$ is an invariant probability vector for P in the sense $pP = p$. If A is irreducible and aperiodic we can use a standard result in probability theory to get the following theorem.

Theorem 0.17. *Let A be an irreducible and aperiodic non-negative matrix. Let*

$$u = (u_1, \dots, u_k), \qquad v = \begin{pmatrix} v_1 \\ \vdots \\ v_k \end{pmatrix}$$

be the strictly positive eigenvectors corresponding to the largest eigenvalue λ as in Theorem 0.16. Then for each pair i, j $\lim_{n \to \infty} \lambda^{-n} a_{ij}^{(n)} = u_j v_i$.

This theorem follows from the renewal theorem which we shall need for another purpose (see Feller [1], p. 291).

Theorem 0.18 (Renewal Theorem). *Let $\{c_n\}_0^\infty$ and $\{u_n\}_0^\infty$ be bounded sequences of real numbers with $0 \le c_n \le 1$ and $d_n \ge 0$ for all n. Suppose the greatest common divisor of all integers n with $c_n > 0$ is one. Suppose $\{u_n\}_0^\infty$ satisfies $u_n = d_n + c_0 u_n + c_1 u_{n-1} + \cdots + c_n u_0$ for all n. If $\sum_{n=0}^\infty c_n = 1$ and $\sum_0^\infty d_n < \infty$ then $\lim_{n \to \infty} u_n = (\sum_0^\infty d_n)(\sum_0^\infty nc_n)^{-1}$ where this is interpreted as zero if $\sum_0^\infty nc_u = \infty$.*

§0.10 Topology

We shall be interested in compact metric spaces. There is the following result.

Theorem 0.19. *Let X be a compact Hausdorff space. The following are equivalent:*

(1) *X is metrisable.*
(2) *X has a countable base.*
(3) *$C(X)$ (the space of all complex-valued continuous functions on X) has a countable dense subset.*

See Kelley [1] for the proof.

We shall usually denote the metric on a space X by d, and the open ball of centre x_0 and radius r by $B(x_0; r)$. The diameter of a subset A of X will be denoted by diam(A), the closure of A by $\bar{A}$, the interior of A by int(A) and the boundary of A by ∂A. We have $\partial(A \backslash B) \subset \partial A \cup \partial B$, $\partial(A \cap B) \subset \partial A \cup \partial B$ and $\partial(A \cup B) \subset \partial A \cup \partial B$.

When X is compact the space $C(X)$ is a Banach algebra with norm $\|f\| = \sup\{|f(x)| : x \in X\}$. We shall use $C(X, R)$ to denote the real Banach algebra of all continuous real-valued functions on X.

We shall be interested in open covers of X and we shall use the following basic result.

Theorem 0.20 (Lebesgue Covering Lemma). *If (X, d) is a compact metric space and α is an open cover of X then there exists a $\delta > 0$ such that each subset of X of diameter less than or equal to δ lies in some member of α. (Such a δ is called a Lebesgue number for α.)*

PROOF. We may as well suppose α is a finite cover.

Let $\alpha = \{A_1, \ldots, A_p\}$. Assume the theorem is false. Then for each $n \geq 1$ there exists $B_n \subseteq X$ such that diam(B_n) $\leq 1/n$ and B_n is not contained in any A_i. Choose $x_n \in B_n$ and select a subsequence $\{x_{n_i}\}$ which converges, say $x_{n_i} \to x$. Suppose $x \in A_j \in \alpha$. Let $a = d(x, X \backslash A_j) > 0$. Choose n_i such that $n_i > 2/a$ and $d(x_{n_i}, x) < a/2$. Then if $y \in B_{n_i}$

$$d(y, x) \leq d(y, x_{n_i}) + d(x_{n_i}, x) \leq \frac{1}{n_i} + \frac{a}{2} < a.$$

Hence $B_{n_i} \subseteq A_j$, a contradiction. $\square$

If α is a cover of X, we write diam(α) for $\sup\{\text{diam}(A) | A \in \alpha\}$.

There is a natural σ-algebra of subsets of a compact metric space X. This is the σ-algebra, $\mathscr{B}(X)$, of Borel subsets of X which is defined to be the smallest σ-algebra containing all the open subsets of X. Every complex-valued continuous function on X is measurable relative to this σ-algebra.

CHAPTER 1
Measure-Preserving Transformations

In this chapter we shall discuss measure-preserving transformations and some of their basic properties.

§1.1 Definition and Examples

Definition 1.1. Suppose $(X_1, \mathscr{B}_1, m_1)$, $(X_2, \mathscr{B}_2, m_2)$ are probability spaces.

(a) A transformation $T: X_1 \to X_2$ is *measurable* if $T^{-1}(\mathscr{B}_2) \subset \mathscr{B}_1$ (i.e. $B_2 \in \mathscr{B}_2 \Rightarrow T^{-1}B_2 \in \mathscr{B}_1$).

(b) A transformation $T: X_1 \to X_2$ is *measure-preserving* if T is measurable and $m_1(T^{-1}(B_2)) = m_2(B_2) \ \forall B_2 \in \mathscr{B}_2$.

(c) We say that $T: X_1 \to X_2$ is an *invertible measure-preserving transformation* if T is measure-preserving, bijective, and T^{-1} is also measure-preserving.

Remarks

(1) We should write $T: (X_1, \mathscr{B}_1, m_1) \to (X_2, \mathscr{B}_2, m_2)$ since the measure-preserving property depends on the $\mathscr{B}$'s and m's.

(2) If $T: X_1 \to X_2$ and $S: X_2 \to X_3$ are measure-preserving so is $S \circ T: X_1 \to X_3$.

(3) Measure-preserving transformations are the structure preserving maps (morphisms) between measure spaces.

(4) Let $(X_i, \bar{\mathscr{B}}_i, \bar{m}_i)$ denote the completion of $(X_i, \mathscr{B}_i, m_i)$ $i = 1, 2$. If $T: (X_1, \mathscr{B}_1, m_1) \to (X_2, \mathscr{B}_2, m_2)$ is measure-preserving then so is $T: (X_1, \bar{\mathscr{B}}_1, \bar{m}_1) \to (X_2, \bar{\mathscr{B}}_2, \bar{m}_2)$. Recall that $\bar{\mathscr{B}}_i$ is the collection of sets $B_i \triangle F_i$ where $B_i \in \mathscr{B}_i$

and F_i is a subset of an element N_i of $\mathcal{B}_i$ of zero measure and $\bar{m}_i(B_i \triangle F_i) = m_i(B_i)$. Then $T^{-1}(B_2 \triangle F_2) = T^{-1}B_2 \triangle T^{-1}F_2$ where $T^{-1}B_2 \in \mathcal{B}_1$, $T^{-1}F_2 \subset T^{-1}N_2$ and $m_1(T^{-1}N_2) = m_2(N_2) = 0$. Therefore $T^{-1}(B_2 \triangle F_2) \in \bar{\mathcal{B}}_1$ and $\bar{m}_1(T^{-1}(B_2 \triangle F_2)) = m_1(T^{-1}B_2) = m_2(B_2) = \bar{m}_2(B_2 \triangle F_2)$.

(5) We shall be mainly interested in the case $(X_1, \mathcal{B}_1, m_1) = (X_2, \mathcal{B}_2, m_2)$ since we wish to study the iterates T^n. When $T: X \to X$ is a measure-preserving transformation of $(X, \mathcal{B}, m)$ we say that T preserves m or that m is T-invariant.

In practice it would be difficult to check, using Definition 1.1, whether a given transformation is measure-preserving or not since one usually does not have explicit knowledge of all the members of $\mathcal{B}$. However we often do have explicit knowledge of a semi-algebra $\mathcal{S}$ generating $\mathcal{B}$ (for example, when X is the unit interval $\mathcal{S}$ may be the semi-algebra of all subintervals of X, and when X is a direct product space $\mathcal{S}$ may be the collection of all measurable rectangles). The following result is therefore desirable in checking whether transformations are measure-preserving or not.

Theorem 1.1. *Suppose* $(X_1, \mathcal{B}_1, m_1)$, $(X_2, \mathcal{B}_2, m_2)$ *are probability spaces and* $T: X_1 \to X_2$ *is a transformation. Let* $\mathcal{S}_2$ *be a semi-algebra which generates* $\mathcal{B}_2$. *If for each* $A_2 \in \mathcal{S}_2$ *we have* $T^{-1}(A_2) \in \mathcal{B}_1$ *and* $m_1(T^{-1}(A_2)) = m_2(A_2)$ *then* T *is measure-preserving.*

PROOF. Let $\mathcal{C}_2 = \{B \in \mathcal{B}_2 : T^{-1}(B) \in \mathcal{B}_1, m_1(T^{-1}(B)) = m_2(B)\}$. We want to show that $\mathcal{C}_2 = \mathcal{B}_2$. However $\mathcal{S}_2 \subseteq \mathcal{C}_2$ and since each member of the algebra $\mathcal{A}(\mathcal{S}_2)$ generated by $\mathcal{S}_2$ is a finite disjoint union of members of $\mathcal{S}_2$ we have $\mathcal{A}(\mathcal{S}_2) \subset \mathcal{C}_2$. Since $\mathcal{C}_2$ is easily seen to be a monotone class, the result follows since the σ-algebra generated by $\mathcal{A}(\mathcal{S}_2)$ is the monotone class generated by $\mathcal{A}(\mathcal{S}_2)$. $\qquad \square$

Examples of Measure-Preserving Transformations

(1) The identity transformation I on $(X, \mathcal{B}, m)$ is obviously measure-preserving.

(2) Let $K = \{z \in \mathbb{C} : |z| = 1\}$, let $\mathcal{B}$ be the σ-algebra of Borel subsets of K and let m be Haar measure. Let $a \in K$ and define $T: K \to K$ by $T(z) = az$. Then T is measure-preserving since m is Haar measure. The transformation T is called a rotation of K.

(3) The transformation $T(x) = ax$ defined on any compact group G (where a is a fixed element of G) preserves Haar measure. Such transformations are called *rotations* of G.

(4) Any continuous endomorphism of a compact group onto itself preserves Haar measure.

PROOF. Let $A: G \to G$ be a continuous endomorphism and let m denote Haar measure on G. Define a probability measure on the Borel subsets of G by $\mu(E) = m(A^{-1}(E))$. The measure μ is regular since m is regular.

Also $\mu(Ax \cdot E) = m(A^{-1}(Ax \cdot E)) = m(x \cdot A^{-1}E) = \mu(E)$. Since A maps G onto G we see that μ is rotation invariant and therefore $\mu = m$ by the uniqueness property of Haar measure. This shows A is measure-preserving. $\square$

For example $T(z) = z^n$ preserves Haar measure on the unit circle if $n \in Z \backslash \{0\}$.

(5) Any affine transformation of a compact group G preserves Haar measure. An *affine transformation* is a map of the form $T(x) = a \cdot A(x)$ where $a \in G$ is fixed and $A: G \to G$ is a surjective endomorphism. It follows that T is measure-preserving because it is the composition of a rotation and an endomorphism. (See remark (2).) When $A = I$ we have example (3) and when a is the identity element of G we have example (4). We shall see that examples 3 and 4 have very different behaviour and so we expect the class of affine transformations to provide examples with a range of interesting behaviours.

When dealing with affine transformations as measure-preserving transformations we always assume the measure involved is normalised Haar measure.

(6) Let $k \geq 2$ be a fixed integer and let $(p_0, p_1, \ldots, p_{k-1})$ be a probability vector with non-zero entries (i.e., $p_i > 0$ each i and $\sum_{i=0}^{k-1} p_i = 1$). Let $(Y, 2^Y, \mu)$ denote the measure space where $Y = \{0, 1, \ldots, k-1\}$ and the point i has measure p_i. Let $(X, \mathcal{B}, m) = \prod_{-\infty}^{\infty} (Y, 2^Y, \mu)$. Define $T: X \to X$ by $T(\{x_n\}) = \{y_n\}$ where $y_n = x_{n+1}$. If $\mathcal{S}$ denotes the semi-algebra of all measurable rectangles then $m(T^{-1}A) = m(A)\ \forall A \in \mathcal{S}$. By Theorem 1.1, T is measure-preserving. We call T the *two-sided* $(p_0, \ldots, p_{k-1})$*-shift*. (The description "two-sided" refers to the fact that the direct product ranges over all integers rather than just the non-negative integers.) This is an example of an invertible measure-preserving transformation. We sometimes use the notation $(\ldots, x_{-1} \overset{*}{x}_0 x_1, \ldots)$ for a point of X (the * indicates the 0-th position in the product) and then T can be written $T((\ldots, x_{-1} \overset{*}{x}_0 x_1, \ldots)) = (\ldots, x_{-1} x_0 \overset{*}{x}_1 x_2, \ldots)$. The set Y is called the state space of the shift.

(7) Let $(Y, 2^Y, \mu)$ be as in example (6) and let $(X, \mathcal{B}, m) = \prod_0^{\infty} (Y, 2^Y, \mu)$. If we write points of X in the form $(x_0, x_1, \ldots)$, $x_i \in Y$, then define $T: X \to X$ by $T(x_0, x_1, x_2, \ldots) = (x_1, x_2, \ldots)$. By considering the semi-algebra of measurable rectangles and using Theorem 1.1 we see that T is measure-preserving. This transformation is called the *one-sided* $(p_0, \ldots, p_{k-1})$*-shift*. It is not an invertible measure-preserving transformation.

(8) Fix an integer $k \geq 2$ and consider the measurable space $(Y, 2^Y)$ where $Y = \{0, 1, \ldots, k-1\}$. Let $(X, \mathcal{B})$ denote the direct product space $(X, \mathcal{B}) = \prod_{-\infty}^{\infty} (Y, 2^Y)$. Let $T: X \to X$ denote the shift transformation defined (as in example (6)) by $T((\ldots, x_{-1} \overset{*}{x}_0 x_1, \ldots)) = (\ldots, x_{-1} x_0 \overset{*}{x}_1 x_2, \ldots)$. We shall

now obtain many probability measures on $(X, \mathcal{B})$ which make T a measure-preserving transformation (in fact all of them are obtained in this way). Whenever $n \geq 0$ and $a_i \in Y(0 \leq i \leq n)$ suppose a real number $p_n(a_0, \ldots, a_n)$ is given such that

(i) $p_n(a_0, \ldots, a_n) \geq 0$;

(ii) $\sum_{a_0 \in Y} p_0(a_0) = 1$;

(iii) $p_n(a_0, \ldots, a_n) = \sum_{a_{n+1} \in Y} p_{n+1}(a_0, \ldots, a_n, a_{n+1})$.

If we define m on the semi-algebra of measurable elementary rectangles by $m(\{(x_i) \in X \,|\, x_q = a_0, \ldots, x_{q+n} = a_n\}) = p_n(a_0, \ldots, a_n)$ then m can be extended to a probability measure on $(X, \mathcal{B})$ by Theorem 0.5 and T preserves the measure m by Theorem 1.1.

To obtain example (6) we take $p_n(a_0, \ldots, a_n) = p_{a_0} p_{a_1} \cdots p_{a_n}$. Another important collection of examples is provided by the Markov chains. Here we are given a probability vector $\mathbf{p} = (p_0, p_1, \ldots, p_{k-1})$ and a stochastic matrix $P = (p_{ij})_{i,j \in Y}$ ($p_{ij} \geq 0$, $\sum_{j=0}^{k-1} p_{ij} = 1$) such that $\sum_{i=0}^{k-1} p_i p_{ij} = p_j$, and we put $p_n(a_0, \ldots, a_n) = p_{a_0} p_{a_0 a_1} p_{a_1 a_2} \cdots p_{a_{n-1} a_n}$. We call this measure-preserving transformation the *two-sided* $(\mathbf{p}, P)$-*Markov shift*. We always assume $p_i > 0$ for all i, for if $p_i = 0$ for some i then we could eliminate i from the state space because all rectangles containing i would have zero measure. The two-sided $(p_0, \ldots, p_{k-1})$-shift can be considered as a Markov shift by taking $\mathbf{p} = (p_0, \ldots, p_{k-1})$ and $p_{ij} = p_j$.

The corresponding one-sided transformations can easily be defined.

(9) Let X be a compact metric space and let $T: X \to X$ be continuous. We shall see later (in Chapter 6) that there is always at least one probability measure m defined on the σ-algebra of Borel subsets, $\mathcal{B}(X)$, of X such that T is a measure-preserving transformation of $(X, \mathcal{B}(X), m)$. In particular any diffeomorphism of a compact smooth manifold has at least one invariant probability measure on the Borel subsets of the manifold.

(10) Given any set X_1, any probability space $(X_2, \mathcal{B}_2, m_2)$ and any map $T: X_1 \to X_2$ of X_1 onto X_2 we can choose a σ-algebra $\mathcal{B}_1$ and a measure m_1 on X_1 to make T measure-preserving. In fact let $\mathcal{B}_1 = T^{-1} \mathcal{B}_2$ and define m_1 by $m_1(T^{-1} B_2) = m_2(B_2)$.

Conversely, if $(X_1, \mathcal{B}_1, m_1)$ is any probability space, X_2 any set and $T: X_1 \to X_2$ any map of X_1 onto X_2, then we can choose a σ-algebra $\mathcal{B}_2$ and a measure m_2 on X_2 so that T is measure-preserving. Put

$$\mathcal{B}_2 = \{B : B \subset X_2 \text{ and } T^{-1} B \in \mathcal{B}_1\}$$

and $m_2(B) = m_1(T^{-1} B)$ for $B \in \mathcal{B}_2$.

In remark (2) we have given one way of obtaining a new measure-preserving transformation from two given ones. Another way of constructing a new measure-preserving transformation from two given ones is by direct products:

Definition 1.2. Let $T_i: X_i \to X_i$ be a measure-preserving transformation of a probability space $(X_i, \mathscr{B}_i, m_i)$ for $i = 1, 2$. The *direct product* $T_1 \times T_2$ is the measure-preserving transformation of $(X_1 \times X_2, \mathscr{B}_1 \times \mathscr{B}_2, m_1 \times m_2)$ given by $(T_1 \times T_2)(x_1, x_2) = (T_1(x_1), T_2(x_2))$.

This transformation is measure-preserving because it clearly preserves the measure of rectangles, and Theorem 1.1 can then be used. In a similar way one can define the direct product of any finite or countable number of measure-preserving transformations.

As we have mentioned before we shall be interested in the study of the powers T^n, $n \in Z$, of a given measure-preserving transformation $T: X \to X$ of a probability space $(X, \mathscr{B}, m)$.

§1.2 Problems in Ergodic Theory

There are two types of problems in measure-theoretic ergodic theory. The first type, which we call internal problems, are concerned with understanding measure-preserving transformations and trying to decide when two of them are isomorphic. In the rest of this chapter we study general properties of measure-preserving transformations and in Chapter 2 we consider the isomorphism problem. The attacks on the isomorphism problem usually involve searching for isomorphism invariants. We say more about this in Chapter 2.

The second type of problem involves applications of measure-theoretic ergodic theory. How can this theory be applied to problems in other branches of mathematics and physics? The problems that can be tackled using ergodic theory have a formulation in terms of a transformation T of a space X and the problem involves studying the asymptotic properties of T (i.e., T^n for large n). The space X usually carries some structure (e.g. X is a topological space or a smooth manifold) and T preserves this structure (e.g. T is a homeomorphism or a diffeomorphism). Then one needs to find a measure on X which is preserved by T and which is a useful measure for the problem. For example if T is a map of the unit interval (or of a smooth manifold) we may like the invariant measure to have the same null sets as Lebesgue measure (or the same null sets as a smooth measure) so that any conclusions we make would hold for almost all points relative to Lebesgue measure. In these applications one needs to find the most suitable measure for the particular problem. As we shall see in Chapter 6, a given transformation of a measurable space may preserve many measures.

As an example suppose we wish to study continued fraction expansions of points in $(0, 1)$. This leads to the transformation

$$T:[0, 1) \to [0, 1) \text{ given by } T(x) = \begin{cases} 0 & \text{if } x = 0 \\ \{1/x\} & \text{if } x \neq 0 \end{cases}$$

where $\{y\}$ denotes the fractional part of a positive real number y. This is because the n-th partial quotient of an irrational number x is $a(T^n x)$ where $a:(0, 1) \to Z$ is given by $a(y) = [1/y]$, the integer part of $1/y$. So the partial quotients are expressable in terms of one function and the iterates of T. It turns out that T preserves the Gauss measure on $[0, 1)$ given by

$$m(A) = \frac{1}{\log 2} \int_A \frac{1}{1 + x} \, dl$$

where l denotes Lebesgue measure. Therefore m has the same null sets as l and so any statement which holds almost everywhere for m will hold almost everywhere for l. For example one can deduce from the ergodic theorem (Theorem 1.14) that for almost every point x of $[0, 1)$ (relative to Lebesgue measure) the relative frequency of the natural number k in the partial quotients of x is

$$\frac{1}{\log 2} \int_{1/k+1}^{1/k} \frac{1}{1 + x} \, dl = \frac{1}{\log 2} \log \frac{(k + 1)^2}{k(k + 2)}.$$

A corresponding situation arises in Hamiltonian mechanical systems. Here we have a one-parameter family $\{T_t\}_{t \in R}$ of diffeomorphisms of a certain manifold M and the classical Lioville theorem gives a smooth measure on M preserved by each transformation T_t.

In Chapters 6, 8, and 9 we shall study invariant probability measures for a given continuous transformation of a compact metric space, and we shall consider some canonical ways of choosing certain invariant measures.

We now continue the study of the general theory of measure-preserving transformations.

§1.3 Associated Isometries

To any measure space $(X, \mathscr{B}, m)$ one can associate the Banach-spaces $L^P(X, \mathscr{B}, \mu) = \{f : X \to \mathbb{C} \mid f \text{ is measurable and } \int |f|^P \, dm < \infty\}$ $(p \geq 1)$. These spaces are some of the most useful tools for dealing with problems taking place on measure-spaces. Since a measure-preserving transformation T is a morphism of measure spaces we would expect to be able to associate to T a morphism of L^P spaces. We now describe this.

Let $L^\circ(X, \mathscr{B}, m)$ denote the space of all complex-valued measurable functions of $(X, \mathscr{B}, m)$ (where two functions are identified if they are equal almost everywhere) and let $L_R^\circ(X, \mathscr{B}, m)$ denote the real-valued members of $L^\circ(X, \mathscr{B}, m)$.

Definition 1.3. Let $(X_i, \mathscr{B}_i, m_i)$ be a probability space, $i = 1, 2$. If $T: X_1 \to X_2$ is measure-preserving the induced operator $U_T: L^0(X_2, \mathscr{B}_2, m_2) \to L^0(X_1, \mathscr{B}_1, m_1)$ is defined by $(U_T f)(x) = f(Tx)$, $f \in L^0(X_2, \mathscr{B}_2, m_2)$, $x \in X_1$.

Clearly, U_T is linear and $U_T L_R^0(X_2, \mathscr{B}_2, m_2) \subset L_R^0(X_1, \mathscr{B}_1, m_1)$. Also, $U_T(f \cdot g) = (U_T f)(U_T g)$ and $U_T c = c$ where c denotes a constant function. If $f \geq 0$ then $U_T f \geq 0$ so U_T is a positive operator. Notice also $U_T \chi_B = \chi_{T^{-1}B}$, $B \in \mathscr{B}_2$. We want to show that $U_T L^p(X_2, \mathscr{B}_2, m_2) \subset L^p(X_1, \mathscr{B}_1, m_1)$. This will follow from the following simple result.

Lemma 1.2. Let $(X_i, \mathscr{B}_i, m_i)$ $i = 1, 2$ be a probability space and let $T: X_1 \to X_2$ be measure-preserving. If $F \in L^0(X_2, \mathscr{B}_2, m_2)$ then $\int U_T F \, dm_1 = \int F \, dm_2$ (where if one side doesn't exist or is infinite, then the other side has the same property).

PROOF. It suffices to prove the result when F is real-valued and, by considering positive and negative parts of F, it suffices to consider non-negative functions. So suppose $F \geq 0$. If F is a simple function then the result is true because T is measure-preserving. Choose simple functions F_n increasing to F. Then $U_T F_n$ are simple functions increasing to $U_T F$ so

$$\int U_T F \, dm_1 = \lim_{n \to \infty} \int U_T F_n \, dm_1 = \lim_{n \to \infty} \int F_n \, dm_2 = \int F \, dm_2. \qquad \square$$

Theorem 1.3. Let $p \geq 1$. With the above notation $U_T L^p(X_2, \mathscr{B}_2, m_2) \subset L^p(X_1, \mathscr{B}_1, m_1)$ and $\|U_T f\|_p = \|f\|_p \forall f \in L^p(X_2, \mathscr{B}_2, m_2)$. Also

$$U_T L_R^p(X_2, \mathscr{B}_2, m_2) \subset L_R^p(X_1, \mathscr{B}_1, m_1).$$

PROOF. Let $f \in L^p(X_2, \mathscr{B}_2, m_2)$. Put $F(x) = |f(x)|^p$ in Lemma 1.2 to get $\|U_T f\|_p = \|f\|_p$. $\qquad \square$

Therefore a measure-preserving transformation $T: X_1 \to X_2$ induces a linear isometry of $L^p(X_2, \mathscr{B}_2, m_2)$ into $L^p(X_1, \mathscr{B}_1, m_1)$ for all $p \geq 1$. In particular if $T: X \to X$ is an invertible measure-preserving transformation of $(X, \mathscr{B}, m)$ then U_T is a unitary operator on $L^2(X, \mathscr{B}, m)$. The study of U_T is called the spectral study of T and we shall see later how this is useful in formulating concepts such as ergodicity and mixing and also in helping with the isomorphism problem for measure-preserving transformations.

If $T_1: X_1 \to X_2$, $T_2: X_2 \to X_3$ are measure-preserving then $U_{T_2 \circ T_1} = U_{T_1} \circ U_{T_2}$ so that if $T: X \to X$ is measure-preserving $U_{T^n} = (U_T)^n$, $n \in Z^+$.

§1.4 Recurrence

One property that is enjoyed by all measure-preserving transformations is recurrence:

Theorem 1.4 (Poincaré's Recurrence Theorem). *Let $T: X \to X$ be a measure-preserving transformation of a probability space $(X, \mathcal{B}, m)$. Let $E \in \mathcal{B}$ with $m(E) > 0$. Then almost all points of E return infinitely often to E under positive iteration by T (i.e., there exists $F \subset E$ with $m(F) = m(E)$ such that for each $x \in F$ there is a sequence $n_1 < n_2 < n_3 < \cdots$ of natural numbers with $T^{n_i}(x) \in F$ for each i).*

PROOF. For $N \geq 0$ let $E_N = \bigcup_{n=N}^{\infty} T^{-n}E$. Then $\bigcap_{N=0}^{\infty} E_N$ is the set of all points of X which enter E infinitely often under positive iteration by T. Hence the set $F = E \cap \bigcap_{N=0}^{\infty} E_N$ consists of all points of E that enter E infinitely often under positive iteration by T. If $x \in F$ then there is a sequence $0 < n_1 < n_2 < \cdots$ of natural numbers with $T^{n_i}(x) \in E$ for all i. For each i we have $T^{n_i}(x) \in F$ because $T^{n_j - n_i}(T^{n_i}x) \in E$ for all j. It remains to show $m(F) = m(E)$.

Since $T^{-1}E_N = E_{N+1}$ we have $m(E_N) = m(E_{N+1})$ and hence $m(E_0) = m(E_N)$ for all N. Since $E_0 \supset E_1 \supset E_2 \supset \cdots$ we have $m(\bigcap_{N=0}^{\infty} E_N) = m(E_0)$. Therefore $m(F) = m(E \cap E_0) = m(E)$ since $E \subset E_0$. ◻

Remarks

1. In the proof the measure-preserving property of T was used only in the weaker form of incompressibility (i.e., if $B \in \mathcal{B}$ and $T^{-1}B \subset B$ then $m(B) = m(T^{-1}B)$). Therefore Theorem 1.4 is true for incompressible transformations.

2. Theorem 1.4 is false if a measure space of infinite measure is used. An example is given by the map $T(x) = x + 1$ defined on Z with the measure on Z which gives each integer unit measure. For this example if E denotes the set $\{0\}$ then the sets E_N all have infinite measure but $\bigcap_{N=0}^{\infty} E_N$ is empty.

§1.5 Ergodicity

Let $(X, \mathcal{B}, m)$ be a probability space and $T: X \to X$ be a measure-preserving transformation. If $T^{-1}B = B$ for $B \in \mathcal{B}$, then also $T^{-1}(X \setminus B) = X \setminus B$ and we could study T by studying the two simpler transformations $T|_B$ and $T|_{X \setminus B}$. If $0 < m(B) < 1$ this has simplified the study of T. If $m(B) = 0$ (or $m(X \setminus B) = 0$) we can ignore B (or $X \setminus B$) and we have not significantly simplified T since neglecting a set of zero measure is allowed in measure theory. This raises the idea of studying those transformations that cannot be decomposed as above and of trying to express every measure-preserving transformation in terms of these indecomposable ones. The indecomposable transformations are called ergodic:

Definition 1.4. Let $(X, \mathcal{B}, m)$ be a probability space. A measure-preserving transformation T of $(X, \mathcal{B}, m)$ is called ergodic if the only members B of $\mathcal{B}$ with $T^{-1}B = B$ satisfy $m(B) = 0$ or $m(B) = 1$.

There are several other ways of stating the ergodicity condition and we present some of them in the next two theorems.

Theorem 1.5. *If* $T: X \to X$ *is a measure-preserving transformation of the probability space* $(X, \mathcal{B}, m)$ *then the following statements are equivalent*:

(i) *T is ergodic.*

(ii) *The only members B of $\mathcal{B}$ with $m(T^{-1}B \triangle B) = 0$ are those with* $m(B) = 0$ *or* $m(B) = 1$.

(iii) *For every $A \in \mathcal{B}$ with $m(A) > 0$ we have* $m(\bigcup_{n=1}^{\infty} T^{-n}A) = 1$.

(iv) *For every $A, B \in \mathcal{B}$ with $m(A) > 0$, $m(B) > 0$ there exists $n > 0$ with* $m(T^{-n}A \cap B) > 0$.

Proof

(i) $\Rightarrow$ (ii). Let $B \in \mathcal{B}$ and $m(T^{-1}B \triangle B) = 0$. We shall construct a set B_∞ with $T^{-1}B_\infty = B_\infty$ and $m(B \triangle B_\infty) = 0$. For each $n \geq 0$ we have $m(T^{-n}B \triangle B) = 0$ because $T^{-n}B \triangle B \subset \bigcup_{i=0}^{n-1} T^{-(i+1)}B \triangle T^{-i}B = \bigcup_{i=0}^{n-1} T^{-i}(T^{-1}B \triangle B)$ and hence $m(T^{-n}B \triangle B) \leq nm(T^{-1}B \triangle B)$. Let $B_\infty = \bigcap_{n=0}^{\infty} \bigcup_{i=n}^{\infty} T^{-i}B$. By the above we know $m(B \triangle \bigcup_{i=n}^{\infty} T^{-i}B) \leq \sum_{i=n}^{\infty} m(B \triangle T^{-i}B) = 0$ for each $n \geq 0$. Since the sets $\bigcup_{i=n}^{\infty} T^{-i}B$ decrease with n and each has measure equal to B we have $m(B_\infty \triangle B) = 0$ and hence $m(B_\infty) = m(B)$. Also $T^{-1}B_\infty = \bigcap_{n=0}^{\infty} \bigcup_{i=n}^{\infty} T^{-(i+1)}B = \bigcap_{n=0}^{\infty} \bigcup_{i=n+1}^{\infty} T^{-i}B = B_\infty$. Therefore we have obtained a set B_∞ with $T^{-1}B_\infty = B_\infty$ and $m(B_\infty \triangle B) = 0$. By ergodicity we must have $m(B_\infty) = 0$ or 1 and hence $m(B) = 0$ or 1.

(ii) $\Rightarrow$ (iii). Let $A \in \mathcal{B}$ and $m(A) > 0$. Let $A_1 = \bigcup_{n=1}^{\infty} T^{-n}A$. We have $T^{-1}A_1 \subset A_1$ and since $m(T^{-1}A_1) = m(A_1)$ we have $m(T^{-1}A_1 \triangle A_1) = 0$. By (ii) we get $m(A_1) = 0$ or 1. We cannot have $m(A_1) = 0$ because $T^{-1}A \subset A_1$ and $m(T^{-1}A) = m(A) > 0$. Therefore $m(A_1) = 1$.

(iii) $\Rightarrow$ (iv). Let $m(A) > 0$ and $m(B) > 0$. By (iii) we have $m(\bigcup_{n=1}^{\infty} T^{-n}A) = 1$ so that $0 < m(B) = m(B \cap \bigcup_{n=1}^{\infty} T^{-n}A) = m(\bigcup_{n=1}^{\infty} B \cap T^{-n}A)$. Therefore $m(B \cap T^{-n}A) > 0$ for some $n \geq 1$.

(iv) $\Rightarrow$ (i). Suppose $B \in \mathcal{B}$ and $T^{-1}B = B$. If $0 < m(B) < 1$ then $0 = m(B \cap (X \setminus B)) = m(T^{-n}B \cap (X \setminus B))$ for all $n \geq 1$, which contradicts (iv). $\square$

Remark. Notice that we could replace (iii) by the statement "For every $A \in \mathcal{B}$ with $m(A) > 0$ and every natural number N we have $m(\bigcup_{n=N}^{\infty} T^{-n}A) = 1$" because $\bigcup_{n=N}^{\infty} T^{-n}A = T^{-N}(\bigcup_{n=0}^{\infty} T^{-n}A)$. Consequently we could replace

(iv) by "for every $A, B \in \mathcal{B}$ with $m(A) > 0, m(B) > 0$ and every natural number N there exists $n > N$ with $m(T^{-n}A \cap B) > 0$". One can think of (iii) and (iv) as saying that the orbit $\{T^{-n}A\}_{n=0}^{\infty}$ of any non-trivial set A sweeps out the whole space X (or that each non-trivial set A has a dense orbit in a measure-theoretical sense).

The next theorem characterises ergodicity in terms of the operator U_T.

Theorem 1.6. *If* $(X, \mathcal{B}, m)$ *is a probability space and* $T: X \to X$ *is measure-preserving then the following statements are equivalent:*

  (i) *T is ergodic.*
  (ii) *Whenever f is measurable and $(f \circ T)(x) = f(x) \ \forall x \in X$ then f is constant a.e.*
  (iii) *Whenever f is measurable and $(f \circ T)(x) = f(x)$ a.e. then f is constant a.e.*
  (iv) *Whenever $f \in L^2(m)$ and $(f \circ T)(x) = f(x) \ \forall x \in X$ then f is constant a.e.*
  (v) *Whenever $f \in L^2(m)$ and $(f \circ T)(x) = f(x)$ a.e. then f is constant a.e.*

PROOF. Trivially we have (iii) $\Rightarrow$ (ii), (ii) $\Rightarrow$ (iv), (v) $\Rightarrow$ (iv), and (iii) $\Rightarrow$ (v). So it remains to show (i) $\Rightarrow$ (iii) and (iv) $\Rightarrow$ (i). We first show (i) $\Rightarrow$ (iii). Let T be ergodic and suppose f is measurable and $f \circ T = f$ a.e. We can assume that f is real-valued for if f is complex-valued we can consider the real and imaginary parts separately. Define, for $k \in Z$ and $n > 0$,

$$X(k, n) = \{x : k/2^n \leq f(x) < (k + 1)/2^n\} = f^{-1}([k/2^n, (k + 1)/2^n)).$$

We have
$$T^{-1}X(k, n) \triangle X(k, n) \subset \{x : (f \circ T)(x) \neq f(x)\}$$

and hence $m(T^{-1}X(k, n) \triangle X(k, n)) = 0$ so that by (ii) of Theorem 1.5 $m(X(k, n)) = 0$ or 1.

For each fixed n $\bigcup_{k \in Z} X(k, n) = X$ is a disjoint union and so there exists a unique k_n with $m(X(k_n, n)) = 1$. Let $Y = \bigcap_{n=1}^{\infty} X(k_n, n)$. Then $m(Y) = 1$ and f is constant on Y so that f is constant a.e.

(iv) $\Rightarrow$ (i). Suppose $T^{-1}E = E$, $E \in \mathcal{B}$. Then $\chi_E \in L^2(m)$ and $(\chi_E \circ T)(x) = \chi_E(x) \ \forall x \in X$ so, by (iv), χ_E is constant a.e. Hence $\chi_E = 0$ a.e. or $\chi_E = 1$ a.e. and $m(E) = \int \chi_E \, dm = 0$ or 1. $\qquad \square$

Remarks

(1) A similar characterization in terms of $L^p(m)$ functions (for any $p \geq 1$) is true, since in the last part of the proof χ_E is in $L^p(m)$ as well as $L^2(m)$. Also we could use real $L^p(m)$ spaces.

(2) Another characterization of ergodicity of T is: whenever $f: X \to R$ is measurable and $f(Tx) \geq f(x)$ a.e. then f is constant a.e.

This is clearly a stronger statement than (iii). To see that ergodicity implies the stated property, let T be ergodic and suppose $f(Tx) \geq f(x)$ a.e. If f is not constant a.e. then there is some $c \in R$ with $B = \{x \in X \mid f(x) \geq c\}$ having $0 < m(B) < 1$. But $T^{-1}B \supset B$ so $m(T^{-1}B \triangle B) = 0$ and hence $m(B) = 0$ or 1. This contradicts $0 < m(B) < 1$.

We want to analyse our examples to see which of them are ergodic. The next result will be useful for this and it also relates the idea of measure-theoretic dense orbits (Theorem 1.5(iii) and (iv)) to the usual notion of dense orbits for continuous maps. We shall say more about this in Chapter 6. Recall that the σ-algebra of Borel subsets of a topological space is the σ-algebra generated by the open sets.

Theorem 1.7. *Let X be a compact metric space, $\mathscr{B}(X)$ the σ-algebra of Borel subsets of X and let m be a probability measure on $(X, \mathscr{B}(X))$ such that $m(U) > 0$ for every non-empty open set U. Suppose $T: X \to X$ is a continuous transformation which preserves the measure m and is ergodic. Then almost all points of X have a dense orbit under T i.e. $\{x \in X \mid (T^n x)_{n=0}^{\infty} \text{ is a dense subset of } X\}$ has m-measure one.*

PROOF. Let $\{U_n\}_{n=1}^{\infty}$ be a base for the topology of X. Then $\{T^n x \mid n \geq 0\}$ is dense in X iff $x \in \bigcap_{n=1}^{\infty} \bigcup_{k=0}^{\infty} T^{-k} U_n$. Since $T^{-1}(\bigcup_{k=0}^{\infty} T^{-k} U_n) \subset \bigcup_{k=0}^{\infty} T^{-k} U_n$ and T is measure-preserving and ergodic we have $m(\bigcup_{k=0}^{\infty} T^{-k} U_n) = 0$ or 1. Since $\bigcup_{k=0}^{\infty} T^{-k} U_n$ is a non-empty open set we have $m(\bigcup_{k=0}^{\infty} T^{-k} U_n) = 1$. The result follows. $\qquad\square$

Note that this result is applicable when m is Haar measure on a compact metric group and T is an affine transformation which is ergodic.

We shall now see when the examples of §1 are ergodic.

(1) Clearly the identity transformation on $(X, \mathscr{B}, m)$ is ergodic iff all members of $\mathscr{B}$ have measure 0 or 1.

(2) We have the following theorem concerning rotations of the unit circle K.

Theorem 1.8. *The rotation $T(z) = az$ of the unit circle K is ergodic (relative to Haar measure m) iff a is not a root of unity.*

PROOF. Suppose a is a root of unity, then $a^p = 1$ for some $p \neq 0$. Let $f(z) = z^p$. Then $f \circ T = f$ and f is not constant a.e. Therefore T is not ergodic by Theorem 1.6(ii). Conversely, suppose a is not a root of unity and $f \circ T = f$, $f \in L^2(m)$. Let $f(z) = \sum_{n=-\infty}^{\infty} b_n z^n$ be its Fourier series. Then $f(az) = \sum_{n=-\infty}^{\infty} b_n a^n z^n$ and hence $b_n(a^n - 1) = 0$ for each n. If $n \neq 0$ then $b_n = 0$, and so f is constant a.e. Theorem 1.6(v) gives that T is ergodic. $\qquad\square$

An equivalent way to say $a \in K$ is not a root of unity is to say that $\{a^n\}_{-\infty}^{\infty}$ is dense in K. (If a is a root of unity then $\{a^n\}_{-\infty}^{\infty}$ is a finite set and so not dense in K. If a is not a root of unity we can obtain an ε-dense subset of K as follows. Since the set $\{a^n\}_{-\infty}^{\infty}$ consists of infinitely many points there are two points a^p, a^q with $d(a^p, a^q) < \varepsilon$ (d is the usual Euclidean distance on K). Then $d(1, a^{q-p}) < \varepsilon$ so that $\{a^{n(q-p)}\}_{n=-\infty}^{\infty}$ is ε-dense.) This formulation is used to generalise Theorem 1.8 to the general case.

(3) We consider a rotation $T(x) = ax$ of a general compact group. The measure involved is normalised Haar measure m.

Theorem 1.9. *Let G be a compact group and $T(x) = ax$ a rotation of G. Then T is ergodic iff $\{a^n\}_{-\infty}^{\infty}$ is dense in G. In particular, if T is ergodic, then G is abelian.*

PROOF. Suppose T is ergodic. Let H denote the closure of the subgroup $\{a^n\}_{-\infty}^{\infty}$ of G. If $H \neq G$ then by (7) of §0.7 there exists $\gamma \in \hat{G}$ with $\gamma \not\equiv 1$ but $\gamma(h) = 1 \ \forall h \in H$. Then $\gamma(Tx) = \gamma(ax) = \gamma(a)\gamma(x) = \gamma(x)$, and this contradicts ergodicity of T. Therefore $H = G$. (If G is metric we could have used Theorem 1.7 instead of the above proof.) Conversely, suppose $\{a^n\}_{n \in Z}$ is dense in G. This implies G is abelian. Let $f \in L^2(m)$ and $f \circ T = f$. By (9) of §0.7 f can be represented as a Fourier series $\sum_i b_i \gamma_i$, where $\gamma_i \in \hat{G}$. Then $\sum_i b_i \gamma_i(a)\gamma_i(x) = \sum_i b_i \gamma_i(x)$ so that if $b_i \neq 0$ then $\gamma_i(a) = 1$ and, since $\gamma_i(a^n) = \gamma_i(a))^n = 1$, $\gamma_i \equiv 1$. Therefore only the constant term of the Fourier series of f can be non-zero, i.e., f is constant a.e. Theorem 1.6(v) gives that T is ergodic. $\qquad\Box$

(4) Let G be a compact group and $A: G \rightarrow G$ be a continuous endomorphism of G onto G. We know that A preserves Haar measure m. We first consider the special case of endomorphisms $A(z) = z^p$ of the unit circle K. We shall show $A(z) = z^p$ is ergodic if $|p| > 1$. Suppose $f \in L^2(m)$ and $f \circ A = f$. If $f(z)$ has Fourier series $f(z) = \sum_{n=-\infty}^{\infty} a_n z^n$ then $f(Az) = \sum_{n=-\infty}^{\infty} a_n z^{pn}$. Therefore $a_n = a_{pn} = a_{p^2n} = a_{p^3n} = \cdots$ so that if $n \neq 0$ we have $a_n = 0$ because the Fourier coefficients must satisfy $\sum_{j=-\infty}^{\infty} |a_j|^2 < \infty$. Only the constant term of the Fourier series can be non-zero so f is constant a.e. Therefore A is ergodic.

In the case of a general compact abelian group we have the following result due independently to Rohlin and Halmos.

Theorem 1.10. *If G is a compact abelian group (equipped with normalized Haar measure) and $A: G \rightarrow G$ is a surjective continuous endomorphism of G then A is ergodic iff the trivial character $\gamma \equiv 1$ is the only $\gamma \in \hat{G}$ that satisfies $\gamma \circ A^n = \gamma$ for some $n > 0$.*

PROOF. Suppose that whenever $\gamma A^n = \gamma$ for some $n \geq 1$ we have $\gamma \equiv 1$. Let $f \circ A = f$ with $f \in L^2(m)$. Let $f(x)$ have the Fourier series $\sum a_n \gamma_n$ where $\gamma_n \in \hat{G}$

and $\sum |a_n|^2 < \infty$. Then $\sum a_n \gamma_n(Ax) = \sum a_n \gamma_n(x)$, so that if $\gamma_n, \gamma_n \circ A, \gamma_n \circ A^2, \ldots,$ are all distinct their coefficients are equal and therefore zero. So if $a_n \neq 0$, $\gamma_n(A^p) = \gamma_n$ for some $p > 0$. Then $\gamma_n \equiv 1$ by assumption and so f is constant a.e. Therefore A is ergodic by Theorem 1.6(v).

Conversely let A be ergodic and $\gamma A^n = \gamma$, $n > 0$. If n is the least such integer, $f = \gamma + \gamma A + \cdots + \gamma A^{n-1}$ is invariant under A and not a.e. constant (being the sum of orthogonal functions), contradicting Theorem 1.6(v). □

We are especially interested in the case of the n-torus K^n. Recall from §0.8 that a surjective endomorphism $A: K^n \to K^n$ is given by an $n \times n$ matrix $[A]$ of integers and that $\hat{K}^n$ can be identified with Z^n and the induced action $\hat{A}: \hat{K}^n \to \hat{K}^n$ corresponds to the action of the transpose matrix $[A]_t$ on Z^n.

Corollary 1.10.1. *Let $A: K^n \to K^n$ be a surjective continuous endomorphism of the n-torus. Then A is ergodic iff the matrix $[A]$ has no roots of unity as eigenvalues.*

PROOF. If A is not ergodic Theorem 1.10 gives the existence of $q \in Z^n$ $q \neq 0$ and $k > 0$ with $[A]_t^k q = q$. Then $[A]_t^k$ has 1 as an eigenvalue so that $[A]_t$, and hence $[A]$, has a k-th root of unity as an eigenvalue.

Conversely if $[A]_t$ has a k-th root of unity as an eigenvalue then $[A]_t^k$ has 1 as an eigenvalue. Therefore $([A]_t^k - I)(y) = 0 \in R^n$ for some $y \in R^n$, and since the matrix $[A]_t$ has integral entries we can find such a $y \in Z^n$. Hence $[A]_t^k y = y$ and A is not ergodic by Theorem 1.10. □

When G is not abelian similar necessary and sufficient conditions for ergodicity of an endomorphism can be stated in terms of the irreducible unitary representations of G. (When G is abelian these representations are the characters of G.) This has been studied by Kaplansky (Kaplansky [1]).

(5) For affine transformations of compact metric groups necessary and sufficient conditions for ergodicity are known. The simplest case is when G is a compact, connected, metric, abelian group.

Theorem 1.11. *If $T(x) = a \cdot A(x)$ is an affine transformation of the compact, connected, metric, abelian group G then the following are equivalent:*

   (i) *T is ergodic (relative to Haar measure).*
   (ii) (a) *whenever $\gamma \circ A^k = \gamma$ for $k > 0$ then $\gamma \circ A = \gamma$; and*
        (b) *the smallest closed subgroup containing a and BG (where $Bx = x^{-1} \cdot A(x)$) is G(i.e., $[a, BG] = G$).*
   (iii) *$\exists x_0 \in G$ with $\{T^n(x_0): n \geq 0\}$ dense in G.*
   (iv) *$m(\{x: \{T^n x: n \geq 0 \text{ is dense}\}\}) = 1$.*

(Note that conditions (a) and (b) reduce to the conditions given in (3) and (4) in the special cases. The equivalence of (i) and (ii) was investigated by Hahn, Hoare and Parry.)

PROOF. First note that B is an endomorphism of G (but maybe nonsurjective) and commutes with A.

(ii) $\Rightarrow$ (i). Suppose (a) and (b) of (ii) hold. If $f \circ T = f$, $f \in L^2(m)$, let $f = \sum b_i \gamma_i$, $\gamma_i \in \hat{G}$, be the Fourier series of f. Then

$$\sum_i b_i \gamma_i(a) \gamma_i(Ax) = \sum_i b_i \gamma_i(x). \tag{*}$$

If γ_i, $\gamma_i \circ A$, $\gamma_i \circ A^2, \ldots,$ are all distinct then $b_i = 0$ or else $\sum |b_i|^2 < \infty$ is violated. Hence, if $b_i \neq 0$ then $\gamma_i \circ A^n = \gamma_i$ for some $n > 0$, and by (a) $\gamma \circ A = \gamma$. But then (*) implies $\gamma_i(a) = 1$ and so $\gamma_i(x) = 1 \; \forall x \in [a, BG]$ and by (b) $\gamma_i = 1$. So f is constant a.e. Therefore T is ergodic by Theorem 1.6(v).

(i) $\Rightarrow$ (iv). This follows by Theorem 1.7.

(vi) $\Rightarrow$ (iii) is trivial.

(iii) $\Rightarrow$ (ii). It remains to show that if $\exists x_0 \in G$ with $\{T^n x_0 : n \geq 0\}$ dense in G then conditions (a) and (b) of (ii) hold. Suppose $\gamma \circ A^k = \gamma$, $k \geq 1$, $\gamma \in \hat{G}$. Let $\gamma_1 = \gamma \circ B$. Then $\gamma_1(T^k x) = \gamma_1(a \cdot Aa \cdot \cdots \cdot A^{k-1}a)\gamma_1(A^k x) = \gamma(a^{-1}A^k a)\gamma_1(x) = \gamma_1(x)$. Hence γ_1 assumes only the finite number of values $\gamma_1(x_0), \gamma_1(Tx_0), \ldots,$ $\gamma_1(T^{k-1}x_0)$ on the dense set $\{T^n x_0 : n \geq 0\}$ and hence assumes only these values on G. Since G is connected γ_1 must be constant, and so $\gamma_1 = 1$. Hence $\gamma A = \gamma$ and condition (a) holds.

If $[a, BG] \neq G \; \exists \gamma \neq 1$, $\gamma \in \hat{G}$, with $\gamma(a) = 1$ and $\gamma(Bx) = 1$ (see (7) of §0.7). Then $\gamma(Tx) = \gamma(x)$ and so γ assumes only the value $\gamma(x_0)$ on the dense set $\{T^n x_0 : n \geq 0\}$ and therefore γ is a constant. Hence $\gamma = 1$, a contradiction, and we have shown that (iii) implies (b). $\qquad\square$

When G is K^n the equivalence of (i) and (ii) becomes: $T = a \cdot A$ is ergodic iff

(a) the matrix $[A]$ has no proper roots of unity (i.e., other than 1) as eigenvalues, and

(b) $[a, BK^n] = K^n$.

This is easily proved by a method similar to the one used in (4) for the endomorphism case.

Conditions for ergodicity of affine transformations of compact non-abelian groups may be found in Chu [1].

(6) We now consider two-sided shift transformations.

Theorem 1.12. *The two-sided $(p_0, \ldots, p_{k-1})$ shift is ergodic.*

PROOF. Let $\mathscr{A}$ denote the algebra of all finite unions of measurable rectangles. Suppose $T^{-1}E = E$, $E \in \mathscr{B}$. Let $\varepsilon > 0$ be given, and choose $A \in \mathscr{A}$ with $m(E \triangle A) < \varepsilon$. Then

$$|m(E) - m(A)| = |m(E \cap A) + m(E \backslash A) - m(A \cap E) - m(A \backslash E)|$$
$$< m(E \backslash A) + m(A \backslash E) < \varepsilon.$$

Choose n_0 so large that $B = T^{-n_0}A$ depends upon different coordinates from A. Then $m(B \cap A) = m(B)m(A) = m(A)^2$ because m is a product measure. We have

$$m(E \triangle B) = m(T^{-n}E \triangle T^{-n}A) = m(E \triangle A) < \varepsilon,$$

and since $E \triangle (A \cap B) \subset (E \triangle A) \cup (E \triangle B)$ we have $m(E \triangle (A \cap B)) < 2\varepsilon$. Hence

$$|m(E) - m(A \cap B)| < 2\varepsilon$$

and

$$
\begin{aligned}
|m(E) - m(E)^2| &\leq |m(E) - m(A \cap B)| + |m(A \cap B) - m(E)^2| \\
&< 2\varepsilon + |m(A)^2 - m(E)^2| \\
&\leq 2\varepsilon + m(A)|m(A) - m(E)| + m(E)|m(A) - m(E)| \\
&< 4\varepsilon.
\end{aligned}
$$

Since ε is arbitrary $m(E) = m(E)^2$ which implies that $m(E) = 0$ or 1. $\square$

(7) By a similar argument, we see that the 1-sided $(p_0, \ldots, p_{k-1})$-shift is ergodic.

(8) We have the following theorem for the $(\mathbf{p}, P)$ Markov shift.

Theorem 1.13. *If T is the $(\mathbf{p}, P)$ Markov shift (either one-sided or two-sided) then T is ergodic iff the matrix P is irreducible (i.e. $\forall i, j \; \exists n > 0$ with $p_{ij}^{(n)} > 0$ where $p_{i,j}^{(n)}$ is the (i,j)-th entry of the matrix P^n).*

We shall give the proof of this theorem in §1.7 (Theorem 1.19) after we have used the ergodic theorem to derive another way of expressing ergodicity. We shall then only have to check a condition on measurable rectangles.

As we mentioned earlier ergodic transformations are the "irreducible" measure-preserving transformations and we would like every measure-preserving transformation to be built out of ergodic ones. To get some idea how this ergodic decomposition of a given transformation may be formulated consider a map T of a cylinder $X = [0, 1] \times K$ given by $T(x, z) = (x, az)$ where $a \in K$ is not a root of unity. In other words T is the direct product of the identity map, I, of $[0, 1]$ and the rotation $Sz = az$ of K. So T maps each circular vertical section of the cylinder to itself and acts on each section by S. We can think of $[0, 1] \times K$ as partitioned into the

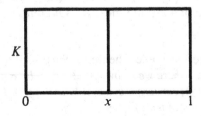

sets $\{x\} \times K$ and we can consider the direct product measure $m_1 \times m_2$, where m_1 is Lebesgue measure on $[0,1]$ and m_2 is Haar measure on K, as being decomposed into copies of m_2 on each element $\{x\} \times K$ of the partition. In other words we have a partition ζ of X on each element of which $m_1 \times m_2$ induces a probability measure and T induces a transformation which is ergodic relative to the induced measure. This is the ergodic decomposition of T. It turns out that this procedure can always be performed when $(X, \mathcal{B}, m)$ is a nice measure space, namely a Lebesgue space. This type of space will be discussed more in Chapter 2. It turns out that if X is a complete separable metric space, m is a probability measure on the σ-algebra $\mathcal{B}(X)$ of Borel subsets of X and $\mathcal{B}$ is the completion of $\mathcal{B}(X)$ by m then $(X, \mathcal{B}, m)$ is a Lebesgue space. Suppose T is a measure-preserving transformation of a Lebesgue space $(X, \mathcal{B}, m)$. Let $\mathcal{I}(T)$ be the σ-algebra consisting of all measurable sets B with $T^{-1}B = B$. The theory of Lebesgue spaces determines a partition ζ of $(X, \mathcal{B}, m)$ such that if $\mathcal{B}(\zeta)$ denotes the smallest σ-algebra containing all members of ζ then $\mathcal{B}(\zeta) \doteq \mathcal{I}(T)$ in the sense that each element of one σ-algebra differs only by a null set from an element of the other. Moreover $TC \subset C$ for each element C of ζ. Also, the measure m can be decomposed into probability measures m_c on the elements C of ζ. The transformation $T|C$ turns out to be ergodic relative to the measure m_c. This decomposition is essentially unique. A full account can be found in Rohlin's article [2].

When $T: K \to K$ is a rotation, $Tz = az$, where a is a root of unity (say $a^p = 1$ and $p > 0$ is the smallest such) then the partition ζ is the partition into cosets of the subgroup $H = \{1, a, a^2, \ldots, a^{p-1}\}$ of K. If $C = zH$ is a typical element of ζ the induced measure on C gives equal weight to each point za^i in C and $T|C$ maps za^i to za^{i+1}.

§1.6 The Ergodic Theorem

The first major result in ergodic theory was proved in 1931 by G. D. Birkhoff. We shall state it for a measure-preserving transformation of a σ-finite measure space. A σ-finite measure on a measurable space $(X, \mathcal{B})$ is a map $m: \mathcal{B} \to R^+ \cup \{\infty\}$ such that $m(\varnothing) = 0$, $m(\bigcup_{n=1}^{\infty} B_n) = \sum_{n=1}^{\infty} m(B_n)$ whenever $\{B_n\}$ is a sequence of members of $\mathcal{B}$ which are pairwise disjoint subsets of X, and there is a countable collection $\{A_n\}_1^{\infty}$ of elements of $\mathcal{B}$ with $m(A_n) < \infty$ for all n and $\bigcup_{n=1}^{\infty} A_n = X$. The Lebesgue measure on $(R^n, \mathcal{B}(R^n))$ provides an example of a σ-finite measure. Of course any probability measure is σ-finite.

Theorem 1.14 (Birkhoff Ergodic Theorem). *Suppose* $T: (X, \mathcal{B}, m) \to (X, \mathcal{B}, m)$ *is measure-preserving (where we allow* $(X, \mathcal{B}, m)$ *to be* σ-*finite) and* $f \in L^1(m)$. *Then* $(1/n) \sum_{i=0}^{n-1} f(T^i(x))$ *converges a.e. to a function* $f^* \in L^1(m)$. *Also* $f^* \circ T = f^*$ *a.e. and if* $m(X) < \infty$, *then* $\int f^* \, dm = \int f \, dm$.

Remark. If T is ergodic then f^* is constant a.e. and so if $m(x) < \infty$ $f^* = (1/m(X)) \int f \, dm$ a.e. If $(X, \mathcal{B}, m)$ is a probability space and T is ergodic we have $\forall f \in L^1(m) \lim_{n \to \infty} (1/n) \sum_{i=0}^{n-1} f(T^i x) = \int f \, dm$ a.e.

We shall give the proof at the end of this section after we have discussed some motivation for the ergodic theorem and some applications of it.

(i) Let $(X, \mathcal{B}, m)$ be a probability space and let $T: X \to X$ be measure-preserving. Let $E \in \mathcal{B}$. For $x \in X$ we could ask with what frequency do the elements of the set $\{x, T(x), T^2(x), \ldots\}$ lie in the set E?

Clearly $T^i(x) \in E$ iff $\chi_E T^i(x) = 1$, so the number of elements of $\{x, T(x), \ldots, T^{n-1}(x)\}$ in E is $\sum_{k=0}^{n-1} \chi_E T^k(x)$. The relative number of elements of $\{x, T(x), \ldots, T^{n-1}(x)\}$ in E equals $(1/n) \sum_{i=0}^{n-1} \chi_E T^k(x)$ and if T is ergodic then $(1/n) \sum_{i=0}^{n-1} \chi_E T^i(x) \to m(E)$ a.e. by the ergodic theorem. Thus the orbit of almost every point of X enters the set E with asymptotic relative frequency $m(E)$.

(ii) Let T be a measure-preserving transformation of the probability space $(X, \mathcal{B}, m)$ and let $f \in L^1(m)$. We define the time mean of f at x to be

$$\lim_{n \to \infty} \frac{1}{n} \sum_{i=0}^{n-1} f(T^i(x)) \text{ if the limit exists.}$$

The phase or space mean of f is defined to be

$$\int_X f(x) \, dm.$$

The ergodic theorem implies these means are equal a.e. for all $f \in L^1(m)$ iff T is ergodic. Since these two means are equated in some arguments in statistical mechanics it is important to verify ergodicity for certain transformations arising in physics. This application to time means and space means is more realistic in the case of a 1-parameter flow $\{T_t\}_{t \in R}$ of measure-preserving transformations. The ergodic theorem then asserts $\lim_{T \to \infty} (1/T) \int_0^T f(T_t x) \, dt$ exists a.e. for $f \in L^1(m)$ and equals $\int_X f \, dm$ if the flow $\{T_t\}$ is ergodic and $(X, \mathcal{B}, m)$ is a probability space.

(iii) We now illustrate how the ergodic theorem gives rise to results in number theory.

Theorem 1.15 (Borel's Theorem on Normal Numbers). *Almost all numbers in $[0, 1)$ are normal to base 2, i.e., for a.e. $x \in [0, 1)$ the frequency of 1's in the binary expansion of x is $\frac{1}{2}$.*

PROOF. Let $T: [0, 1) \to [0, 1)$ be defined by $T(x) = 2x \bmod 1$. We know that T preserves Lebesgue measure m and is ergodic, by Example 4 at the end of §1.5. Let Y denote the set of points of $[0, 1)$ that have a unique binary expansion. Then Y has a countable complement so $m(Y) = 1$.

Suppose $x = a_1/2 + a_2/2^2 + \cdots$ has a unique binary expansion. Then

$$T(x) = T\left(\frac{a_1}{2} + \frac{a_2}{2^2} + \frac{a^3}{2^3} + \cdots\right) = \frac{a_2}{2} + \frac{a_3}{2^2} + \cdots.$$

Let $f(x) = \chi_{[1/2,1)}(x)$. Then

$$f(T^i(x)) = f\left(\frac{a_{i+1}}{2} + \frac{a_{i+2}}{2^2} + \cdots\right) = \begin{cases} 1 & \text{iff } a_{i+1} = 1 \\ 0 & \text{iff } a_{i+1} = 0. \end{cases}$$

Hence if $x \in Y$ the number of 1's in the first n digits of the dyadic expansion of x is $\sum_{i=0}^{n-1} f(T^i(x))$. Dividing both sides of this equality by n and applying the ergodic theorem we see that

$$\frac{1}{n} \sum_{i=0}^{n-1} f(T^i x) \to \int \chi_{[1/2,1)} \, dm = \frac{1}{2} \quad \text{a.e.}$$

This says the frequency of 1's in the binary expansion of almost all points is $\frac{1}{2}$. □

The ergodic theorem can be applied to give other number theoretic results. Some are obtained in Billingsley [1] and Avez-Arnold [1].

Before moving to the proof of Theorem 1.14 we present some corollaries of it.

Corollary 1.14.1 (L^p Ergodic Theorem of Von Neumann). *Let $1 \le p < \infty$ and let T be a measure-preserving transformation of the probability space $(X, \mathcal{B}, m)$. If $f \in L^p(m)$ there exists $f^* \in L^p(m)$ with $f^* \circ T = f^*$ a.e. and $\left\|(1/n)\sum_{i=0}^{n-1} f(T^i x) - f^*(x)\right\|_p \to 0$.*

PROOF. If g is bounded and measurable then $g \in L^p$ and by the ergodic theorem we have that $(1/n)\sum_{i=0}^{n-1} g(T^i x) \to g^*(x)$ a.e. Clearly $g^* \in L^\infty(m)$ and hence $g^* \in L^p(m)$. Also $|(1/n)\sum_{i=0}^{n-1} g(T^i x) - g^*(x)|^p \to 0$ a.e. and by the bounded convergence theorem $\left\|(1/n)\sum_{i=0}^{n-1} g(T^i x) - g^*(x)\right\|_p \to 0$. If $\varepsilon > 0$ we can choose $N(\varepsilon, g)$ such that if $n > N(\varepsilon, g)$ and $k > 0$ then

$$\left\| \frac{1}{n} \sum_{i=0}^{n-1} g(T^i x) - \frac{1}{n+k} \sum_{i=0}^{n+k-1} g(T^i x) \right\|_p < \varepsilon.$$

Let $f \in L^p(m)$ and $S_n(f)(x) = (1/n)\sum_{i=0}^{n-1} f(T^i x)$. We must show that $\{S_n(f)\}$ is a Cauchy sequence in $L^p(m)$. Note that $\|S_n(f)\|_p \le \|f\|_p$. Let $\varepsilon > 0$ and choose $g \in L^\infty(m)$ such that $\|f - g\|_p < \varepsilon/4$. Then

$$\|S_n f - S_{n+k} f\|_p \le \|S_n f - S_n g\|_p + \|S_n g - S_{n+k} g\|_p + \|S_{n+k} g - S_{n+k} f\|_p$$
$$\le \varepsilon/4 + \varepsilon/2 + \varepsilon/4 = \varepsilon$$

if $n > N(\varepsilon/2, g)$ and $k > 0$. Therefore $\{S_n(f)\}$ is a Cauchy sequence in $L^p(m)$ and hence $\|S_n f - f^*\|_p \to 0$ for some $f^* \in L^p(m)$. We have $f^* \circ T = f^*$ a.e. because

$$\left(\frac{n+1}{n}\right)(S_{n+1} f)(x) - (S_n f)(Tx) = \frac{f(x)}{n}. \qquad \square$$

The next corollary gives another form of the definition of ergodicity. It illustrates some of the power of the ergodic theorem because by assuming T is ergodic (i.e., $\forall A, B \in \mathscr{B}$ with $m(A) > 0, m(B) > 0$ there exists some $n \geq 1$ with $m(T^{-n}A \cap B) > 0$ (Theorem 1.5(iv)) we can actually conclude that $m(T^{-n}A \cap B)$ converges, in a Cesaro sense, to $m(A)m(B)$.

Corollary 1.14.2. *Let $(X, \mathscr{B}, m)$ be a probability space and let $T : X \to X$ be a measure-preserving transformation. Then T is ergodic iff $\forall A, B \in \mathscr{B}$*

$$\frac{1}{n} \sum_{i=0}^{n-1} m(T^{-i}A \cap B) \to m(A)m(B).$$

PROOF. Suppose T is ergodic. Putting $f = \chi_A$ in Theorem 1.14 gives $(1/n) \sum_{i=0}^{n-1} \chi_A(T^i(x)) \to m(A)$ a.e. Multiplying by χ_B gives

$$\frac{1}{n} \sum_{i=0}^{n-1} \chi_A(T^i(x))\chi_B \to m(A)\chi_B \text{ a.e.,}$$

and the dominated convergence theorem implies

$$\frac{1}{n} \sum_{i=0}^{n-1} m(T^{-i}A \cap B) \to m(A)m(B).$$

Conversely, suppose the convergence property holds. Let $T^{-1}E = E$, $E \in \mathscr{B}$. Put $A = B = E$ in the convergence property to get $(1/n) \sum_{i=0}^{n-1} m(E) \to m(E)^2$. Hence $m(E) = m(E)^2$ and $m(E) = 0$ or 1. $\qquad \square$

We now turn to the proof of Theorem 1.14. If T is a measure-preserving transformation of the probability space $(X, \mathscr{B}, m)$ then the operator U_T is defined and $U_T L^1(m) \subset L^1(m)$, $U_T L_R^1(m) \subset L_R^1(m)$ and $\|U_T f\|_1 = \|f\|_1 \forall f \in L^1(m)$. We shall need the following result which we shall apply to the operator U_T. Recall that an operator $U : L_R^1(m) \to L_R^1(m)$ is called positive if whenever $f \geq 0$ then also $Uf \geq 0$.

Theorem 1.16 (Maximal Ergodic Theorem). *Let $U : L_R^1(m) \to L_R^1(m)$ be a positive linear operator with $\|U\| \leq 1$. Let $N > 0$ be an integer and let $f \in L_R^1(m)$. Define $f_0 = 0$, $f_n = f + Uf + U^2f + \cdots + U^{n-1}f$ for $n \geq 1$, and $F_N = \max_{0 \leq n \leq N} f_n \geq 0$. Then $\int_{\{x : F_N(x) > 0\}} f \, dm \geq 0$.*

PROOF. (due to A. Garsia) Clearly $F_N \in L_R^1(m)$. For $0 \leq n \leq N$ we have $F_N \geq f_n$ so $UF_N \geq Uf_n$ by positivity, and hence $UF_N + f \geq f_{n+1}$. Therefore

$$UF_N(x) + f(x) \geq \max_{1 \leq n \leq N} f_n(x)$$

$$= \max_{0 \leq n \leq N} f_n(x) \quad \text{when } F_N(x) > 0$$

$$= F_N(x).$$

Thus $f \geq F_N - UF_N$ on $A = \{x : F_N(x) > 0\}$, so

$$\int_A f \, dm \geq \int_A F_N \, dm - \int_A UF_N \, dm$$

$$= \int_X F_N \, dm - \int_A UF_N \, dm \quad \text{since } F_N = 0 \text{ on } X \backslash A.$$

$$\geq \int_X F_N \, dm - \int_X UF_N \, dm \quad \text{since } F_N \geq 0 \text{ and hence } UF_N \geq 0.$$

$$\geq 0 \text{ since } \|U\| \leq 1. \qquad \square$$

Corollary 1.16.1. *Let* $T : X \to X$ *be measure-preserving. If* $g \in L^1_R(m)$ *and*

$$B_\alpha = \left\{ x \in X : \sup_{n \geq 1} \frac{1}{n} \sum_{i=0}^{n-1} g(T^i(x)) > \alpha \right\}$$

then

$$\int_{B_\alpha \cap A} g \, dm \geq \alpha m(B_\alpha \cap A)$$

if $T^{-1}A = A$ *and* $m(A) < \infty$.

PROOF. We first prove this result under the assumptions $m(X) < \infty$ and $A = X$. Let $f = g - \alpha$, then $B_\alpha = \bigcup_{N=0}^{\infty} \{x : F_N(x) > 0\}$ so that $\int_{B_\alpha} f \, dm > 0$ by Theorem 1.16 and therefore $\int_{B_\alpha} g \, dm \geq \alpha m(B_\alpha)$. In the general case we apply the above to $T \mid A$ to get $\int_{A \cap B_\alpha} g \, dm \geq \alpha m(A \cap B_\alpha)$. $\qquad \square$

PROOF OF THEOREM 1.14. We first assume $m(X) < \infty$. By considering real and imaginary parts it suffices to consider $f \in L^1_R(m)$. For such an f let $f^*(x) = \limsup_{n \to \infty} (1/n) \sum_{i=0}^{n-1} f(T^i x)$ and $f_*(x) = \liminf_{n \to \infty} (1/n) \sum_{i=0}^{n-1} f(T^i x)$. We have $f^* \circ T = f^*$ and $f_* \circ T = f_*$ because if $a_n(x) = (1/n) \sum_{i=0}^{n-1} f(T^i x)$ then $((n+1)/n) a_{n+1}(x) - a_n(Tx) = f(x)/n$. We have to show that $f^* = f_*$ a.e. and that they belong to $L^1(m)$.

For real numbers α, β let $E_{\alpha,\beta} = \{x \in X \mid f_*(x) < \beta \text{ and } \alpha < f^*(x)\}$. Since $\{x \mid f_*(x) < f^*(x)\} = \bigcup \{E_{\alpha,\beta} \mid \beta < \alpha \text{ and } \alpha, \beta \text{ both rational}\}$ we shall show $m(E_{\alpha,\beta}) = 0$ if $\beta < \alpha$ because then we shall have $f^* = f_*$ a.e. Clearly $T^{-1}E_{\alpha,\beta} = E_{\alpha,\beta}$ and if we put $B_\alpha = \{x \in X \mid \sup_{n \geq 1} (1/n) \sum_{i=0}^{n-1} f(T^i x) > \alpha\}$ then $E_{\alpha,\beta} \cap B_\alpha = E_{\alpha,\beta}$. From Corollary 1.16.1 we get

$$\int_{E_{\alpha,\beta}} f \, dm = \int_{E_{\alpha,\beta} \cap B_\alpha} f \, dm \geq \alpha m(E_{\alpha,\beta} \cap B_\alpha) = \alpha m(E_{\alpha,\beta}).$$

Therefore $\int_{E_{\alpha,\beta}} f \, dm \geq \alpha m(E_{\alpha,\beta})$.

If we replace f, α, β by $-f, -\beta, -\alpha$, respectively, then since $(-f)^* = -f_*$ and $(-f)_* = -f^*$ we get

$$\int_{E_{\alpha,\beta}} f \, dm \leq \beta m(E_{\alpha,\beta}).$$

Therefore $\alpha m(E_{\alpha,\beta}) \leq \beta m(E_{\alpha,\beta})$, so if $\beta < \alpha$ then $m(E_{\alpha,\beta}) = 0$. This gives $f^* = f_*$ a.e. as we explained above. Therefore $(1/n) \sum_{i=0}^{n-1} f(T^i x) \to f^*$ a.e.

To show $f^* \in L^1(m)$ we use the part of Faton's lemma that asserts $\lim g_n \in L^1(m)$ if $\{g_n\}$ is a (pointwise) convergent sequence of non-negative integrable functions with $\lim \inf \int g_n \, dm < \infty$. Let $g_n(x) = |(1/n) \sum_{i=0}^{n-1} f(T^i(x))|$. Then $\int g_n \, dm \leq \int |f| \, dm$ so we can apply the Faton lemma to assert $\lim_{n \to \infty} g_n(x) = |\lim_{n \to \infty} (1/n) \sum_{i=0}^{n-1} f(T^i x)| = |f^*|$ belongs to $L^1(m)$.

It remains to show that $\int f \, dm = \int f^* \, dm$ if $m(X) < \infty$. Let $D_k^n = \{x \in X : (k/n) \leq f^*(x) < (k+1)/n\}$ where $k \in Z$, $n \geq 1$. For each small $\varepsilon > 0$ we have $D_k^n \cap B_{(k/n)-\varepsilon} = D_k^n$ and by Corollary 1.16.1

$$\int_{D_k^n} f \, dm \geq \left(\frac{k}{n} - \varepsilon\right) m(D_k^n) \quad \text{so that} \quad \int_{D_k^n} f \, dm \geq \frac{k}{n} m(D_k^n).$$

Then

$$\int_{D_k^n} f^* \, dm \leq \frac{k+1}{n} m(D_k^n) \leq \frac{1}{n} m(D_k^n) + \int_{D_k^n} f \, dm$$

by the above inequality. Summing over k gives $\int_X f^* \, dm \leq (m(X)/n) + \int_X f \, dm$. Since this holds for all $n \geq 1$ we have $\int_X f^* \, dm \leq \int_X f \, dm$. Applying this to $-f$ instead of f gives $\int_X (-f)^* \, dm \leq \int_X -f \, dm$ so that $\int_X f_* \, dm \geq \int_X f \, dm$. But $f_* = f^*$ a.e. so $\int_X f^* \, dm = \int_X f \, dm$. This finishes the proof when $m(X) < \infty$.

When $m(X) = \infty$ the above proof is valid once we have shown that $m(E_{\alpha,\beta}) < \infty$ when $\beta < \alpha$. (We need to show this to apply Corollary 1.16.1.) Suppose firstly that $\alpha > 0$. Let $C \in \mathcal{B}$ be any set with $C \subset E_{\alpha,\beta}$ and $m(C) < \infty$. (Such a set exists because X is σ-finite.) Then $h = f - \alpha\chi_c \in L^1(m)$ so by the maximal ergodic theorem

$$\int_{\{x : H_N(x) > 0\}} (f - \alpha\chi_c) \, dm \geq 0 \qquad \text{for all } N \geq 1.$$

(The function H_N is associated to h by Theorem 1.16.) But $C \subset E_{\alpha,\beta} \subset \bigcup_{N=0}^{\infty} \{x : H_N(x) > 0\}$ and therefore $\int_X |f| \, dm \geq \alpha m(C)$. Hence $m(C) \leq (1/\alpha) \int_X |f| \, dm$ for each $C \in \mathcal{B}$ with $C \subset E_{\alpha,\beta}$ and $m(C) < \infty$. Since X is σ-finite we have $m(E_{\alpha,\beta}) < \infty$. If $\alpha \leq 0$ then $\beta < 0$ so we can apply the above with $-f$ and $-\beta$ instead of f and α to get $m(E_{\alpha,\beta}) < \infty$. $\square$

§1.7 Mixing

If T is a measure-preserving transformation of a probability space we have deduced from the ergodic theorem that T is ergodic iff $\forall A, B \in \mathcal{B}$,

$$\lim_{n \to \infty} \frac{1}{n} \sum_{i=0}^{n-1} m(T^{-i}A \cap B) = m(A)m(B).$$

We can make changes in the method of this convergence to give the following notions.

Definition 1.5. Let T be a measure-preserving transformation of a probability space $(X, \mathscr{B}, m)$.

  (i) *T is weak-mixing if* $\forall A, B \in \mathscr{B}$

$$\lim_{n \to \infty} \frac{1}{n} \sum_{i=0}^{n-1} |m(T^{-i}A \cap B) - m(A)m(B)| = 0.$$

  (ii) *T is strong-mixing if* $\forall A, B \in \mathscr{B}$

$$\lim_{n \to \infty} m(T^{-n}A \cap B) = m(A)m(B).$$

Remarks

  (1) Every strong-mixing transformation is weak-mixing and every weak-mixing transformation is ergodic. This is because if $\{a_n\}$ is a sequence of real numbers then $\lim_{n \to \infty} a_n = 0$ implies

$$\lim_{n \to \infty} \frac{1}{n} \sum_{i=0}^{n-1} |a_i| = 0$$

and this second condition implies

$$\lim_{n \to \infty} \frac{1}{n} \sum_{i=0}^{n-1} a_i = 0.$$

  (2) An example of an ergodic transformation which is not weak-mixing is given by a rotation $T(z) = az$ on the unit circle K. This will be proved at the end of this section but one can see it roughly as follows. If A and B are two small intervals on K then $T^{-i}A$ will be disjoint from B for at least half of the values of i so that $(1/n) \sum_{i=0}^{n-1} |m(T^{-i}A \cap B) - m(A)m(B)| \geq \frac{1}{2}m(A)m(B)$ for large n. From this one sees that, intuitively, a weak-mixing transformation has to do some "stretching."

  (3) There are examples of weak-mixing T which are not strong-mixing. Kakutani [1] has an example constructed by combinatorial methods and Maruyama [1] constructed an example using Gaussian processes. Chacon and Katok and Stepin ([1] p. 94) also have examples. If $(X, \mathscr{B}, m)$ is a probability space let $\tau(X)$ denote the collection of all invertible measure-preserving transformations of $(X, \mathscr{B}, m)$. If we topologize $\tau(X)$ with the "weak" topology (see Halmos [1]), the class of weak-mixing transformations is of second category while the class of strong-mixing transformations is of first category. So from the point of view of this topology most transformations are weak-mixing but not strong-mixing.

  (4) Intuitive descriptions of ergodicity and strong-mixing can be given as follows. To say T is strong-mixing means that for any set A the sequence of sets $T^{-n}A$ becomes, asymptotically, independent of any other set B. Ergodicity means $T^{-n}A$ becomes independent of B on the average, for each pair of sets $A, B \in \mathscr{B}$. We shall give a similar description of weak-mixing after Theorem 1.22.

The following theorem gives a way of checking the mixing properties for examples by reducing the computations to a class of sets we can manipulate with. For example, it implies we need only consider measurable rectangles when dealing with the mixing properties of shifts.

Theorem 1.17. *Let $(X, \mathscr{B}, m)$ be a measure space and let $\mathscr{S}$ be a semi-algebra that generates $\mathscr{B}$. Let $T: X \to X$ be a measure-preserving transformation. Then*

(i) *T is ergodic iff $\forall A, B \in \mathscr{S}$*

$$\lim_{n \to \infty} \frac{1}{n} \sum_{i=0}^{n-1} m(T^{-i}A \cap B) = m(A)m(B),$$

(ii) *T is weak-mixing iff $\forall A, B \in \mathscr{S}$*

$$\lim_{n \to \infty} \frac{1}{n} \sum_{i=0}^{n-1} \left| m(T^{-i}A \cap B) - m(A)m(B) \right| = 0, \quad and$$

(iii) *T is strong-mixing iff $\forall A, B \in \mathscr{S}$*

$$\lim_{n \to \infty} m(T^{-n}A \cap B) = m(A)m(B).$$

PROOF. Since each member of the algebra, $\mathscr{A}(\mathscr{S})$, generated by $\mathscr{S}$ can be written as a finite disjoint union of members of $\mathscr{S}$ it follows that if any of the three convergence properties hold for all members of $\mathscr{S}$ then they hold for all members of $\mathscr{A}(\mathscr{S})$.

Let $\varepsilon > 0$ be given and let $A, B \in \mathscr{B}$. Choose $A_0, B_0 \in \mathscr{A}(\mathscr{S})$ with $m(A \triangle A_0) < \varepsilon$ and $m(B \triangle B_0) < \varepsilon$. For any $i \geq 0, (T^{-i}A \cap B) \triangle (T^{-i}A_0 \cap B_0) \subset (T^{-i}A \triangle T^{-i}A_0) \cup (B \triangle B_0)$, so we have $m((T^{-n}A \cap B) \triangle (T^{-i}A_0 \cap B_0)) < 2\varepsilon$, and therefore $\left| m(T^{-i}A \cap B) - m(T^{-i}A_0 \cap B_0) \right| < 2\varepsilon$. Therefore

$$\begin{aligned}
\left| m(T^{-i}A \cap B) - m(A)m(B) \right| &\leq \left| m(T^{-i}A \cap B) - m(T^{-i}A_0 \cap B_0) \right| \\
&+ \left| m(T^{-i}A_0 \cap B_0) - m(A_0)m(B_0) \right| \\
&+ \left| m(A_0)m(B_0) - m(A)m(B_0) \right| \\
&+ \left| m(A)m(B_0) - m(A)m(B) \right| \\
&< 4\varepsilon + \left| m(T^{-i}A_0 \cap B_0) - m(A_0)m(B_0) \right|.
\end{aligned}$$

This inequality together with the known behaviour of the righthand term proves (ii) and (iii). To prove (i) one can easily obtain

$$\left| \frac{1}{n} \sum_{i=0}^{n-1} m(T^{-i}A \cap B) - m(A)m(B) \right| < 4\varepsilon$$

$$+ \left| \frac{1}{n} \sum_{i=0}^{n-1} m(T^{-i}A_0 \cap B_0) - m(A_0)m(B_0) \right|$$

and then use the known behaviour of the right-hand side. □

As an application of this result we shall prove the result about ergodicity of Markov shifts mentioned in §1.5. To do this we shall use the following

Lemma 1.18. *Let P be a stochastic matrix, having a strictly positive probability vector $\mathbf{p}$ with $\mathbf{p}P = \mathbf{p}$. Then $Q = \lim_{N \to \infty} (1/N) \sum_{n=0}^{N-1} P^n$ exists. The matrix Q is also stochastic and $QP = PQ = Q$. Any eigenvector of P for the eigenvalue 1 is also an eigenvector of Q. Also $Q^2 = Q$.*

PROOF. Let m denote the $(\mathbf{p}, P)$ Markov measure and T be two-sided $(\mathbf{p}, P)$ Markov shift. Let χ_i denote the characteristic function of the cylinder $_0[i]_0 = \{(x_n)_{-\infty}^\infty \mid x_0 = i\}$. By Birkhoff's ergodic theorem $(1/N) \sum_{n=0}^{N-1} \chi_j(T^n x) \to \chi_j^*(x)$ a.e., and by multiplying by $\chi_i(x)$ and using the dominated convergence theorem we have $(1/N) \sum_{n=0}^{N-1} p_i p_{ij}^{(n)} \to \int \chi_j^*(x)\chi_i(x)\,dm(x)$. So $Q = (q_{ij})$ is given by $q_{ij} = (1/p_i)\int \chi_j^*(x)\chi_i(x)\,dm(x)$. The other properties are clear. ☐

Theorem 1.19. *Let T denote the $(\mathbf{p}, P)$ Markov shift (either one-sided or two-sided). We can assume $p_i > 0$ for each i where $\mathbf{p} = (p_0, \ldots, p_{k-1})$. Let Q be the matrix obtained in Lemma 1.18. The following are equivalent:*

(i) *T is ergodic.*
(ii) *All rows of the matrix Q are identical.*
(iii) *Every entry in Q is strictly positive.*
(iv) *P is irreducible.*
(v) *1 is a simple eigenvalue of P.*

PROOF. Let m denote the $(\mathbf{p}, P)$ Markov measure.

(i) $\Rightarrow$ (ii). As in the proof of Lemma 1.18 Birkhoff's ergodic theorem gives $(1/N)\sum_{n=0}^{N-1} m(_0[i]_0 \cap {}_n[j]_n) = p_i q_{ij}$. Since T is ergodic the limit is $m(_0[i]_0)$. $m(_n[j]_n) = p_i p_j$. Therefore $q_{ij} = p_j$ and so the rows of Q are identical.

(ii) $\Rightarrow$ (iii). If the rows of Q are identical then $\mathbf{p}Q = \mathbf{p}$ implies $q_{ij} = p_j$ and so $q_{ij} > 0$.

(iii) $\Rightarrow$ (iv). Fix i, j. Since $(1/N) \sum_{i=0}^{N-1} p_{ij}^{(n)} \to q_{ij} > 0$, then $p_{ij}^{(n)} > 0$ for some n.

(iv) $\Rightarrow$ (iii). Fix i and let S_i denote the collection of states j with $q_{ij} > 0$. Since $Q = QP$ we have $q_{ij} \geq q_{il}p_{lj}$ for each l. Therefore if $l \in S_i$ and $p_{lj} > 0$ then $q_{ij} > 0$ so $j \in S_i$. This implies that if $l \in S_i$ then $\sum_{j \in S_i} p_{lj} = 1$. Since P is irreducible we must have that S_i is the whole state space and so $q_{ij} > 0$ for all j.

(iii) $\Rightarrow$ (ii). Fix j and put $q_j = \max_i q_{ij}$. We know $Q^2 = Q$. If $q_{ij} < q_j$ for some i then

$$q_{lj} = \sum_i q_{li}q_{ij} < \sum_i q_{li}q_j = q_j \quad \text{for all } l,$$

and this contradicts the definition of q_j. Hence $q_{ij} = q_j$ for all i.

(ii) $\Rightarrow$ (i). To show T is ergodic it suffices (by Theorem 1.17) to show that for any two blocks $A = {}_a[i_0, \ldots, i_r]_{a+r}, B = {}_b[j_0, \ldots, j_s]_{b+s}$ we have

$\lim_{N \to \infty} (1/N) \sum_{n=0}^{N-1} m(T^{-n}A \cap B) = m(A)m(B)$. For $n > b + s - a$ we have

$$m(T^{-n}A \cap B) = p_{j_0} p_{j_0 j_1} \cdots p_{j_s j_{s-1}} p_{j_s i_0}^{(a+n-b-s)} p_{i_0 i_1} \cdots p_{i_{r-1} i_r}$$

and since we know that (ii) implies $q_{ij} = p_j$ this gives

$$\lim_{N \to \infty} \frac{1}{N} \sum_{n=0}^{N-1} m(T^{-n}A \cap B) = p_{j_0} p_{j_0 j_1} \cdots p_{j_{s-1} j_s} p_{i_0} p_{i_0 i_1} \cdots p_{i_{r-1} i_r}$$

$$= m(A)m(B).$$

(ii) $\Rightarrow$ (v). We know (ii) implies $q_{ij} = p_j$ so that the only left eigenvectors of Q for the eigenvalue 1 are multiples of **p**. By Lemma 1.18 there are also the only left eigenvectors of P for the eigenvalue 1.

(v) $\Rightarrow$ (ii). Suppose 1 is a simple eigenvalue of P. Since $Q = QP$ each row of Q is a left eigenvector and so they are identical. $\qquad \square$

The condition (v) gives a practical way to test ergodicity. We shall use Theorem 1.17 at the end of this section to find necessary and sufficient conditions for a Markov shift to have the mixing properties (Theorem 1.31).

We shall use the following result, about sequences of real numbers, to obtain other formulations of weak-mixing.

Theorem 1.20. *If $\{a_n\}$ is a bounded sequence of real numbers then the following are equivalent*:

(i)
$$\lim_{n \to \infty} \frac{1}{n} \sum_{i=0}^{n-1} |a_i| = 0.$$

(ii) *There exists a subset J of Z^+ of density zero (i.e.,*

$$\left(\frac{\text{cardinality } (J \cap \{0, 1, \ldots, n-1\})}{n} \right) \to 0),$$

such that $\lim_n a_n = 0$ provided $n \notin J$.

(iii) $\displaystyle \lim_{n \to \infty} \frac{1}{n} \sum_{i=0}^{n-1} |a_i|^2 = 0.$

PROOF. If $M \subset Z^+$ let $\alpha_M(n)$ denote the cardinality of $\{0, 1, \ldots, n-1\} \cap M$.

(i) $\Rightarrow$ (ii). Let $J_k = \{n \in Z^+ : |a_n| \geq 1/k\}$ $(k > 0)$. Then $J_1 \subset J_2 \subset \cdots$. Each J_k has density zero since

$$\frac{1}{n} \sum_{i=0}^{n-1} |a_i| \geq \frac{1}{n} \frac{1}{k} \alpha_{J_k}(n).$$

Therefore there exist integers $0 = l_0 < l_1 < l_2 < \cdots$ such that for $n \geq l_k$,

$$\frac{1}{n} \alpha_{J_{k+1}}(n) < \frac{1}{k+1}.$$

Set $J = \bigcup_{k=0}^{\infty} [J_{k+1} \cap [l_k, l_{k+1})]$. We now show that J has density zero. Since $J_1 \subset J_2 \subset \cdots$, if $l_k \leq n < l_{k+1}$ we have

$$J \cap [0, n) = [J \cap [0, l_k)] \cup [J \cap [l_k, n)] \subset [J_k \cap [0, l_k)] \cup [J_{k+1} \cap [0, n)],$$

and therefore

$$\frac{1}{n} \alpha_J(n) \leq \frac{1}{n} [\alpha_{J_k}(l_k) + \alpha_{J_{k+1}}(n)] \leq \frac{1}{n} [\alpha_{J_k}(n) + \alpha_{J_{k+1}}(n)] < \frac{1}{k} + \frac{1}{k+1}.$$

Hence $(1/n)\alpha_J(n) \to 0$ as $n \to \infty$, and so J has density zero. If $n > l_k$ and $n \notin J$ then $n \notin J_{k+1}$ and therefore $|a_n| < 1/(k+1)$. Hence

$$\lim_{J \not\ni n \to \infty} |a_n| = 0.$$

(ii) $\Rightarrow$ (i). Suppose $|a_n| \leq K$ $\forall n$. Let $\varepsilon > 0$. There exists N_ε such that $n \geq N_\varepsilon$, $n \notin J$ imply $|a_n| < \varepsilon$, and such that $N \geq N_\varepsilon$ implies $(\alpha_J(n)/n) < \varepsilon$. Then $n \geq N_\varepsilon$ implies

$$\frac{1}{n} \sum_{i=0}^{n-1} |a_i| = \frac{1}{n} \left[\sum_{i \in J \cap \{0,1,\ldots,n-1\}} |a_i| + \sum_{i \notin J \cap \{0,1,\ldots,n-1\}} |a_i| \right]$$

$$< \frac{K}{n} \alpha_J(n) + \varepsilon < (K+1)\varepsilon.$$

(i) $\Rightarrow$ (iii). By the above it suffices to note that $\lim_{J \not\ni n \to \infty} |a_n| = 0$ iff $\lim_{J \not\ni n \to \infty} |a_n|^2 = 0$. $\square$

Theorem 1.21. *If T is a measure-preserving transformation of a probability space $(X, \mathscr{B}, m)$ the following are equivalent:*

(i) *T is weak-mixing.*

(ii) *For every pair of elements A, B of $\mathscr{B}$ there is a subset $J(A, B)$ of Z^+ of density zero such that*

$$\lim_{J(A,B) \not\ni n \to \infty} m(T^{-n}A \cap B) = m(A)m(B).$$

(iii) *For every pair of elements A, B of $\mathscr{B}$ we have*

$$\lim_{n \to \infty} \frac{1}{n} \sum_{i=0}^{n-1} |m(T^{-i}A \cap B) - m(A)m(B)|^2 = 0.$$

PROOF. Apply Theorem 1.20 with $a_n = m(T^{-n}A \cap B) - m(A)m(B)$. $\square$

We now show that for most useful measure spaces we can strengthen statement (ii) to obtain a set of density zero that works for all pairs of sets A and B.

Definition 1.6. The probability space $(X, \mathscr{B}, m)$ has a countable basis if there is a sequence $\{B_k\}_{k=1}^{\infty}$ of members of $\mathscr{B}$ such that for each $\varepsilon > 0$ and each $B \in \mathscr{B}$ there exists some B_k with $m(B \triangle B_k) < \varepsilon$.

This condition is equivalent to the condition that $L^2(m)$ has a countable dense subset (see §0.5).

If X is a metric space with a countable topological base and $\mathscr{B}$ is the σ-algebra of Borel subsets of X then $(X, \mathscr{B}, m)$ has a countable basis for any probability measure m on $(X, \mathscr{B})$. This follows from Theorem 6.1. This is also true if $\mathscr{B}$ is the completion, under m, of the σ-algebra of Borel subsets of X.

Theorem 1.22. *Let $(X, \mathscr{B}, m)$ be a probability space with a countable basis and let $T: X \to X$ be a measure-preserving transformation. Then T is weak-mixing if there is a subset J of Z^+ of density zero such that for all $A, B \in \mathscr{B}$* $\lim_{J \not\ni n \to \infty} m(T^{-n}A \cap B) = m(A)m(B)$.

PROOF. It suffices to prove that the stated condition holds if T is weak-mixing. Let $\{B_k\}^\infty$ be a countable basis for $(X, \mathscr{B}, m)$. Put

$$a_n = \sum_{i,j=1}^{\infty} \frac{\left| m(T^{-n}B_i \cap B_j) - m(B_i)m(B_j) \right|}{2^{i+j}}.$$

Since T is weak-mixing we have $(1/n)\sum_{l=0}^{n-1} a_l \to 0$ so by Theorem 1.20 there is a subset J of Z^+ of density zero such that $\lim_{J \not\ni n \to \infty} a_n = 0$. Therefore $\lim_{J \not\ni n \to \infty} m(T^{-n}B_i \cap B_j) = m(B_i)m(B_j)$ for all i, j, and the result follows by a simple approximation argument. $\qquad\square$

Remark. We can use Theorem 1.21 to give an intuitive description of weak-mixing. It means that for each set $A \in \mathscr{B}$ the sequence $T^{-n}A$ becomes independent of any other set $B \in \mathscr{B}$ provided we neglect a few instants of time.

The next result expresses the mixing concepts in functional form. This will be useful for checking whether examples have the mixing properties. Recall that U_T is defined on functions by $U_T f = f \circ T$.

Theorem 1.23. *Suppose $(X, \mathscr{B}, m)$ is a probability space and $T: X \to X$ is measure-preserving.*

(i) *The following are equivalent:*
  (1) *T is ergodic.*
  (2) *For all $f, g \in L^2(m)$ $\lim_{n \to \infty} (1/n)\sum_{i=0}^{n-1} (U_T^i f, g) = (f, 1)(1, g)$.*
  (3) *For all $f \in L^2(m)$ $\lim_{n \to \infty} (1/n)\sum_{i=0}^{n-1} (U_T^i f, f) = (f, 1)(1, f)$.*
(ii) *The following are equivalent:*
  (1) *T is weak-mixing.*
  (2) *For all $f, g \in L^2(m)$ $\lim_{n \to \infty} (1/n)\sum_{i=0}^{n-1} \left| (U_T^i f, g) - (f, 1)(1, g) \right| = 0$.*
  (3) *For all $f \in L^2(m)$ $\lim_{n \to \infty} (1/n)\sum_{i=0}^{n-1} \left| (U_T^i f, f) - (f, 1)(1, f) \right| = 0$.*
  (4) *For all $f \in L^2(m)$ $\lim_{n \to \infty} (1/n)\sum_{i=0}^{n-1} \left| (U_T^i f, f) - (f, 1)(1, f) \right|^2 = 0$.*
(iii) *The following are equivalent:*
  (1) *T is strong-mixing.*
  (2) *For all $f, g \in L^2(m)$, $\lim_{n \to \infty} (U_T^n f, g) = (f, 1)(1, g)$.*
  (3) *For all $f \in L^2(m)$, $\lim_{n \to \infty} (U_T^n f, f) = (f, 1)(1, f)$.*

PROOF. (i), (ii) and (iii) are proved using similar methods. We shall prove (iii) to illustrate the ideas. Slight modification of this proof will prove (i) and (ii).

(2) $\Rightarrow$ (1). This follows by putting $f = \chi_A$, $g = \chi_B$, for $A, B \in \mathscr{B}$.

(1) $\Rightarrow$ (3). We easily get that for any $A, B \in \mathscr{B}$, $(U_T^n \chi_A, \chi_B) \to (\chi_A, 1)(1, \chi_B)$. Fixing B, we get that $(U_T^n h, \chi_B) \to (h, 1)(1, \chi_B)$ for any simple function h. Then, fixing h, we get that $(U_T^n h, h) \to (h, 1)(1, h)$. So (3) is true for all simple functions.

Suppose $f \in L^2(m)$, and let $\varepsilon < 0$. Choose a simple function h with $\|f - h\|_2 < \varepsilon$, and choose $N(\varepsilon)$ so that $n \geq N(\varepsilon)$ implies

$$|(U_T^n h, h) - (h, 1)(1, h)| < \varepsilon.$$

Then if $n \geq N(\varepsilon)$

$$
\begin{aligned}
|(U_T^n f, f) - (f, 1)(1, f)| &\leq |(U_T^n f, f) - (U_T^n h, f)| + |(U_T^n h, f) \\
&\quad - (U_T^n h, h)| + |(U_T^n h, h) - (h, 1)(1, h)| + |(h, 1)(1, h) \\
&\quad - (f, 1)(1, h)| + |(f, 1)(1, h) - (f, 1)(1, f)| \\
&\leq |(U_T^n (f - h), f)| + |(U_T^n h, f - h)| \\
&\quad + \varepsilon + |(1, h)| |(h - f, 1)| + |(f, 1)| |(1, h - f)| \\
&\leq \|f - h\|_2 \|f\|_2 + \|f - h\|_2 \|h\|_2 + \varepsilon + \|h\|_2 \|f - h\|_2 \\
&\quad + \|f\|_2 \|h - f\|_2 \quad \text{by the Schwartz inequality} \\
&\leq \varepsilon \|f\|_2 + \varepsilon(\|f\|_2 + \varepsilon) + \varepsilon + (\|f\|_2 + \varepsilon)\varepsilon + \varepsilon \|f\|_2.
\end{aligned}
$$

Therefore $\lim_{n \to \infty} (U_T^n f, f) = (f, 1)(1, f)$.

(3) $\Rightarrow$ (2). Let $f \in L^2(m)$ and let $\mathscr{H}_f$ denote the smallest (closed) subspace of $L^2(m)$ containing f and the constant functions and satisfying $U_T \mathscr{H}_f \subset \mathscr{H}_f$. The set

$$\mathscr{F}_f = \left\{ g \in L^2(m) \colon \lim_{n \to \infty} (U_T^n f, g) = (f, 1)(1, g) \right\}$$

is a closed subspace of $L^2(m)$ and contains f and the constant functions. Since it is U_T invariant it contains $\mathscr{H}_f$. If $g \in \mathscr{H}_f^\perp$ then $(U_T^n f, g) = 0$ for $n \geq 0$ and $(1, g) = 0$ and therefore $\mathscr{H}_f^\perp \subset \mathscr{F}_f$. Hence $\mathscr{F}_f = L^2(m)$. $\square$

Remark. Another form of weak-mixing is in terms of sets of density zero. Let us suppose that $(X, \mathscr{B}, m)$ has a countable basis. Then T is weak-mixing iff there exists a subset J of Z^+ of density zero such that $\lim_{J \not\ni n \to \infty} (U_T^n f, g) = (f, 1)(1, g)$ for all $f, g \in L^2(m)$.

The next result connects weak-mixing of T with the ergodicity of $T \times T$.

Theorem 1.24. *If T is a measure-preserving transformation on a probability space $(X, \mathscr{B}, m)$ then the following are equivalent:*

(i)   *T is weak-mixing.*
(ii) *$T \times T$ is ergodic.*
(iii) *$T \times T$ is weak-mixing.*

PROOF. We first show (i) $\Rightarrow$ (iii). Let $A, B, C, D \in \mathscr{B}$. There exist subsets J_1, J_2 of Z^+ of density zero such that

$$\lim_{J_1 \not\ni n \to \infty} m(T^{-n}A \cap B) = m(A)m(B)$$

$$\lim_{J_2 \not\ni n \to \infty} m(T^{-n}C \cap D) = m(C)m(D).$$

Then

$$\lim_{J_1 \cup J_2 \not\ni n \to \infty} (m \times m)\{(T \times T)^{-n}(A \times C) \cap (B \times D)\}$$

$$= \lim_{J_1 \cup J_2 \not\ni n \to \infty} m(T^{-n}A \cap B)m(T^{-n}C \cap D)$$

$$= m(A)m(B)m(C)m(D)$$

$$= (m \times m)(A \times C)(m \times m)(B \times D).$$

By Theorem 1.20 we know $\lim_{n \to \infty} (1/n) \sum_{i=0}^{n-1} |(m \times m)\{(T \times T)^{-n}(A \times C) \cap (B \times D)\} - (m \times m)(A \times C)(m \times m)(B \times D)| = 0$. Since the measurable rectangles form a semi-algebra that generates $\mathscr{B} \times \mathscr{B}$ Theorem 1.17 asserts that $T \times T$ is weak-mixing.

It is clear that (iii) implies (ii).

We now show (ii) $\Rightarrow$ (i). Let $A, B \in \mathscr{B}$. We shall show $\lim_{n \to \infty} (1/n) \sum_{i=0}^{n-1} (m(T^{-i}A \cap B) - m(A)m(B))^2 = 0$. We have

$$\frac{1}{n} \sum_{i=0}^{n-1} m(T^{-i}A \cap B) = \frac{1}{n} \sum_{i=0}^{n-1} (m \times m)((T \times T)^{-i}(A \times X) \cap (B \times X))$$

$$\to (m \times m)(A \times X)(m \times m)(B \times X) \quad \text{by (ii)}$$

$$= m(A)m(B).$$

Also

$$\frac{1}{n} \sum_{i=0}^{n-1} (m(T^{-i}A \cap B))^2 = \frac{1}{n} \sum_{i=0}^{n-1} (m \times m)((T \times T)^{-i}(A \times A) \cap (B \times B))$$

$$\to (m \times m)(A \times A)(m \times m)(B \times B) \quad \text{by (ii)}$$

$$= m(A)^2 m(B)^2.$$

Thus

$$\frac{1}{n} \sum_{i=0}^{n-1} \{m(T^{-i}A \cap B) - m(A)m(B)\}^2$$

$$= \frac{1}{n} \sum_{i=0}^{n-1} \{m(T^{-i}A \cap B)^2 - 2m(T^{-i}A \cap B)m(A)m(B) + m(A)^2 m(B)^2\}$$

$$\to 2m(A)^2 m(B)^2 - 2m(A)^2 m(B)^2 = 0.$$

Therefore T is weak-mixing by Theorem 1.21. $\qquad\qquad\qquad\qquad \square$

Remark. It is easy to show that T is strong-mixing iff $T \times T$ is strong-mixing.

We now relate the weak-mixing of T to a spectral property of the operator U_T on $L^2(m)$.

Definition 1.7. Let T be a measure-preserving transformation of the probability space $(X, \mathcal{B}, m)$. We say a complex number λ is an eigenvalue of T if it is an eigenvalue of the isometry $U_T : L^2(m) \to L^2(m)$ i.e. if there exists $f \in L^2(m)$ $f \neq 0$ with $U_T f = \lambda f$ (or $f(Tx) = \lambda f(x)$ a.e.). Such an f is called an eigenfunction corresponding to λ.

Remarks

(i) If λ is an eigenvalue of T then $|\lambda| = 1$ since

$$\|f\|^2 = \|U_T f\|^2 = (U_T f, U_T f) = (\lambda f, \lambda f) = |\lambda|^2 \|f\|^2.$$

(ii) A measure-preserving transformation always has $\lambda = 1$ as an eigenvalue and any non-zero constant function is a corresponding eigenfunction.

Definition 1.8. We say that a measure-preserving transformation T of a probability space $(X, \mathcal{B}, m)$ has *continuous spectrum* if 1 is the only eigenvalue of T and the only eigenfunctions are the constants.

Observe that T has continuous spectrum iff $\lambda = 1$ is the only eigenvalue and T is ergodic.

We shall need the following result from spectral theory to prove the next Theorem. The proof can be found in Halmos [2].

Theorem 1.25 (Spectral Theorem for Unitary Operators). *Suppose U is a unitary operator on a complex Hilbert space $\mathcal{H}$. Then for each $f \in \mathcal{H}$ there exists a unique finite Borel measure μ_f on K such that*

$$(U^n f, f) = \int_K \lambda^n \, d\mu_f(\lambda) \quad \forall n \in Z.$$

If T is an invertible measure-preserving transformation then U_T is unitary, and if T has continuous spectrum and $(f, 1) = 0$ then μ_f has no atoms (i.e. each point of K has zero μ_f-measure).

Theorem 1.26. *If T is an invertible measure-preserving transformation of a probability space $(X, \mathcal{B}, m)$ then T is weak-mixing iff T has continuous spectrum.*

PROOF. Suppose T is weak-mixing and let $U_T f = \lambda f, f \in L^2(m)$. If $\lambda \neq 1$ then integration gives $(f, 1) = 0$ and by the weak-mixing property we have

$$\frac{1}{n} \sum_{i=0}^{n-1} |(U_T^i f, f)| \to 0$$

and hence

$$\frac{1}{n} \sum_{i=0}^{n-1} |(\lambda^i f, f)| \to 0.$$

Since $|\lambda| = 1$ this gives $(f, f) = 0$ and therefore $f = 0$ a.e. If $\lambda = 1$ then $f = $ constant a.e. by the ergodicity of T. (This part of the proof did not use the spectral theorem.)

Now suppose T has continuous spectrum. We show that if $f \in L^2(m)$ then

$$\frac{1}{n} \sum_{i=0}^{n-1} |(U_T^i f, f) - (f, 1)(1, f)|^2 \to 0.$$

If f is constant a.e. this is true. Hence all we need to show is that $(f, 1) = 0$ implies

$$\frac{1}{n} \sum_{i=0}^{n-1} |(U_T^i f, f)|^2 \to 0.$$

By the spectral theorem it suffices to show that if μ_f is a continuous (non-atomic) measure on K then

$$\frac{1}{n} \sum_{i=0}^{n-1} \left| \int \lambda^i \, d\mu_f(\lambda) \right|^2 \to 0.$$

We have

$$\frac{1}{n} \sum_{i=0}^{n-1} \left| \int \lambda^i \, d\mu_f(\lambda) \right|^2 = \frac{1}{n} \sum_{i=0}^{n-1} \left(\int \lambda^i \, d\mu_f(\lambda) \cdot \int \lambda^{-i} \, d\mu_f(\lambda) \right)$$

$$= \frac{1}{n} \sum_{i=0}^{n-1} \left(\int \lambda^i \, d\mu_f(\lambda) \cdot \int \tau^{-i} \, d\mu_f(\tau) \right)$$

$$= \frac{1}{n} \sum_{i=0}^{n-1} \iint_{K \times K} (\lambda \bar\tau)^i \, d(\mu_f \times \mu_f)(\lambda, \tau) \quad \text{(by Fubini's Theorem)}$$

$$= \iint_{K \times K} \left(\frac{1}{n} \sum_{i=0}^{n-1} (\lambda \bar\tau)^i \right) d(\mu_f \times \mu_f)(\lambda, \tau).$$

If (λ, τ) is not in the diagonal of $K \times K$ then

$$\frac{1}{n} \sum_{i=0}^{n-1} (\lambda \bar\tau)^i = \frac{1}{n} \left[\frac{1 - (\lambda \bar\tau)^n}{1 - (\lambda \bar\tau)} \right] \to 0$$

as $n \to \infty$. Since μ_f has no atoms the diagonal has measure 0 for $\mu_f \times \mu_f$ and therefore $(1/n) \sum_{i=0}^{n-1} (\lambda \bar\tau)^i \to 0$ a.e. $(\mu_f \times \mu_f)$. The modulus of the integrand is bounded by 1, so we can apply the bounded convergence theorem to obtain the result. $\square$

We now investigate the mixing properties of the examples mentioned in §1.1.

Examples

(1) Clearly the identity transformation I of $(X, \mathscr{B}, m)$ is ergodic iff all the elements of $\mathscr{B}$ have measure 0 or 1 iff I is strong-mixing.

(2) A rotation $T(z) = az$ of the unit circle K is never weak-mixing. This follows because because if $f(z) = z$ then $f(Tz) = f(az) = af(z)$ and we can apply the easy part of Theorem 1.26.

(3) **Theorem 1.27.** *No rotation $Tx = ax$ on a compact group is weak-mixing.*

PROOF. We know that if T is ergodic then the group G is abelian, and if γ is any character of G we have $\gamma(Tx) = \gamma(a)\gamma(x)$, which shows that T is not weak-mixing by the easy part of Theorem 1.26.   □

(4) **Theorem 1.28.** *For an endomorphism of a compact group strong-mixing, weak-mixing and ergodicity are all equivalent. (The condition for ergodicity was given in Theorem 1.10.)*

PROOF. We shall give the proof when G is abelian. It suffices to show that if the endomorphism $A : G \to G$ is ergodic then A is strong-mixing. If $\gamma, \delta \in \hat{G}$ then $(U_A^n \gamma, \delta) = 0$ eventually unless $\gamma = \delta \equiv 1$. So always $(U_A^n \gamma, \delta) \to (\gamma, 1)(1, \delta)$. Fix $\delta \in \hat{G}$. The collection

$$\mathscr{H}_\delta = \{ f \in L^2(m) : (U_A^n f, \delta) \to (f, 1)(1, \delta) \}$$

is a subspace of $L^2(m)$ which is closed. (To check $\mathscr{H}_\delta$ is closed, suppose $f_k \in \mathscr{H}$ and $f_k \to f \in L^2(m)$. For $\delta \equiv 1$ it is clear that $\mathscr{H}_\delta = L^2(m)$, so suppose $(1, \delta) = 0$. Then

$$
\begin{aligned}
|(U_A^n f, \delta)| &\leq |(U_A^n f, \delta) - (U_A^n f_k, \delta)| + |(U_A^n f_k, \delta)| \\
&\leq \|f - f_k\|_2 \|\delta\|_2 + |(U_A^n f_k, \delta)| \quad \text{(by the Schwarz inequality)} \\
&= \|f - f_k\|_2 + |(U_A^n f_k, \delta)|.
\end{aligned}
$$

If $\varepsilon > 0$ is given choose k so that $\|f - f_k\|_2 < \varepsilon/2$ and then choose $N(\varepsilon)$ so that $n \geq N(\varepsilon)$ implies $|(U_A^n f_k, \delta)| < \varepsilon/2$.) Since $\mathscr{H}_\delta$ contains $\hat{G}$ it is equal to $L^2(m)$. Fix $f \in L^2(m)$ and consider $\mathscr{L}_f = \{ g \in L^2(m) : (U_A^n f, g) \to (f, 1)(1, g) \}$. Then $\mathscr{L}_f$ is a closed subspace of $L^2(m)$, contains $\hat{G}$ by the above, and so equals $L^2(m)$. Hence A is strong-mixing.   □

(5) **Theorem 1.29.** *For an affine transformation $T = a \cdot A$ on a compact metric abelian group the following are equivalent:*

(i)   *T is strong-mixing.*
(ii) *T is weak-mixing.*
(iii) *A is ergodic.*

PROOF. We shall give the proof in the case when G is connected. Let $Bx = x^{-1}A(x)$ and recall that T is ergodic iff

(a) $\gamma \circ A^k = \gamma$, $k > 0$, implies $\gamma \circ A = \gamma$, and
(b) $[a, BG] = G$.

If A is ergodic then $BG = G$ since the endomorphism $\hat{B}$ of $\hat{G}$ is one-to-one. Choose $b \in G$ so that $B(b) = a$. Define $\phi: G \to G$ by $\phi(x) = bx$. Then $\phi T = A\phi$ and ϕ preserves Haar measure m. By (4) above A is strong-mixing and so if C, $D \in \mathcal{B}$ we have $m(T^{-n}C \cap B) = m(\phi(T^{-n}C \cap D) = m(\phi T^{-n}C \cap \phi D) = m(A^{-n}\phi C \cap \phi D) \to m(\phi C)m(\phi D) = m(C)m(D)$. Therefore T is strong-mixing.

Conversely if T is weak-mixing and A is not ergodic then by (a) $\gamma \circ A = \gamma$ for some $\gamma \not\equiv 1$. But then

$$|(U_T^n \gamma, \gamma)| = |(\gamma(a)\gamma(Aa) \cdots \gamma(A^{n-1}a)\gamma, \gamma)| = \|\gamma\|_2^2 = 1$$

for all n contradicting the weak-mixing of T. So if T is weak-mixing then A is ergodic. $\square$

(6) **Theorem 1.30.** *The two-sided* $(p_0, \ldots, p_{k-1})$-*shift is strong-mixing.*

PROOF. Use Theorem 1.17 after verifying the correct behaviour for measurable rectangles. $\square$

(7) Similarly, the one-sided $(p_0, \ldots, p_{k-1})$-shift is strong-mixing.
(8) We have the following theorem for the $(\mathbf{p}, P)$ Markov shift. Recall that we can always assume each $p_i > 0$.

Theorem 1.31. *If T is the $(\mathbf{p}, P)$ Markov shift (either one-sided or two-sided) the following are equivalent:*

(i)   *T is weak-mixing.*
(ii) *T is strong-mixing.*
(iii) *The matrix P is irreducible and aperiodic (i.e. $\exists N > 0$ such that the matrix P^N has no zero entries).*
(iv) *For all states i, j $p_{ij}^{(n)} \to p_j$.*

PROOF. That (iii) and (iv) are equivalent is a standard use of the renewal theorem in probability theory.

(i) $\Rightarrow$ (iii). Since $(1/N) \sum_{i=0}^{N-1} |m(_0[i]_0 \cap T_0^{-n}[j]_0) - m(_0[i]_0)m(_0[j]_0)| \to 0$ we have $(1/N) \sum_{n=0}^{N-1} |p_{ij}^{(n)} - p_j| \to 0$. By Theorem 1.20 we get a set J of density zero in Z^+ such that

$$\lim_{\substack{J \not\ni n \to \infty}} p_{ij}^{(n)} = p_j \quad \text{for all } i, j.$$

Therefore P is irreducible and aperiodic.

(iv) $\Rightarrow$ (ii). By Theorem 1.17 it suffices to show that for two blocks $A = {}_a[i_0 \cdots i_r]_{a+r}$, $B = {}_b[j_0 \cdots j_s]_{b+s}$ we have $m(T^{-n}A \cap B) \to m(A)m(B)$. This is a straightforward calculation (see proof of Theorem 1.19 for a similar one).

□

Remark. We know that T is ergodic iff $\forall f \in L^1(m)$ $(1/n) \sum_{i=0}^{n-1} U_T^i f \to \int f \, dm$ a.e. There is the following connection between strong-mixing and convergence of ergodic averages. The measure-preserving transformation T is strong-mixing iff for every increasing sequence $k_1 < k_2 < \cdots$ of natural numbers and every $f \in L^2(m)$ we have $\left\| (1/n) \sum_{i=0}^{n-1} U_T^{k_i} f - \int f \, dm \right\|_2 \to 0$ (Blum and Hanson [2]).

CHAPTER 2
Isomorphism, Conjugacy, and Spectral Isomorphism

So far we have been studying measure-preserving transformations on probability spaces. We now want to consider the notion of isomorphism for measure-preserving transformations; in other words, when should we consider two measure-preserving transformations as being "the same" or being equivalent? We must bear in mind that in measure theory a set of measure zero can be ignored. We first consider ways of ignoring sets of measure zero before considering isomorphism of measure-preserving transformations.

§2.1 Point Maps and Set Maps

One of the most important notions in measure theory is that of neglecting sets of measure zero. With this in mind let us consider what we should mean by saying two probability spaces $(X_1, \mathscr{B}_1, m_1)$, $(X_2, \mathscr{B}_2, m_2)$ are isomorphic. One way to view this is to require that the spaces be connected by an invertible measure-preserving transformation after removing sets of measure zero from each space.

Definition 2.1. The probability spaces $(X_1, \mathscr{B}_1, m_1)$, $(X_2, \mathscr{B}_2, m_2)$ are said to be *isomorphic* if there exist $M_1 \in \mathscr{B}_1, M_2 \in \mathscr{B}_2$ with $m_1(M_1) = 1 = m_2(M_2)$ and an invertible measure-preserving transformation $\phi : M_1 \to M_2$. (The space M_i is assumed to be equipped with the σ-algebra $M_i \cap \mathscr{B}_i = \{M_i \cap B \,|\, B \in \mathscr{B}_i\}$ and the restriction of the measure m_i to this σ-algebra.)

There is the following theorem on isomorphism of measure spaces.

Theorem 2.1. *Let X be a complete separable metric space, let $\mathscr{B}(X)$ be its σ-algebra of Borel subsets and let m be a probability measure on $\mathscr{B}(X)$ with*

$m(\{x\}) = 0$ *for each set* $\{x\}$ *consisting of a single point* $x \in X$. *Let* $([0,1],$ $\mathscr{B}([0,1]), l)$ *denote the closed unit interval with its σ-algebra of Borel sets and Lebesgue measure* l. *Then* $(X, \mathscr{B}(X), m)$ *and* $([0, 1], \mathscr{B}([0, 1]), l)$ *are isomorphic. If* $(X, \mathscr{B}_m(X), m)$ *denotes the completion of* $(X, \mathscr{B}(X), m)$ *then* $(X, \mathscr{B}_m(X), m)$ *is isomorphic to* $([0, 1], \mathscr{L}, l)$ *where* $\mathscr{L}$ *is the σ-algebra of Lebesgue measurable sets* (*which is the completion* $\mathscr{B}_l([0, 1])$).

A proof is given in Theorem 9, page 327 of Royden's book (Royden [1]).

If the condition that m have no points of positive measure is omitted then there are at most a countable collection of points $\{x_n\}_1^\infty$ of X with positive measure and then $(X, \mathscr{B}(X), m)$ is isomorphic to a measure space consisting of points $\{y_n\}_1^\infty$ with measures $\{m(x_n)\}$ together with $([0, s],$ $\mathscr{B}([0, s]), l)$ where $s = 1 - \sum_{n=1}^\infty m(x_n)$. The corresponding statement for completed spaces is true. Therefore all the probability spaces on which our examples act come under these results.

There is another way to handle the omission of sets of measure zero, and this other way is, perhaps, mathematically more natural but less practical in applications. This is the idea of using measure algebras.

Let $(X, \mathscr{B}, m)$ be a probability space. Define an equivalence relation on $\mathscr{B}$ by saying A and B are equivalent $(A \sim B)$ iff $m(A \triangle B) = 0$. Let $\tilde{\mathscr{B}}$ denote the collection of equivalence classes. Then $\tilde{\mathscr{B}}$ is a Boolean σ-algebra under the operations of complementation, union and intersection inherited from $\mathscr{B}$. The measure m induces a measure $\tilde{m}$ on $\tilde{\mathscr{B}}$ by $\tilde{m}(\tilde{B}) = m(B)$. (Here $\tilde{B}$ is the equivalence class to which B belongs.) The pair $(\tilde{\mathscr{B}}, \tilde{m})$ is called a measure algebra.

From this point of view one says $(X_1, \mathscr{B}_1, m_1)$ and $(X_2, \mathscr{B}_2, m_2)$ are "equivalent" if their corresponding measure algebras are isomorphic:

Definition 2.2. Let $(X_1, \mathscr{B}_1, m_1)$, $(X_2, \mathscr{B}_2, m_2)$ be probability spaces with measure algebras $(\tilde{\mathscr{B}}_1, \tilde{m}_1)$, $(\tilde{\mathscr{B}}_2, \tilde{m}_2)$. The measure algebras are isomorphic if there is a bijection $\Phi: \tilde{\mathscr{B}}_2 \to \tilde{\mathscr{B}}_1$ which preserves complements, countable unions and intersections and satisfies $\tilde{m}_1(\Phi\tilde{B}) = \tilde{m}_2(\tilde{B}) \; \forall \tilde{B} \in \mathscr{B}_2$. The probability spaces are said to be *conjugate* if their measure algebras are isomorphic.

Conjugacy of measure spaces is weaker than isomorphism because if $(X_1, \mathscr{B}_1, m_1)$ and $(X_2, \mathscr{B}_2, m_2)$ are isomorphic as in Definition 2.1 then they are conjugate via $\Phi: \tilde{\mathscr{B}}_2 \to \tilde{\mathscr{B}}_1$ defined by $\Phi(\tilde{B}) = (\phi^{-1}(M_2 \cap B))^{\tilde{}}$. It is easy to give examples of conjugate measure spaces which are not isomorphic. Let $(X_1, \mathscr{B}_1, m_1)$ be a space of one point and let $(X_2, \mathscr{B}_2, m_2)$ be a space with two points and $\mathscr{B}_2 = \{\phi, X_2\}$. The two spaces are conjugate but they are not isomorphic because a set of zero measure cannot be omitted from X_2 so that the remaining set is mapped bijectively with X_1. The main reason this example works is that $\mathscr{B}_2$ does not separate the points of X_2. When some conditions are placed on the probability spaces the notions of conjugacy and isomorphism coincide:

Theorem 2.2. *Let* X_1, X_2 *be complete separable metric spaces, let* $\mathscr{B}(X_1)$, $\mathscr{B}(X_2)$ *be their σ-algebras of Borel subsets and let* m_1, m_2 *be probability measures on* $\mathscr{B}(X_1)$, $\mathscr{B}(X_2)$. *Let* $\Phi : \tilde{\mathscr{B}}(X_2) \to \tilde{\mathscr{B}}(X_1)$ *be an isomorphism of measure algebras. Then there exist* $M_1 \in \mathscr{B}(X_1)$, $M_2 \in \mathscr{B}(X_2)$ *with* $m_1(M_1) = 1 = m_2(M_2)$ *and an invertible measure-preserving transformation* $\phi : M_1 \to M_2$ *such that* $\Phi(\tilde{B}) = (\phi^{-1}(B \cap M_2))^{\tilde{}}$ $\forall B \in \mathscr{B}(X_2)$. *If* ψ *is any other isomorphism from* $(X_1, \mathscr{B}(X_1), m_1)$ *to* $(X_2, \mathscr{B}(X_2), m_2)$ *which induces* Φ *then*

$$m_1(\{x \in X_1 \,|\, \phi(x) \neq \psi(x)\}) = 0.$$

The proof is given in Theorem 12, page 329 of Royden [1].

The corresponding statement for measure-algebra homomorphisms holds (they are induced by (not necessarily invertible) measure-preserving transformations).

Therefore set maps are always induced by point maps for the probability spaces used in our examples.

Often in the ergodic theory literature all probability spaces used are assumed to be Lebesgue spaces:

Definition 2.3. A probability space $(X, \mathscr{B}, m)$ is a *Lebesgue space* if it is isomorphic to a probability space which is the disjoint union of a countable (or finite) number of points $\{y_1, y_2, \ldots\}$ each of positive measure and the space $([0, s], \mathscr{L}([0, s], l)$ where $\mathscr{L}([0, s])$ is the σ-algebra of Lebesgue measurable subsets of $[0, s]$ and l is Lebesgue measure. Here $s = 1 - \sum_{n=1}^{\infty} p_n$ where p_n is the measure of the point y_n.

The theory of Lebesgue spaces is given in Rohlin [1]. In particular the analogue of Theorem 2.2 is true for Lebesgue spaces (i.e. set maps are always induced by point maps) and so the two ways of dealing with sets of measure zero coincide. Notice that the remarks following Theorem 2.1 show that if X is a complete separable metric space and $\mathscr{B}_m(X)$ denotes the completion of $\mathscr{B}(X)$ under a probability measure m then $(X, \mathscr{B}_m(X), m)$ is a Lebesgue space.

A third way to study a probability space $(X, \mathscr{B}, m)$ is to study the Hilbert space $L^2(m)$. If $(X_1, \mathscr{B}_1, m_1)$, $(X_2, \mathscr{B}_2, m_2)$ are both spaces with countable basis then $L^2(m_1)$, $L^2(m_2)$ are separable (see §0.5) and hence unitarily isomorphic (i.e. there is a bijective linear map $W : L^2(m_2) \to L^2(m_1)$ such that $(Wf, Wg) = (f, g)$ $\forall f, g \in L^2(m_2)$). However, $L^2(m)$ has some extra structure because one can multiply certain members of $L^2(m)$ to obtain a function in $L^2(m)$. It turns out that conjugacy of measure spaces is equivalent to their L^2 spaces being equivalent in a sense involving this multiplication.

Let $(X, \mathscr{B}, m)$ be a probability space. Since $L^2(m)$ consists of equivalence classes of functions (f and g are equivalent if $f = g$ a.e.) we see that if $\tilde{B} \in \tilde{\mathscr{B}}$ then χ_B is a well-defined member of $L^2(m)$. There is the following simple result.

Theorem 2.3. *Let* $(X_1, \mathscr{B}_1, m_1)$, $(X_2, \mathscr{B}_2, m_2)$ *be probability spaces and let* $\Phi : (\tilde{\mathscr{B}}_2, \tilde{m}_2) \to (\tilde{\mathscr{B}}_2, \tilde{m}_1)$ *be a measure algebra isomorphism. Then* Φ *induces a* *bijective linear map* $V:L^2(m_2) \to L^2(m_1)$ *with the properties*

(a) $(Vf, Vg) = (f, g) \; \forall f, g \in L^2(m_2)$
(b) V *and* V^{-1} *map bounded functions to bounded functions*
(c) $V(fg) = (Vf)(Vg)$ *whenever* f, g *are bounded.*

The map V *is defined on characteristic functions by* $V\chi_{\tilde{B}} = \chi_{\Phi(\tilde{B})}$.

PROOF. For a characteristic function $\chi_{\tilde{B}}$, $\tilde{B} \in \tilde{\mathscr{B}}_2$ define $V\chi_{\tilde{B}} = \chi_{\Phi(\tilde{B})}$. Notice $\|V\chi_{\tilde{B}}\|_2 = \|\chi_{\tilde{B}}\|_2$. Extend V to simple functions to preserve linearity i.e. $V(\sum_{i=1}^{n} a_i X_{\tilde{B}_i}) = \sum_{i=1}^{n} a_i V X_{\tilde{B}_i}$ whenever $\{\tilde{B}_1, \ldots, \tilde{B}_n\}$ are mutually disjoint members of $\tilde{\mathscr{B}}_2$. This definition is independent of the representation of a simple function as a linear combination of characteristic functions. We have $\|Vf\|_2 = \|f\|_2$ for all simple functions f. Now let f be any element of $L^2(m_2)$. Choose simple functions f_n with $\|f_n - f\|_2 \to 0$. Since $\{f_n\}$ is a Cauchy sequence and V is an isometry on simple functions we know $\{Vf_n\}$ is a Cauchy sequence. Denote the limit of this sequence by Vf. If $\{g_n\}$ is another sequence of simple functions with $\|g_n - f\|_2 \to 0$ then $\|g_n - f_n\|_2 \to 0$ so that $\|Vf_n - Vg_n\|_2 \to 0$. This shows that Vf is well defined. The inverse of V is constructed in a similar way from Φ^{-1}. The linearity of V is clear from the definition. Property (a) holds because it is equivalent to V preserving norm. A function $f \in L^2(m_2)$ is bounded iff it is the limit of a sequence of simple functions which are uniformly bounded. By the definition of V on simple functions we see that Vf has the same bounds as f. Property (c) is easily seen to hold for simple functions and hence for bounded functions. □

If Φ is not necessarily surjective then V is still defined but may be non-surjective.

It turns out that the converse of Theorem 2.3 is true.

Theorem 2.4. *Let* $(X_1, \mathscr{B}_1, m_1)$, $(X_2, \mathscr{B}_2, m_2)$ *be probability spaces. Suppose* $V:L^2(m_2) \to L^2(m_1)$ *is a bijective linear map with the properties* (a), (b), (c) *listed in Theorem 2.3. Then* V *is induced by an isomorphism* $\Phi:(\tilde{\mathscr{B}}_2, \tilde{m}_2) \to (\tilde{\mathscr{B}}_1, \tilde{m}_1)$ *of measure algebras in the sense that* $V\chi_{\tilde{B}} = \chi_{\Phi(\tilde{B})} \; \forall \tilde{B} \in \tilde{\mathscr{B}}_2$.

PROOF. Let $\tilde{B}_2 \in \tilde{\mathscr{B}}_2$. We have $\chi_{\tilde{B}_2}^2 = \chi_{\tilde{B}_2}$ so that

$$V(\chi_{\tilde{B}_2}^2) = V(\chi_{\tilde{B}_2})V(\chi_{\tilde{B}_2}) = V(\chi_{\tilde{B}_2}),$$

and we see that $V(\chi_{\tilde{B}_2})$ takes 1 and 0 as its only values. Thus there exists $\tilde{B}_1 \in \tilde{\mathscr{B}}_1$ such that $V(\chi_{\tilde{B}_2}) = \chi_{\tilde{B}_1}$. We define $\Phi:(\tilde{\mathscr{B}}_2, \tilde{m}_2) \to (\tilde{\mathscr{B}}_1, \tilde{m}_1)$ by by $\Phi(\tilde{B}_2) = \tilde{B}_1$.

We now show that Φ is an isomorphism of measure algebras. By doing the above procedure for V^{-1} we obtain an inverse for Φ so Φ is invertible. Also

$$\tilde{m}_2(\tilde{B}_2) = (\chi_{\tilde{B}_2}, \chi_{\tilde{B}_2})$$
$$= (V\chi_{\tilde{B}_2}, V\chi_{\tilde{B}_2}) = (\chi_{\Phi(\tilde{B}_2)}, \chi_{\Phi(\tilde{B}_2)}) = \tilde{m}_1(\Phi\tilde{B}_2).$$

It remains to show that Φ preserves complements and countable unions. First note that since V is norm-preserving and maps characteristic functions to characteristic functions we have $V(1) = 1$. Since $\chi_{\tilde{B}_2} + \chi_{\tilde{X}_2 \setminus \tilde{B}_2} = 1$ in $L^2(m_2)$, applying V to both sides gives $\chi_{\Phi\tilde{B}_2} + \chi_{\Phi(\tilde{X}_2 \setminus \tilde{B}_2)} = 1$ so $\tilde{X}_1 \setminus \Phi\tilde{B}_2 = \Phi(\tilde{X}_2 \setminus \tilde{B}_2)$. Therefore Φ preserves complements.

Suppose $\tilde{B}, \tilde{C} \in \mathscr{B}_2$. Then

$$\chi_{\tilde{B} \cup \tilde{C}} = \chi_{\tilde{B}} + \chi_{\tilde{C}} - \chi_{\tilde{B} \cap \tilde{C}} = \chi_{\tilde{B}} + \chi_{\tilde{C}} - \chi_{\tilde{B}} \chi_{\tilde{C}}.$$

Applying V to both sides gives

$$\chi_{\Phi(\tilde{B} \cup \tilde{C})} = \chi_{\Phi(\tilde{B})} + \chi_{\Phi(\tilde{C})} - \chi_{\Phi(\tilde{B})} \chi_{\Phi(\tilde{C})} = \chi_{\Phi(\tilde{B}) \cup \Phi(\tilde{C})}.$$

Thus $\Phi(\tilde{B} \cup \tilde{C}) = \Phi(\tilde{B}) \cup \Phi(\tilde{C})$ and hence (by induction) Φ preserves all finite unions.

Now let $B_n \in \mathscr{B}_n$, $n \geq 1$. We have

$$\chi_{\bigcup_{i=1}^{n} B_i} \to \chi_{\bigcup_{i=1}^{\infty} B_i} \quad \text{a.e.}$$

and also in $L^2(m_2)$ by the bounded convergence theorem. Since V is an isometry it is continuous, so

$$V\chi_{\bigcup_{i=1}^{n} \tilde{B}_i} \to V\chi_{\bigcup_{i=1}^{\infty} \tilde{B}_i} = \chi_{\Phi\left(\bigcup_{i=1}^{\infty} \tilde{B}_i\right)} \quad \text{in } L^2(m_1).$$

On the other hand,

$$V\chi_{\bigcup_{i=1}^{n} \tilde{B}_i} = \chi_{\Phi\left(\bigcup_{i=1}^{n} \tilde{B}_i\right)} = \chi_{\bigcup_{i=1}^{n} \Phi\tilde{B}_i}$$

converges to

$$\chi_{\bigcup_{i=1}^{\infty} \Phi\tilde{B}_i}$$

in $L^2(m_1)$. Therefore $\Phi(\bigcup_{i=1}^{\infty} \tilde{B}_i) = \bigcup_{i=1}^{\infty} \Phi\tilde{B}_i$. $\qquad\qquad \square$

We have shown that $(X_1, \mathscr{B}_1, m_1)$ and $(X_2, \mathscr{B}_2, m_2)$ are conjugate iff $L^2(m_1)$ and $L^2(m_2)$ are connected by a bijective linear map satisfying the properties a, b, c of Theorem 2.3.

§2.2 Isomorphism of Measure-Preserving Transformations

What should we mean by saying that two measure-preserving transformations are the "same"? We must bear in mind that sets of measure 0 do not matter from the point of view of measure theory. Let us consider two examples.

(1) Let T be the transformation $Tz = z^2$ on the unit circle K with Borel sets and Haar measure, and let S be given by $Sx = 2x \bmod 1$ on $[0, 1)$ with

Borel sets and Lebesgue measure. Consider the map $\phi:[0,1) \to K$ defined by $x \to e^{2\pi i x}$. The map ϕ is a bijection and preserves measure (check on intervals and use Theorem 1.1). Also $\phi S = T\phi$. We want to regard T and S as the "same", because an isomorphism ϕ of the measure spaces $[0,1)$ and K maps S to $\phi S \phi^{-1}$ which is T.

(2) Again let S be the transformation $Sx = 2x \bmod 1$ on $[0,1)$ with Borel sets and Lebesgue measure, and let $T_2:X \to X$ be the 1-sided $(\frac{1}{2},\frac{1}{2})$-shift. Define $\psi:X \to [0,1)$ by

$$\psi(a_1, a_2, a_3, \ldots) = \frac{a_1}{2} + \frac{a_2}{2^2} + \frac{a_3}{2^3} + \cdots.$$

The map ψ is one-to-one on the complement of the set of all points $(a_1, a_2, \ldots)$ whose coordinates are constant eventually. However ψ is onto and $\psi T_2 = S\psi$. Also ψ preserves measure; we can check this on dyadic intervals and apply Theorem 1.1.

Suppose D_2 is the set of points of the space X of the 1-sided $(\frac{1}{2}, \frac{1}{2})$-shift which have constant coordinates eventually. Then $T_2^{-1}D_2 = D_2$ and so $T_2^{-1}(X\backslash D_2) = X\backslash D_2$. Let D consist of the dyadic rationals in $[0,1)$. Then $S^{-1}D = D$, so that $S^{-1}([0,1)\backslash D) = [0,1)\backslash D$. We see that D_2 and D both have zero measure and ψ maps $X\backslash D_2$ bijectively to $[0,1)\backslash D$. Also $\psi T_2(x) = S\psi(x)$ $\forall x \in X\backslash D_2$. We would like to consider S and T_2 as isomorphic since, after removing sets of measure zero, we can change one to the other by an invertible measure-preserving transformation.

We arrive at the following definition of isomorphism.

Definition 2.4. Suppose $(X_1, \mathcal{B}_1, m_1)$ and $(X_2, \mathcal{B}_2, m_2)$ are probability spaces together with measure-preserving transformations

$$T_1:X_1 \to X_1, \quad T_2:X_2 \to X_2.$$

We say that T_1 is *isomorphic* to T_2 if there exist $M_1 \in \mathcal{B}_1$, $M_2 \in \mathcal{B}_2$ with $m_1(M_1) = 1$, $m_2(M_2) = 1$ such that

(i) $T_1 M_1 \subseteq M_1$, $T_2 M_2 \subseteq M_2$, and
(ii) there is an invertible measure-preserving transformation

$$\phi:M_1 \to M_2 \text{ with } \phi T_1(x) = T_2\phi(x) \, \forall x \in M_1.$$

We write $T_1 \simeq T_2$. (In (ii) the set M_i $(i = 1, 2)$ is assumed to be equipped with the σ-algebra $M_i \cap \mathcal{B}_i = \{M_i \cap B | B \in \mathcal{B}_i\}$ and the restriction of the measure m_i to this σ-algebra.)

Remarks

(1) Isomorphism is an equivalence relation.
(2) If $T_1 \simeq T_2$ then $T_1^n \simeq T_2^n$ $\forall n > 0$.
(3) If T_1 and T_2 are invertible we can take M_1, M_2 so that $T_1 M_1 = M_1$, $T_2 M_2 = M_2$: we just take $\bigcap_{-\infty}^{\infty} T_1^k M_1$, $\bigcap_{-\infty}^{\infty} T_2^k M_2$ as the new sets.

(4) Isomorphism of measure spaces (Definition 2.1) means that the identity transformations of these spaces are isomorphic.

Because of the principle of neglecting sets of measure zero we should consider two measure-preserving transformations T_1, T_2 of $(X, \mathscr{B}, m)$ as identical if we can remove a set N of zero measure from X so that T_1 and T_2 agree on $X \backslash N$. (We can always suppose $T_i^{-1} N \subset N$ and hence $T_i^{-1}(X \backslash N) \subset X \backslash N$, $i = 1, 2$, by replacing N by

$$\bigcup_{n=0}^{\infty} \bigcup_{i=1}^{n} \cdots \bigcup_{r_i, s_i = 0}^{\infty} T_1^{-r_1} T_2^{-s_1} T_1^{-r_2} T_2^{-s_2} \cdots T_1^{-r_n} T_2^{-s_n} N.)$$

This is the same as T_1 being isomorphic to T_2 via the identity isomorphism. We shall write $T_1 = T_2 \bmod 0$ when this occurs. Also we shall say a measure-preserving transformation T of $(X, \mathscr{B}, m)$ is invertible mod 0 if there is a set N of measure zero such that $T(X \backslash N) = X \backslash N$ and $T|_{X \backslash N}$ is an invertible measure-preserving transformation. Since the transformation $S: X \to X$ given by $S(x) = x$ if $x \in N$, $S(x) = T(x)$ if $x \in X \backslash N$ is then an invertible measure-preserving transformation the requirement is equivalent to saying that the identity provides an isomorphism between T and an invertible measure-preserving transformation.

§2.3 Conjugacy of Measure-Preserving Transformations

Although the concept of isomorphism of measure-preserving transformations, introduced above, is useful in practice the procedure of studying what happens on measure algebras is perhaps mathematically more natural.

Let $(X_1, \mathscr{B}_1, m_1)$, $(X_2, \mathscr{B}_2, m_2)$ be probability spaces with corresponding measure algebras $(\tilde{\mathscr{B}}_1, \tilde{m}_1)$, $(\tilde{\mathscr{B}}_2, \tilde{m}_2)$. If $\phi: X_1 \to X_2$ is measure-preserving then we have a map $\tilde{\phi}^{-1}: (\tilde{\mathscr{B}}_2, \tilde{m}_2) \to (\tilde{\mathscr{B}}_1, \tilde{m}_1)$ defined by $(\tilde{\phi}^{-1}(\tilde{B}) = (\phi^{-1}(B))^{\tilde{}}$. This map is well-defined since ϕ is measure-preserving. The map $\tilde{\phi}^{-1}$ preserves complements and countable unions (and hence countable intersections). Also $\tilde{m}_1(\tilde{\phi}^{-1}(\tilde{B})) = \tilde{m}_2(\tilde{B}) \; \forall \tilde{B} \in \tilde{\mathscr{B}}_2$. Therefore $\tilde{\phi}^{-1}$ can be considered a homomorphism of measure algebras. Note that $\tilde{\phi}^{-1}$ is injective.

Recall that an isomorphism $\Phi: (\tilde{\mathscr{B}}_2, \tilde{m}_2) \to (\tilde{\mathscr{B}}_1, \tilde{m}_1)$ of measure-algebras is a bijection which preserves complements and countable unions and satisfies $\tilde{m}_1(\Phi(\tilde{B})) = \tilde{m}_2(\tilde{B}) \; \forall \tilde{B} \in \tilde{\mathscr{B}}_2$.

We shall call two measure-preserving transformations conjugate if they induce isomorphic maps on measure-algebras:

Definition 2.5. Let T_i be a measure-preserving transformation of the probability space $(X_i, \mathscr{B}_i, m_i)$, $i = 1, 2$. We say that T_1 is *conjugate* to T_2 if there is a measure-algebra isomorphism $\Phi:(\tilde{\mathcal{B}}_2, \tilde{m}_2) \to (\tilde{\mathcal{B}}_1, \tilde{m}_1)$ such that $\Phi\tilde{T}_2^{-1} = \tilde{T}_1^{-1}\Phi$.

It is clear that conjugacy is an equivalence relation on the set of all measure-preserving transformations.

Isomorphism and conjugacy are connected by the following simple result.

Theorem 2.5. *For* $i = 1, 2$ *let* T_i *be a measure-preserving transformation of the probability space* $(X_i, \mathcal{B}_i, m_i)$. *If* T_1 *is isomorphic to* T_2 *then* T_1 *is conjugate to* T_2.

PROOF. Let $\phi: M_1 \to M_2$ be the isomorphism as in Definition 2.4. Define $\Phi:(\tilde{\mathcal{B}}_2, \tilde{m}_2) \to (\tilde{\mathcal{B}}_1, \tilde{m}_1)$ by $\Phi(\tilde{B}) = (\phi^{-1}(B \cap M_2))\tilde{}$. Then Φ satisfies the requirements of Definition 2.5. $\qquad\qquad\qquad\qquad\qquad\qquad\qquad\qquad\square$

The converse of this theorem is not true in general because we pointed out in §2.1 that identity maps on two measure spaces can be conjugate without being isomorphic. However the converse is true for some classes of measure spaces.

Theorem 2.6. *Let* $(X_1, \mathcal{B}_1, m_1)$, $(X_2, \mathcal{B}_2, m_2)$ *be probability spaces which are either both Lebesgue spaces or where each* X_i *is a complete separable metric space and* $\mathcal{B}_i$ *is its σ-algebra of Borel sets. Let* $T_i: X_i \to X_i$ *be a measure-preserving transformation,* $i = 1, 2$. *If* T_1 *is conjugate to* T_2 *then* T_1 *is isomorphic to* T_2.

PROOF. Suppose $\Phi:(\tilde{\mathcal{B}}_2, \tilde{m}_2) \to (\tilde{\mathcal{B}}_1, \tilde{m}_1)$ is the isomorphism of measure-algebras with $\Phi\tilde{T}_2^{-1} = \tilde{T}_1^{-1}\Phi$. By Theorem 2.2 (or the corresponding theorem for Lebesgue spaces) there exist $X_1' \in \mathcal{B}_1, X_2' \in \mathcal{B}_2$ with $m_1(X_1') = 1, m_2(X_2') = 1$, and an invertible measure-preserving transformation $\phi: X_1' \to X_2'$ such that $\Phi(\tilde{B}) = (\phi^{-1}(B \cap X_2'))\tilde{}$. Then $\tilde{\phi}^{-1}\tilde{T}_2^{-1} = \tilde{T}_1^{-1}\tilde{\phi}^{-1}$ which can be written $(\widetilde{T_2\phi})^{-1} = (\widetilde{\phi T_1})^{-1}$. It follows that $T_2\phi = \phi T_1$ a.e. Let

$$A_1 = \{x \in X_1 \mid T_2\phi(x) = \phi T_1(x)\}$$

and $M_1 = \bigcap_{n=0}^{\infty} T_1^{-n} A_1$. Then $m_1(M_1) = 1$ and $T_1^{-1}M_1 \supset M_1$ so $T_1 M_1 \subset M_1$. Let $M_2 = \phi M_1$ and then $T_2 M_2 \subset M_2$ and the conditions of Definition 2.4. hold. $\qquad\qquad\qquad\qquad\qquad\qquad\qquad\qquad\qquad\qquad\square$

On these nice measure spaces we can also reformulate the notion of a measure-preserving transformation being invertible mod 0.

Theorem 2.7. *Let* $(X, \mathcal{B}, m)$ *be a Lebesgue space or a probability space where* X *is a complete separable metric space and* $\mathcal{B}$ *its σ-algebra of Borel subsets. Let* $T: X \to X$ *be measure-preserving. Then* T *is invertible mod* 0 *iff* $\tilde{T}^{-1}\tilde{\mathcal{B}} = \tilde{\mathcal{B}}$.

PROOF. We know $\tilde{T}^{-1}:\tilde{\mathcal{B}} \to \tilde{\mathcal{B}}$ is always injective. If T is invertible mod 0 it induces the same map on the measure algebra as that induced by the invertible map. Therefore $\tilde{T}^{-1}\tilde{\mathscr{B}} = \tilde{\mathscr{B}}$. (No assumption on the measure space was needed for this part.) If $\tilde{T}^{-1}\tilde{\mathscr{B}} = \tilde{\mathscr{B}}$ then $\tilde{T}^{-1}:(\tilde{\mathscr{B}}, \tilde{m}) \to (\tilde{\mathscr{B}}, \tilde{m})$ is a bijection and so is induced by an invertible measure-preserving transformation defined on a subset of X of measure one (Theorem 2.2). Therefore T is equal to this invertible transformation a.e. $\qquad\qquad\square$

We can weaken the definitions of isomorphism and conjugacy to say what it means for one measure-preserving transformation to be a factor (or homomorphic image) of another.

Definition 2.6. Suppose $(X_i, \mathscr{B}_i, m_i)$ is a probability space and $T_i: X_i \to X_i$ is measure-preserving, $i = 1, 2$. We say T_2 is a *factor* of T_1 if there exist $M_i \in \mathscr{B}_i$ with $m_i(M_i) = 1$ and $T_i M_i \subset M_i$ $(i = 1, 2)$ and there exists a measure-preserving transformation $\phi: M_1 \to M_2$ with $\phi T_1(x) = T_2 \phi(x)$ $\forall x \in M_1$.

The difference between this and isomorphism is that the transformation ϕ may not be invertible. For example, if S, T are measure-preserving transformations then S is a factor of $S \times T$, the role of ϕ being played by the natural projection.

The corresponding weakening in the notion of conjugacy leads to the following.

Definition 2.7. Let T_i be a measure-preserving transformation of a probability space $(X_i, \mathscr{B}_i, m_i)$, $i = 1, 2$. We say T_2 is a *semi-conjugate image* of T_1 if there is a measure-algebra homomorphism $\Phi:(\tilde{\mathscr{B}}_2, \tilde{m}_2) \to (\tilde{\mathscr{B}}_1, \tilde{m}_1)$ such that $\Phi \tilde{T}_2^{-1} = \tilde{T}_1^{-1}\Phi$.

A map $\Phi:(\tilde{\mathscr{B}}_2, \tilde{m}_2) \to (\tilde{\mathscr{B}}_1, \tilde{m}_1)$ is a measure-algebra homomorphism if $\Phi(\tilde{Y} \backslash \tilde{B}) = \tilde{X} \backslash \Phi(\tilde{B})$ $\forall B \in \mathscr{B}_2$, $\Phi(\bigcup_{n=1}^{\infty} \tilde{B}_n) = \bigcup_{n=1}^{\infty} \Phi(\tilde{B}_n)$, whenever each $B_n \in \mathscr{B}_2$, and $\tilde{m}_1(\Phi(\tilde{B})) = \tilde{m}_2(\tilde{B})$ $\forall B \in \mathscr{B}_2$.

If T_2 is a factor of T_1 (as in Definition 2.6) then T_2 is a semi-conjugate image of T_1 by taking $\Phi(\tilde{B}) = (\phi^{-1}(B \cap M_2))^{\tilde{}}$. If the probability spaces are such that X_2 is a complete separable metric space and $\mathscr{B}_2$ is its σ-algebra of Borel sets then every measure-algebra homomorphism $\Phi:(\tilde{\mathscr{B}}_2, \tilde{m}_2) \to (\tilde{\mathscr{B}}_1, \tilde{m}_1)$ is induced by a measure-preserving transformation defined on a subset of X_1 of measure one (Royden, Theorem 11, p. 329). The same is true if $(X_1, \mathscr{B}_1, m_1)$ and $(X_2, \mathscr{B}_2, m_2)$ are both Lebesgue spaces (Rohlin [1]). Therefore in most measure spaces we encounter the notions of factor and semi-conjugate image coincide.

When we deal only with Lebesgue spaces there is a nice description of all the factors of a given measure-preserving transformation. If $(X, \mathscr{B}, m)$ is a Lebesgue space and $\mathscr{F}$ is a sub-σ-algebra of $\mathscr{B}$ then there is a partition ζ of $(X, \mathscr{B}, m)$ into measurable sets such that if $\mathscr{B}(\zeta)$ denotes the collection of all measurable sets that are unions of members of ζ then $\overline{\mathscr{B}(\zeta)} = \mathscr{F}$. For example, if we take $\mathscr{F} = \mathscr{B}$ then ζ would be the partition of X into individual points. We can form a new set X_ζ by taking the points of X_ζ to be the elements of ζ.

There is a natural map $\pi: X \to X_\zeta$ and we can consider $\mathscr{B}(\zeta)$ as a σ-algebra of subsets of X_ζ. Also $m_\zeta = m \circ \pi^{-1}$ is a probability measure on $(X_\zeta, \mathscr{B}(\zeta))$ and $(X_\zeta, \mathscr{B}(\zeta), m_\zeta)$ turns out to be a Lebesgue space. If $T: X \to X$ is measure-preserving and $T^{-1}\mathscr{F} \subset \mathscr{F}$ then $T^{-1}\zeta \leq \zeta$ (i.e. each set $T^{-1}C$, $C \in \zeta$, is a union of elements of ζ) and we get a measure-preserving transformation $T_\zeta: X_\zeta \to X_\zeta$ defined by $T_\zeta(D) = C$ if $D \subset T^{-1}C$. We have $\pi T = T_\zeta \pi$ so that T_ζ is a factor of T. Therefore each sub-σ-algebra $\mathscr{F}$ of $\mathscr{B}$ with $T^{-1}\mathscr{F} \subset \mathscr{F}$ leads to a factor of T. It turns out that every other factor is isomorphic to one of this type. Suppose $T_2: X_2 \to X_2$ is a factor of $T_1: X_1 \to X_1$, and both probability spaces are Lebesgue spaces. Suppose $\phi: M_1 \to M_2$ is a measure-preserving transformation with $\phi T_1 = T_2\phi$, as in Definition 2.6. Consider the partition ζ of M_1 into sets of the form $\phi^{-1}(x)$, $x \in M_2$. Let $\mathscr{F}$ be the σ-algebra generated by this partition ζ. Since $T_1^{-1}\phi^{-1} = \phi^{-1}T_2^{-1}$ we have $T_1^{-1}\zeta \leq \zeta$ and hence $T_1^{-1}\mathscr{F} \subset \mathscr{F}$. Also ϕ induces an isomorphism between $(T_1)_\zeta: (X_1)_\zeta \to (X_1)_\zeta$ and $T_2: X_2 \to X_2$. So, when dealing with Lebesgue spaces, the factors of a given measure-preserving transformation $T: (X, \mathscr{B}, m) \to (X, B, m)$ are determined by the sub-σ-algebras $\mathscr{F}$ of $\mathscr{B}$ with $T^{-1}\mathscr{F} \subset \mathscr{F}$.

§2.4 The Isomorphism Problem

The main internal problem in ergodic theory is deciding when two measure-preserving transformations are isomorphic (or when they are conjugate). The usual way to tackle such an isomorphism problem is to look for isomorphism invariants. These invariants are usually of two types. The first type of invariant is a property (e.g. ergodicity, weak-mixing) that some measure-preserving transformations have and some do not have, such that two isomorphic measure-preserving transformations either both have the property or both do not have the property. In order for the property to be useful one should be able to check if naturally occurring examples have the property or not. Ergodicity, weak-mixing and strong-mixing are examples of such properties. The second type of invariant is the assignment of some object (e.g. a number, a group) in some mathematical category to each measure-preserving transformation such that the objects associated to two isomorphic measure-preserving transformations are isomorphic in their own category (e.g. equal if they are numbers, isomorphic groups if they are groups). Again, in order for such an invariant to be useful it should be calculable for interesting examples of measure-preserving transformations. Also it will be a good invariant if there is some collection of measure-preserving transformations for which the invariant is complete i.e. any two transformations from this class with isomorphic objects are isomorphic measure-preserving transformations.

We shall discuss two invariants of this type. The first is the group of eigenvalues of a measure-preserving transformation (See Chapter 3). This will be a subgroup of the unit circle in the complex plane, and isomorphic measure-preserving transformations have the same group of eigenvalues. We shall see in Chapter 3 that in the collection of all ergodic measure-preserving transformations with discrete spectrum this invariant is complete, i.e., if two ergodic measure-preserving transformations with discrete spectrum have the same group of eigenvalues then the transformations are isomorphic.

The other invariant we shall consider is entropy. Entropy theory assigns to each measure-preserving transformation T a non-negative number $h(T)$ (which could be $+\infty$) and if T_1 is isomorphic to T_2 then $h(T_1) = h(T_2)$. In 1969, D. S. Ornstein proved the deep result that in the collection of all Bernoulli shifts this invariant is complete i.e. two Bernoulli shifts with the same entropy are isomorphic. We discuss this more fully in Chapter 4.

Before moving to these questions we shall see that some properties of measure-preserving transformations only depend on the unitary isomorphism class of the induced operators on L^2 spaces.

§2.5 Spectral Isomorphism

Let $T: X \to X$ be a measure-preserving transformation on the probability space $(X, \mathscr{B}, m)$. We have defined the operator $U_T: L^2(m) \to L^2(m)$ by $U_T f = f \circ T$. The operator U_T is linear and $(U_T f, U_T g) = (f, g)$; $f, g \in L^2(m)$. Since this last property says U_T is norm-preserving we know that U_T is injective. There is the following simple result.

Theorem 2.8 *Let T be a measure-preserving transformation of the probability space $(X, \mathscr{B}, m)$. Then $U_T: L^2(m) \to L^2(m)$ is surjective iff $\tilde{T}^{-1}: (\tilde{\mathscr{B}}, \tilde{m}) \to (\tilde{\mathscr{B}}, \tilde{m})$ is surjective (i.e. U_T is a unitary operator iff $\tilde{T}^{-1}$ is an automorphism of the measure-algebra).*

PROOF. Let $B \in \mathscr{B}$. Then $U_T \chi_B = \chi_{\tilde{T}^{-1}B}$. If $\tilde{T}^{-1}$ is surjective then the image of U_T contains all characteristic functions and so U_T is surjective. Suppose now U_T is surjective and hence bijective. Let $A \in \mathscr{B}$. If $U_T f = \chi_{\tilde{A}}$ then $U_T(f \cdot f) = U_T f \cdot U_T f = \chi_{\tilde{A}}$ so $f \cdot f = f$. Therefore $f = \chi_{\tilde{C}}$ for some $C \in \mathscr{B}$. Therefore $\tilde{A} = \tilde{T}^{-1}\tilde{C}$ and $\tilde{T}^{-1}$ is surjective.   □

On decent measure spaces Theorem 2.8 asserts that the surjectivity of U_T is equivalent to T being invertible mod 0 (see Theorem 2.7).

We now consider equivalence in the category of isometric operators in L^2 spaces.

Definition 2.8. Measure-preserving transformations T_1 on $(X_1, \mathscr{B}_1, m_1)$, and T_2 on $(X_2, \mathscr{B}_2, m_2)$ are *spectrally isomorphic* if there is a linear operator

$W : L^2(m_2) \to L^2(m_1)$ such that

  (i) W is invertible
  (ii) $(Wf, Wg) = (f, g) \; \forall f, g \in L^2(m_2)$
  (iii) $U_{T_1} W = W U_{T_2}$.

(The conditions (i), (ii) just say that W is an isomorphism of Hilbert spaces.)
  The following shows spectral isomorphism is weaker than conjugacy.

Theorem 2.9. *Let* T_i *($i = 1, 2$) be a measure-preserving transformation of a probability space* $(X_i, \mathscr{B}_i, m_i)$. *If* T_1 *and* T_2 *are conjugate then they are spectrally isomorphic.*

PROOF. Suppose $\Phi : (\tilde{\mathscr{B}}_2, \tilde{m}_2) \to (\tilde{\mathscr{B}}_1, \tilde{m}_1)$ is an isomorphism of measure algebras such that $\Phi \tilde{T}_2^{-1} = \tilde{T}_1^{-1} \Phi$. Let V be defined as in Theorem 2.3. It remains to show $V U_{T_2} = U_{T_1} V$.
  For $B_2 \in \mathscr{B}_2$ we have

$$U_{T_1} V(\chi_{\tilde{B}_2}) = U_{T_1}(\chi_{\Phi \tilde{B}_2}) = \chi_{\tilde{T}_1^{-1} \Phi \tilde{B}_2} = \chi_{\Phi \tilde{T}_2^{-1} \tilde{B}_2} = V(\chi_{\tilde{T}_2^{-1} \tilde{B}_2}) = V U_{T_2}(\chi_{\tilde{B}_2}).$$

Therefore $U_{T_1} V$ and $V U_{T_2}$ agree on characteristic functions and hence on linear combinations of characteristic functions. By their continuity we have $U_{T_1} V = V U_{T_2}$.   □

  The following tells us when spectral isomorphism implies conjugacy.

Theorem 2.10. *If* T_i *($i = 1, 2$) is a measure-preserving transformation of a probability space* $(X_i, \mathscr{B}_i, m_i)$ *and if* $V : L^2(m_2) \to L^2(m_1)$ *is an invertible linear isometry satisfying the conditions of Theorem 2.4 and* $U_{T_1} V = V U_{T_2}$, *then* T_1 *and* T_2 *are conjugate.*

PROOF. By Theorem 2.4 V is induced by an isomorphism of measure-algebras $\Phi : (\tilde{\mathscr{B}}_2, \tilde{m}_2) \to (\tilde{\mathscr{B}}_1, \tilde{m}_1)$ in the sense $V(\chi_{\tilde{B}}) = \chi_{\Phi(\tilde{B})}$, $B \in \mathscr{B}_2$. The equation $U_{T_1} V(\chi_{\tilde{B}}) = V U_{T_2}(\chi_{\tilde{B}})$ becomes $\chi_{\tilde{T}_1^{-1} \Phi(\tilde{B})} = \chi_{\Phi \tilde{T}_2^{-1}(\tilde{B})}$ and so $\tilde{T}_1^{-1} \Phi = \Phi \tilde{T}_2^{-1}$.
                                                                    □

  Spectral isomorphism is much weaker than conjugacy as we shall see in the following. For one class of transformations spectral isomorphism implies conjugacy. This class consists of ergodic transformations with discrete spectrum and will be discussed fully in Chapter 3. An ergodic measure-preserving transformation T of a probability space $(X, \mathscr{B}, m)$ has discrete spectrum if $L^2(m)$ has an orthonormal basis $\{f_i\}$ consisting of eigenfunctions of U_T (i.e. $U_T f_i = \lambda_i f_i$ for some complex number λ_i). For this class of transformations a basis member f_i is mapped by U_T to a constant multiple of itself. We now briefly discuss transformations which have the "opposite" type of behaviour, in the sense that there is a basis where each basis member is mapped by U_T to another basis member.

Definition 2.9. Let $(X, \mathscr{B}, m)$ be a probability space $(X, \mathscr{B}, m)$ with a countable basis (because we want $L^2(m)$ to be separable). An invertible measure-preserving transformation T of $(X, \mathscr{B}, m)$ is said to have *countable Lebesgue spectrum* if there is a sequence $\{f_j\}_{j=0}^{\infty}$, with $f_0 \equiv 1$, of members of $L^2(m)$ such that $\{f_0\} \cup \{U_T^n f_j | j \geq 1, n \in Z\}$ is an orthonormal basis of $L^2(m)$.

Diagramatically the basis has the form

$$f_0 \equiv 1$$
$$\ldots, U_T^{-2} f_1, U_T^{-1} f_1, f_1, U_T f_1, U_T^2 f_1, \ldots$$
$$\ldots, U_T^{-2} f_2, U_T^{-1} f_2, f_2, U_T f_2, U_T^2 f_2, \ldots$$
$$\vdots \qquad \vdots \qquad \vdots \qquad \vdots \qquad \vdots$$

We shall show in §4.9 that every Kolmogorov automorphism has countable Lebesgue spectrum. This implies that every Bernoulli shift has countable Lebesgue spectrum. We shall now indicate why the two-sdied $(\frac{1}{2}, \frac{1}{2})$-shift T has countable Lebesgue spectrum. Here the state space, $\{0, 1\}$ consists of two points each with measure $\frac{1}{2}$. A basis for the L^2-space of the state space consists of the constant function 1 and the map $\{0, 1\} \to \mathbb{C}$ given by $t \to e^{\pi i t}$, $t \in \{0, 1\}$. The transformation T acts on the direct product space $X = \{0, 1\}^{\mathbb{Z}}$ equipped with the product measure m. Since $L^2(m)$ is the tensor product the L^2-space of the state space there is a basis for $L^2(m)$ of the form $\{g_{n_1, \ldots, n_r} | r \geq 1, n_1 < n_2 \cdots < n_r\} \cup \{1\}$ where

$$g_{n_1, \ldots, n_r}(\{x_i\}) = e^{\pi i (x_{n_1} + x_{n_2} + \cdots + x_{n_r})}.$$

Note that $U_T g_{n_1, \ldots, n_r} = g_{1+n_1, \ldots, 1+n_r}$. It is now clear that we can rename the basis so that it has the form $\{U_T^n f_i | i \geq 1, n \in Z\} \cup \{1\}$.

One can use this method to show directly that the two-sided $(p_0, \ldots, p_{k-1})$-shift has countable Lebesgue spectrum. It is also easy to give a direct proof that an ergodic automorphism A of a compact abelian metric group G has countable Lebesgue spectrum. The elements of the character group $\hat{G}$ form an orthonormal basis. The ergodicity of A implies that if $\gamma \in \hat{G}$ and $\gamma \not\equiv 1$ then the collection $\{\hat{A}^n \gamma | n \in Z\}$ consists of district characters. (Here $\hat{A}$ is the dual automorphism to A.) Therefore A has countable Lebesgue spectrum once it is shown that there are infinitely many district sets of the form $\{\hat{A}^n \gamma | n \in Z\}$ in $\hat{G}$. This can be done by a simple group theory argument (Halmos [1] p, 54).

The following results are elementary.

Theorem 2.11. *Any two invertible measure-preserving transformations with countable Lebesgue spectrum are spectrally isomorphic.*

PROOF. Let $(X_i, \mathscr{B}_i, m_i)$ $i = 1, 2$ be a probability space and let $T_i : X_i \to X_i$ be an invertible measure-preserving transformation. Suppose $L^2(m_1)$ has a basis $\{f_0\} \cup \{U_{T_1}^n f_j | j \geq 1, n \in Z\}$ where $f_0 \equiv 1$ and $L^2(m_2)$ has a basis $\{g_0\} \cup$

$\{U_{T_2}^n g_j | j \geq 1, n \in Z\}$ where $g_0 \equiv 1$. Define $W: L^2(m_2) \to L^2(m_1)$ by $W(g_0) = f_0$, $W(U_{T_2}^n g_j) = U_{T_1}^n f_j$ and extend by linearity. Then $WU_{T_2} = U_{T_1} W$ and T_1 and T_2 are spectrally isomorphic.   □

It follows from this and the discussion above that any two Bernoulli shifts are spectrally isomorphic. This was known in 1943 and only when entropy was introduced by A. N. Kolmogorov in 1958 was it shown that there are non-isomorphic Bernoulli shifts (see Chapter 4).

Theorem 2.12. *If a measure-preserving transformation T of a probability space $(X, \mathcal{B}, m)$ has countable Lebesgue spectrum it is strong-mixing.*

PROOF. Let $\{f_0\} \cup \{U_T^n f_j, j \geq 1, n \in Z\}$ be a basis of $L^2(m)$ where $f_0 \equiv 1$. Then if j, $q \geq 0$ $\lim_{p \to \infty} (U_T^p \circ U_T^n f_j, U_T^k f_q) = (U_T^n f_m, 1)(1, U_T^k f_q)$ $\forall k$, $n \in Z$, since both sides are zero unless $j = q = 0$ and then both sides equal one. Fix k and q and consider

$$\mathcal{H}_{k,q} = \left\{ f \in L^2(m): \lim_{p \to \infty} (U_T^p f, U_T^k f_q) = (f, 1)(1, U_T^k f_q) \right\}.$$

Then $\mathcal{H}_{k,q}$ is a closed subspace of $L^2(m)$ and contains the basis by the above calculation. Hence $\mathcal{H}_{k,q} = L^2(m)$. Fix $f \in L^2(m)$ and let

$$\mathcal{L}_f = \left\{ g \in L^2(m): \lim_{p \to \infty} (U_T^p f, g) = (f, 1)(1, g) \right\}.$$

Then $\mathcal{L}_f$ is a closed subspace of $L^2(m)$, contains the basis by the above, and therefore is equal to $L^2(m)$. Hence

$$\lim_{p \to \infty} (U_T^p f, g) = (f, 1)(1, g) \quad \forall f, g \in L^2(m). \qquad \square$$

§2.6 Spectral Invariants

Definition 2.10. A property P of measure-preserving transformations is a
$$\begin{cases} \text{isomorphism} \\ \text{conjugacy} \\ \text{spectral} \end{cases} \text{invariant if the following holds:}$$

Given T_1 has P and T_2 is $\begin{cases} \text{isomorphic} \\ \text{conjugate} \\ \text{spectrally isomorphic} \end{cases}$ to T_1

then T_2 has property P.

Remark. A spectral invariant is a conjugacy invariant, and a conjugacy invariant is an isomorphism invariant.

The following shows that the properties we have considered up to now are spectral invariants.

Theorem 2.13. *The following are spectral invariants of measure-preserving transformations:* (1) *ergodicity,* (2) *weak-mixing,* (3) *strong-mixing.*

PROOF

(1) We know T is ergodic iff $\{f \in L^2(m): U_T f = f\}$ is a one-dimensional subspace, and the latter condition is preserved under spectral isomorphism.

(2) We know T is weak-mixing iff 1 is the only eigenvalue and T is ergodic, and this is preserved under spectral isomorphism.

(3) Suppose $WU_{T_2} = U_{T_1}W$ and T_1 is strong-mixing. We have to show that

$$(U_{T_2}^n h, k) \to (h, 1)(1, k) \quad \forall h, k \in L^2(m_2).$$

Since this is true if h is constant or if k is constant, it suffices to consider the cases when $(h, 1) = 0 = (k, 1)$. Since T_1 is ergodic then T_2 is ergodic by (1) and since W sends the invariant functions for T_2 onto those for T_1 W maps the subspace of constants in $L^2(m_2)$ onto the subspace of constants in $L^2(m_1)$. So $(Wh, 1) = 0 = (1, Wk)$. Since W preserves the inner product,

$$(U_{T_2}^n h, k) = (WU_{T_2}^n h, Wk) = (U_{T_1}^n Wh, Wk) \to 0$$

since T_1 is strong-mixing. Therefore T_2 is strong-mixing. $\qquad\square$

This theorem allows us to easily display non-spectrally isomorphic transformations. For example a non-ergodic transformation (such as a rotation of K by a root of unity) cannot be spectrally isomorphic to an ergodic transformation (such as a rotation of K by a non root of unity). Also a rotation of a compact group, which is not weak-mixing, cannot be spectrally isomorphic to an ergodic automorphism of a compact group because such automorphisms are weak-mixing.

CHAPTER 3

Measure-Preserving Transformations with Discrete Spectrum

In this chapter we study a class of measure-preserving transformations for which the conjugacy problem is solved and for which spectral isomorphism implies conjugacy.

§3.1 Eigenvalues and Eigenfunctions

Definition 3.1. Let T be a measure-preserving transformation of the probability space $(X, \mathscr{B}, m)$, and let U_T be the induced linear isometry of $L^2(m)$. The eigenvalues and eigenfunctions of U_T are called the eigenvalues and eigenfunctions of T. So a complex number λ is called an *eigenvalue* of T if there is $f \in L^2(m)$, with f not the zero function, satisfying $U_T f = \lambda f$. The function f is called an *eigenfunction* of T corresponding to the eigenvalue λ.

The main properties are as follows.

Theorem 3.1. *Let T be a measure-preserving transformation of a probability space $(X, \mathscr{B}, m)$ and suppose T is ergodic. Then the following are true.*

(i) *If $U_T f = \lambda f$, $f \in L^2(m)$, $f \not\equiv 0$, then $|\lambda| = 1$ and $|f|$ is constant a.e.*

(ii) *Eigenfunctions corresponding to different eigenvalues are orthogonal.*

(iii) *If f and g are both eigenfunctions corresponding to the eigenvalue λ then $f = cg$ a.e. for some constant c.*

(iv) *The eigenvalues of T form a subgroup of the unit circle K.*

PROOF

(i) We have $\|U_T f\| = |\lambda| \|f\|$ so that $\|f\| = |\lambda| \|f\|$. Since $\|f\| \neq 0$ we have $|\lambda| = 1$. Then we have $|U_T f| = |\lambda| |f| = |f|$ so by ergodicity $|f|$ is constant a.e.

(ii) Suppose $\lambda \neq \mu$, $U_T f = \lambda f$, $U_T g = \mu g$. Then

$$(f, g) = (U_T f, U_T g) = (\lambda f, \mu g) = \lambda \bar{\mu}(f, g)$$

and $\lambda \bar{\mu} \neq 1$ implies $(f, g) = 0$.

(iii) Since g is not the zero element of $L^2(m)$ and $|g|$ is constant a.e. by (i) we have $g(x) \neq 0$ a.e. Then f/g is an invariant function for T and hence constant a.e.

(iv) If λ, μ are both eigenvalues of T then $f \circ T = \lambda f$, $g \circ T = \mu g$ for some non-zero f, $g \in L^2(m)$. By taking complex conjugates we have $\bar{g} \circ T = \bar{\mu} \bar{g}$ and so $(f\bar{g}) \circ T = (\lambda \bar{\mu})(f\bar{g})$. Therefore $\lambda \mu^{-1} = \lambda \bar{\mu}$ is an eigenvalue of T and the eigenvalues of T form a subgroup of K.   □

Note that (ii) and (iii) say that eigenspaces are one-dimensional and mutually orthogonal. Because eigenspaces are orthogonal we know that if $L^2(m)$ is separable then the group of eigenvalues of T is countable.

Remark. It is clear that if T_1 is spectrally isomorphic to T_2 then T_1 and T_2 have the same eigenvalues.

§3.2 Discrete Spectrum

Definition 3.2. An ergodic measure-preserving transformation T on a probability space $(X, \mathcal{B}, m)$ has *discrete spectrum* (*pure-point spectrum*) if there exists an orthonormal basis for $L^2(m)$ which consists of eigenfunctions of T.

Remark. If T has discrete spectrum then $U_T : L^2(m) \to L^2(m)$ is clearly surjective so that $\tilde{T}^{-1}\tilde{\mathcal{B}} = \tilde{\mathcal{B}}$ (Theorem 2.8). Therefore if $(X, \mathcal{B}, m)$ is a Lebesgue space or arises from a complete separable metric space then T is invertible mod 0.

We shall need the following results for the proof of the main theorem of this section.

Lemma 3.2. *Let* $(X, \mathcal{B}, m)$ *be a probability space. Let* $h \in L^2(m)$. *Then* h *is bounded (i.e.* $\exists c \in R$ *such that* $m(\{x \mid |h(x)| > c\}) = 0$*) iff* $h \cdot f \in L^2(m)$ *for all* $f \in L^2(m)$.

PROOF. If h is bounded then clearly $h \cdot f \in L^2(m)$ when $f \in L^2(m)$. Now suppose h is such that $h \cdot f \in L^2(m)$ whenever $f \in L^2(m)$. Let

$$X_n = \{x \in X \mid n - 1 \leq |h(x)| < n\}$$

for $n \geq 1$. Then $\{X_n\}_1^\infty$ partitions X. Let $f(x) = \sum_{i=1}^\infty i^{-1} m(X_i)^{-1/2} \chi_{X_i}(x)$, where it is understood that the i-term is omitted if $m(X_i) = 0$.

Then

$$\int |f|^2 \, dm \le \sum_1^\infty \frac{1}{i^2} < \infty \quad \text{but} \quad \int |hf|^2 \, dm \ge \sum_{i \in F} \left(\frac{i-1}{i} \right)^2$$

where $F = \{i \mid m(X_i) \ne 0\}$. Since $hf \in L^2(m)$ we have that F is finite and therefore h is bounded.   □

The following abstract group theoretic result is needed.

Lemma 3.3. *Let H be a discrete abelian group and K a divisible subgroup of H (i.e., $\forall k \in K$ and $\forall n > 0$ $\exists a \in K$ such that $a^n = k$). Then there exists a homomorphism $\phi : H \to K$ such that $\phi|_K = $ identity (i.e., K is an algebraic retract of H).*

PROOF. Let $\mathscr{R}$ consist of all retracts onto K from supergroups of K in H, i.e., $\mathscr{R}$ consists of all pairs (M, ϕ) where $K \subset M \subset H$ and $\phi : M \to K$ is a homomorphism such that $\phi|_K = $ identity. We know $\mathscr{R}$ is non-empty as $(K, id_K) \in \mathscr{R}$. We order $\mathscr{R}$ by extension, i.e., $(M_1, \phi_1) < (M_2, \phi_2)$ if $M_1 \subset M_2$ and $\phi_2|_{M_1} = \phi_1$. This is a partial ordering and every linearly ordered subset has an upper bound. So by Zorn's Lemma there exists a maximal element, say (L, p), of $\mathscr{R}$.

We claim that $L = H$. Suppose not, then consider $g \in H \backslash L$ and let M be the group generated by g and L.

Case 1. If no power of g lies in L then every element of M can be uniquely written in the form $g^i a$ where $a \in L$, $i \in Z$. We define $\psi : M \to K$ by $\psi(g^i a) = p(a)$. We can easily check that ψ is a homomorphism and that $\psi|_K = id_K$. This then contradicts the maximality of (L, p).

Case 2. Let n be the least positive integer such that $g^n \in L$. Each element of M can be uniquely written as $g^i a$, where $a \in L$, $0 \le i \le n - 1$. Since K is divisible, let $g_0 \in K$ be such that $p(g^n) = g_0^n$. Then $\psi(g^i a) = g_0^i p(a)$ defines a homomorphism of M into K such that $\psi|_K = id_k$. Again, we have contradicted the maximality of (L, p).

Therefore $L = H$.   □

The following theorem due to Halmos and von Neumann (1942) shows that the eigenvalues determine completely whether two transformations with discrete spectrum are conjugate or not.

Theorem 3.4 (Discrete Spectrum Theorem). *Let T_i be an ergodic measure-preserving transformation of a probability space $(X_i, \mathscr{B}_i, m_i)$ and suppose T_i has discrete spectrum, $i = 1, 2$. The following are equivalent:*

   (i) *T_1 and T_2 are spectrally isomorphic.*
   (ii) *T_1 and T_2 have the same eigenvalues.*
   (iii) *T_1 and T_2 are conjugate.*

PROOF

(i) $\Rightarrow$ (ii) is trivial.

(iii) $\Rightarrow$ (i) is always true (Theorem 2.9).

(ii) $\Rightarrow$ (i). For each eigenvalue λ, choose $f_\lambda \in L^2(m_1)$, $g_\lambda \in L^2(m_2)$ such that $U_{T_1} f_\lambda = \lambda f_\lambda$, $U_{T_2} g_\lambda = \lambda g_\lambda$ and $|f_\lambda| = |g_\lambda| = 1$.

We define $W: L^2(m_2) \to L^2(m_1)$ by $W(g_\lambda) = f_\lambda$ and extending by linearity. We readily see that W is a bijective isometry. Moreover $WU_{T_2} = U_{T_1} W$ by checking this on the g_λ.

We now prove that (ii) $\Rightarrow$ (iii). Let Λ denote the group of eigenvalues of T_1 which we are assuming to equal the group of eigenvalues of T_2. For each $\lambda \in \Lambda$ choose $f_\lambda \in L^2(m_1)$ so that $|f_\lambda| = 1$ and $U_{T_1} f_\lambda = \lambda f_\lambda$. We know that $\{f_\lambda : \lambda \in \Lambda\}$ is a basis for $L^2(m_1)$. Also choose $g_\lambda \in L^2(m_2)$ so that $|g_\lambda| = 1$ and $U_{T_2} g_\lambda = \lambda g_\lambda$. We know that $\{g_\lambda : \lambda \in \Lambda\}$ is a basis for $L^2(m_2)$. We have for every $\lambda, \mu \in \Lambda$

$$U_{T_1} f_{\lambda\mu} = \lambda\mu f_{\lambda\mu}$$

and also

$$U_{T_1}(f_\lambda \cdot f_\mu) = f_\lambda(T) \cdot f_\mu(T) = (\lambda\mu)(f_\lambda \cdot f_\mu).$$

By (iii) of Theorem 3.1 there exists a constant $r(\lambda, \mu) \in K$ such that $f_\lambda(x) f_\mu(x) = r(\lambda, \mu) f_{\lambda\mu}(x)$ a.e. We shall use Lemma 3.3 to show that we can suppose $r(\lambda, \mu) = 1$.

Let H denote the collection of all functions $X \to K$. Clearly H is an abelian group under pointwise multiplication. Moreover K is a subgroup of H if we identify constant functions with their values.

By Lemma 3.3 there exists a homomorphism $\phi: H \to K$ such that $\phi|_K = id_K$. Let $f_\lambda^* = \overline{\phi(f_\lambda)} f_\lambda$. Then $|f_\lambda^*| = 1$, $U_T f_\lambda^* = \lambda f_\lambda^*$ and $\{f_\lambda^* : \lambda \in \Lambda\}$ is a basis for $L^2(m_1)$. Also,

$$
\begin{aligned}
f_\lambda^* f_\mu^* &= \overline{\phi(f_\lambda)}\, \overline{\phi(f_\mu)} f_\lambda f_\mu = \overline{\phi(f_\lambda f_\mu)} f_\lambda f_\mu \\
&= \overline{\phi(r(\lambda, \mu))}\, \overline{\phi(f_{\lambda\mu})} r(\lambda, \mu) f_{\lambda\mu} \\
&= \overline{r(\lambda, \mu)}\, \overline{\phi(f_{\lambda\mu})} r(\lambda, \mu) f_{\lambda\mu} \\
&= f_{\lambda\mu}^*.
\end{aligned}
$$

Thus without loss of generality we can assume that $f_\lambda f_\mu = f_{\lambda\mu}$. Similarly we may as well assume $g_\lambda g_\mu = g_{\lambda\mu}$ $\forall \lambda, \mu \in \Lambda$.

Define $W: L^2(m_2) \to L^2(m_1)$ by $W(g_\lambda) = f_\lambda$ and extended by linearity. The operator W is bijective, linear and preserves the inner product. Also $WU_{T_2} = U_{T_1} W$. If we can show that W satisfies the conditions of Theorem 2.4 then by Theorem 2.10 T_1 and T_2 are conjugate. But

$$W(g_\lambda g_\mu) = W(g_{\lambda\mu}) = f_{\lambda\mu} = f_\lambda f_\mu = W(g_\lambda) W(g_\mu).$$

Let $h, k \in L^2(m_2)$ and let k be bounded. If we fix g_μ and let a finite linear combination of g_λ's converge to h in $L^2(m_2)$ we obtain that $W(hg_\mu) = W(h) W(g_\mu)$. Then if we let a finite linear combination of g_μ's converge to k in $L^2(m_2)$ we get that $W(hk) = W(h) W(k)$. It follows from this and Lemma 3.2

that W maps bounded functions to bounded functions because $hk \in L^2(m_2)$ so $W(hk) \in L^2(m_1)$ $\forall h \in L^2(m_2)$ and so $W(k)f \in L^2(m_1)$ for all $f \in L^2(m_1)$. □

Corollary 3.4.1. *If T is an invertible ergodic measure-preserving transformation with discrete spectrum then T and T^{-1} are conjugate.*

PROOF. They have the same eigenvalues. □

Remark. When the spaces $(X_1, \mathcal{B}_1, m_1)$, $(X_2, \mathcal{B}_2, m_2)$ are both Lebesgue spaces or both complete separate metric spaces then the statements in Theorem 3.4 are also equivalent to T_1 being isomorphic to T_2.

§3.3 Group Rotations

We now discuss a class of examples of ergodic measure-preserving transformations with discrete spectrum.

Let $T: K \to K$ be defined by $T(z) = az$ where a is not a root of unity. We know that T is ergodic. Let $f_n: K \to \mathbb{C}$ be defined by $f_n(z) = z^n$ where $n \in Z$. Then

$$f_n(Tz) = f_n(az) = a^n z^n = a^n f_n(z).$$

Thus f_n is an eigenfunction with eigenvalue a^n. Since the $\{f_n\}$ form a basis for $L^2(K)$ we see that T is ergodic and has discrete spectrum.

These ideas carry over to ergodic rotations on any compact abelian group. Recall that $\hat{G}$ denotes the character group of a compact abelian group G (see §0.7), and that we always use normalised Haar measure, m, on such a group G if no other measure is mentioned. If G is not metrisable then $\hat{G}$ is not countable.

Theorem 3.5. *Let T, given by $T(g) = ag$, be an ergodic rotation of a compact abelian group G. Then T has discrete spectrum. Every eigenfunction of T is a constant multiple of a character, and the eigenvalues of T are $\{\gamma(a) : \gamma \in \hat{G}\}$.*

PROOF. Let $\gamma \in \hat{G}$. Then

$$\gamma(Tg) = \gamma(ag) = \gamma(a)\gamma(g).$$

Therefore each character is an eigenfunction and so T has discrete spectrum since the characters are an orthonormal basis of $L^2(m)$. If there is another eigenvalue besides the members of $\{\gamma(a) : \gamma \in \hat{G}\}$ then the corresponding eigenfunction would be orthogonal to all members of $\hat{G}$, by (iv) of Theorem 3.1, and so is zero. Hence $\{\gamma(a) : \gamma \in \hat{G}\}$ is the group of all eigenvalues of T and the only eigenfunctions are constant multiples of characters, using (iii) of Theorem 3.1. □

It turns out that such rotations are the canonical examples of measure-preserving transformations with discrete spectrum.

Theorem 3.6 (Representation Theorem). *An ergodic measure-preserving transformation T with discrete spectrum on a probability space $(X, \mathcal{B}, m)$ is conjugate to an ergodic rotation on some compact abelian group. The group will be metrisable iff $(X, \mathcal{B}, m)$ has a countable basis.*

PROOF. Let Λ be the group of all eigenvalues of T and give Λ the discrete topology. So Λ is an algebraic subgroup of K but has the discrete topology. (If $L^2(m)$ is separable then Λ is countable). Let $G = \hat{\Lambda}$, the character group of Λ. Then G is compact and abelian. By the duality theorem (3 of §0.7) $\hat{G} = \hat{\hat{\Lambda}}$ is naturally isomorphic to Λ. For $\lambda \in \Lambda$ we shall let $\underline{\lambda}$ denote the corresponding element of $\hat{G}$ i.e. $\underline{\lambda}(g) = g(\lambda) \ \forall g \in G = \hat{\Lambda}$. The map $a: \Lambda \to K$ given by $a(\lambda) = \lambda$ is a homomorphism of the discrete group Λ into K and so belongs to $\hat{\Lambda} = G$. Therefore $\underline{\lambda}(a) = a(\lambda) = \lambda \ \forall \lambda \in \Lambda$.

Define $S: G \to G$ by $S(g) = ag$. We claim that S is ergodic. Let μ denote Haar measure on G and suppose $f \circ S = f, f \in L^2(\mu)$. Then f has a Fourier series $f = \sum_j b_j \underline{\lambda}_j, \ \lambda_j \in \Lambda$. From $f \circ S = f$ we have $\sum_j b_j \underline{\lambda}_j(a) \underline{\lambda}_j(g) = \sum_j b_j \underline{\lambda}_j(g)$ so that $b_j \underline{\lambda}_j(a) = b_j$. Since $\underline{\lambda}_j(a) = \lambda_j$ this gives $b_j \lambda_j = b_j$. If $b_j \neq 0$ then $\lambda_j = 1$ so that $\underline{\lambda}_j \equiv 1$. Therefore the only non-vanishing term in the Fourier series is the constant term and so f is constant a.e. We now know S is ergodic and, by Theorem 3.5, it has discrete spectrum.

Again by Theorem 3.5 the group of eigenvalues of S is $\{\gamma(a): \gamma \in \hat{G}\} = \{a(\lambda): \lambda \in \Lambda\} = \{\lambda: \lambda \in \Lambda\} = \Lambda$. Therefore S and T have the same eigenvalues and both have discrete spectrum. By the Discrete Spectrum Theorem they are conjugate.

The group G is metrisable iff Λ is countable and this is equivalent to $L^2(m)$ being separable. $\square$

Theorem 3.7 (Existence Theorem). *Every subgroup Λ of K is the group of eigenvalues of an ergodic measure-preserving transformation with discrete spectrum.*

PROOF. The desired transformation is the rotation S constructed in the proof of Theorem 3.6. $\square$

The conjugacy problem for ergodic measure-preserving transformations with discrete spectrum is completely solved. We have some very simple invariants, namely the eigenvalues, which determine when two such transformations are conjugate. Each conjugacy class of ergodic measure-preserving transformations with discrete spectrum is characterized by a subgroup of K, and each subgroup of K corresponds to a conjugacy class. So for this class of transformations we get a very satisfying solution to the conjugacy problem. Also there are some simple examples, namely group rotations, such that each ergodic measure-preserving transformation with discrete spectrum is conjugate to one of these examples.

An extension of these results to a wider class of transformations has been carried out by Abramov [1], Hoare and Parry [1] and Hahn and Parry [1], [2]. This class is the collection of all transformations with quasi-discrete spectrum. The invariant is not just one group but a sequence of groups connected by homomorphisms. The canonical examples in this class are certain affine transformations of compact abelian groups.

CHAPTER 4
Entropy

We are searching for conjugacy and/or isomorphism invariants. In 1958 Kolmogorov introduced the concept of entropy into ergodic theory, and this has been the most successful invariant so far. For example, in 1943 it was known that the two-sided $(\frac{1}{2}, \frac{1}{2})$-shift and the two-sided $(\frac{1}{3}, \frac{1}{3}, \frac{1}{3})$-shift both have countable Lebesgue spectrum and hence are spectrally isomorphic, but it was not known whether they are conjugate. This was resolved in 1958 when Kolmogorov showed that they had entropies log 2 and log 3, respectively, and hence are not conjugate. The notion of entropy now used is slightly different from that used by Kolmogorov—the improvement was made by Sinai in 1959.

The definition of the entropy of a measure-preserving transformation T of $(X, \mathscr{B}, m)$ is in three stages: the entropy of a finite sub-σ-algebra of $\mathscr{B}$, the entropy of the transformation T relative to a finite sub-σ-algebra, and, finally, the entropy of T. Each stage of the definition is quite simple to state. Before giving the definition we shall study finite sub-σ-algebras of $\mathscr{B}$ and give some motivation for the definition.

The definitions involve logarithms and we shall use natural logarithms. This is because it will be more natural in Chapter 9, to tie in with some ideas from statistical mechanics. Some authors use logarithms of base 2.

§4.1 Partitions and Subalgebras

Throughout this chapter $(X, \mathscr{B}, m)$ will denote a probability space.

Definition 4.1. A *partition* of $(X, \mathscr{B}, m)$ is a disjoint collection of elements of $\mathscr{B}$ whose union is X.

We shall be interested in finite partitions. They will be denoted by Greek letters, e.g., $\xi = \{A_1, \ldots, A_k\}$.

If ξ is a finite partition of $(X, \mathscr{B}, m)$ then the collection of all elements of $\mathscr{B}$ which are unions of elements of ξ is a finite sub-σ-algebra of $\mathscr{B}$. We denote it by $\mathscr{A}(\xi)$. Conversely, if $\mathscr{C}$ is a finite sub-σ-algebra of $\mathscr{B}$, say $\mathscr{C} = \{\mathscr{C}_i : i = 1, \ldots, n\}$, then the non-empty sets of the form $B_1 \cap \cdots \cap B_n$, where $B_i = C_i$ or $X \backslash C_i$, form a finite partition of $(X, \mathscr{B}, m)$. We denote it by $\xi(\mathscr{C})$. We have $\mathscr{A}(\xi(\mathscr{C})) = \mathscr{C}$ and $\xi(\mathscr{A}(\eta)) = \eta$. Thus we have a one-to-one correspondence between finite partitions and finite sub-σ-algebras of $\mathscr{B}$.

Definition 4.2. Suppose ξ and η are two finite partitions of $(X, \mathscr{B}, m)$. We write $\xi \leq \eta$ to mean that each element of ξ is a union of elements of η. (i.e. η is a *refinement* of ξ). We have $\xi \leq \eta \Leftrightarrow \mathscr{A}(\xi) \subseteq \mathscr{A}(\eta)$, and $\mathscr{A} \subseteq \mathscr{C} \Leftrightarrow \xi(\mathscr{A}) \leq \xi(\mathscr{C})$.

Definition 4.3. Let $\xi = \{A_1, \ldots, A_n\}$, $\eta = \{C_1, \ldots, C_k\}$ be two finite partitions of $(X, \mathscr{B}, m)$. Their *join* is the partition

$$\xi \vee \eta = \{A_i \cap C_j : 1 \leq i \leq n, 1 \leq j \leq k\}.$$

If $\mathscr{A}$ and $\mathscr{C}$ are finite sub-σ-algebras of $\mathscr{B}$ then $\mathscr{A} \vee \mathscr{C}$ denotes the smallest sub-σ-algebra of $\mathscr{B}$ containing $\mathscr{A}$ and $\mathscr{C}$.

Clearly $\mathscr{A} \vee \mathscr{C}$ consists of all sets which are unions of sets of the form $A \cap C$, $A \in \mathscr{A}$, $C \in \mathscr{C}$. We have $\xi(\mathscr{A} \vee \mathscr{C}) = \xi(\mathscr{A}) \vee \xi(\mathscr{C})$, and $\mathscr{A}(\xi \vee \eta) = \mathscr{A}(\xi) \vee \mathscr{A}(\eta)$.

Definition 4.4. Suppose $T : X \to X$ is a measure-preserving transformation. If $\xi = \{A_1, \ldots, A_k\}$, then $T^{-n}\xi$ denotes the partition $\{T^{-n}A_1, \ldots, T^{-n}A_k\}$ and if $\mathscr{A}$ is a sub-σ-algebra of $\mathscr{B}$ then $T^{-n}(\mathscr{A})$ denotes the sub-σ-algebra $\{T^{-n}A : A \in \mathscr{A}\}$ $(n \geq 0)$.

If $n \geq 0$, then, since T^{-n} preserves set theoretic operations, we have

$$\xi(T^{-n}\mathscr{A}) = T^{-n}\xi(\mathscr{A})$$

$$\mathscr{A}(T^{-n}\xi) = T^{-n}\mathscr{A}(\xi)$$

$$T^{-n}(\mathscr{A} \vee \mathscr{C}) = T^{-n}\mathscr{A} \vee T^{-n}\mathscr{C}$$

$$T^{-n}(\xi \vee \eta) = T^{-n}\xi \vee T^{-n}\eta$$

$$\xi \leq \eta \Rightarrow T^{-n}\xi \leq T^{-n}\eta$$

$$A \subseteq \mathscr{C} \Rightarrow T^{-n}\mathscr{A} \subseteq T^{-n}\mathscr{C}.$$

Definition 4.5. If $\mathscr{C}, \mathscr{D}$ are (not necessarily finite) sub-σ-algebras of $\mathscr{B}$ we write $\mathscr{C} \doteq \mathscr{D}$ if for every $C \in \mathscr{C}$ there exists $D \in \mathscr{D}$ with $m(D \triangle C) = 0$. In terms of measure algebras (see §2.1) this is equivalent to $\tilde{\mathscr{C}} \subset \tilde{\mathscr{D}}$. We write $\mathscr{C} \doteq \mathscr{D}$ if $\mathscr{C} \doteq \mathscr{D}$ and $\mathscr{D} \doteq \mathscr{C}$. This is equivalent to $\tilde{\mathscr{C}} = \tilde{\mathscr{D}}$. If ξ, η are finite partitions then $\xi \doteq \eta$ means $\mathscr{A}(\xi) \doteq \mathscr{A}(\eta)$.

If $\mathscr{C}$, $\mathscr{D}$ are finite and $\mathscr{C} \doteq \mathscr{D}$, then if $\zeta(\mathscr{C}) = \{C_1, \ldots, C_p, C_{p+1}, \ldots, C_q\}$, where $m(C_i) > 0$ for $1 \leq i \leq p$ and $m(C_i) = 0$ for $p + 1 \leq i \leq q$, we have $\zeta(\mathscr{D}) = \{D_1, \ldots, D_p, D_{p+1}, \ldots, D_s\}$ where $m(C_i \triangle D_i) = 0$ for $1 \leq i \leq p$ and $m(D_i) = 0$ for $p + 1 \leq i \leq s$.

§4.2 Entropy of a Partition

We shall use natural logarithms, and the expression $0 \log 0$ will be considered to be 0. As in probability theory let us consider a partition $\xi = \{A_1, \ldots, A_k\}$ of $(X, \mathscr{B}, m)$ as listing the possible outcomes of an experiment, where the probability of the outcome A_i is $m(A_i)$. We want to associate to this experiment a number $H(\xi)$ that describes the amount of uncertainty about the outcome of the experiment. In other words, $H(\xi)$ will measure the uncertainty removed (or information gained) by performing the experiment represented by ξ. Suppose we want $H(\xi)$ to depend only on the numbers $\{m(A_1), \ldots, m(A_k)\}$. We shall also denote $H(\xi)$ by $H(m(A_1), \ldots, m(A_k))$. What should this function be? It turns out that the expression for $H(\xi)$ is determined if one requires it to satisfy reasonable properties. To do this we shall derive another function from $H(\xi)$.

Suppose $\xi = \{A_1, \ldots, A_k\}$ and $\eta = \{B_1, \ldots, B_l\}$ are two partitions of $(X, \mathscr{B}, m)$ representing two experiments, and suppose we want to measure the uncertainty about the outcome of ξ if we are to be told the outcome of η. If we know the outcome B_j occurs then A_i occurs with probability $m(A_i \cap B_j)/m(B_j)$, so the uncertainty about the outcome of ξ given B_j occurs is

$$H\left(\left(\frac{m(A_1 \cap B_j)}{m(B_j)}, \frac{m(A_2 \cap B_j)}{m(B_j)}, \ldots, \frac{m(A_k \cap B_j)}{m(B_j)}\right)\right).$$

Therefore the uncertainty about the outcome of ξ, given that we will be told the outcome of η, is

$$(*) \quad H(\xi/\eta) = \sum_{j=1}^{l} m(B_j) H\left(\left(\frac{m(A_1 \cap B_j)}{m(B_j)}, \frac{m(A_2 \cap B_j)}{m(B_j)}, \ldots, \frac{m(A_k \cap B_j)}{m(B_j)}\right)\right).$$

The function $H(\xi)$ is determined by the following result (where $H(\xi)$ means the same as $H((m(A_1), \ldots, m(A_k)))$).

Theorem 4.1. *Let* $\Delta_k = \{(p_1, \ldots, p_k) \in R^k \,|\, p_i \geq 0, \; \sum_{i=1}^{k} p_i = 1\}$. *Suppose* $H: \bigcup_{k=1}^{\infty} \Delta_k \to R$ *has the following properties:*

(i) $H(p_1, \ldots, p_k) \geq 0$, *and* $H(p_1, \ldots, p_k) = 0$ *iff some* $p_i = 1$.
(ii) *For each* $k \geq 1$, $H|_{\Delta_k}$ *is continuous.*
(iii) *For each* $k \geq 1$, $H|_{\Delta_k}$ *is symmetric.*
(iv) *For each* $k \geq 1$, $H|_{\Delta_k}$ *has its largest value at* $(1/k, \ldots, 1/k)$.

(v) $H(\xi \vee \eta) = H(\xi) + H(\eta/\xi)$, where $H(\xi/\eta)$ is defined from H by (*).
(vi) $H((p_1, \ldots, p_k, 0)) = H((p_1, \ldots, p_k))$.

Then there exists $\lambda > 0$ such that $H(p_1, \ldots, p_k) = -\lambda \sum_{i=1}^{k} p_i \log p_i$.

The properties listed in Theorem 4.1 are reasonable properties for $H(\xi)$ to satisfy. Property (i) says the only experiments that give no information are those with only one possible outcome. Property (iv) means that among the experiments with k outcomes the ones having most uncertainty about their outcomes are those with equiprobable outcomes. Property (v) says that the total information gained from performing two experiments ξ and η is the information obtained by performing ξ plus the information gained by performing η knowing that ξ has been performed.

The (elementary) proof of this theorem can be found on page 9 of Khinchine's book [1]. This theorem motivates our definition of the entropy of a partition. We shall prove that entropy has the properties listed in Theorem 4.1.

Definition 4.6. Let $\mathscr{A}$ be a finite sub-algebra of $\mathscr{B}$ with $\xi(\mathscr{A}) = \{A_1, \ldots, A_k\}$. The entropy of $\mathscr{A}$ (or of $\xi(\mathscr{A})$) is the number $H(\mathscr{A}) = H(\xi(\mathscr{A})) = -\sum_{i=1}^{k} m(A_i) \log m(A_i)$.

As mentioned above, $H(\mathscr{A})$ is a measure of the uncertainty removed (or information gained) by performing the experiment with outcomes $\{A_1, \ldots, A_k\}$.

Remarks

(1) If $\mathscr{A} = \{X, \phi\}$ then $H(\mathscr{A}) = 0$. Here $\mathscr{A}$ represents the outcomes of a "certain" experiment so there is no uncertainty about the outcome.
(2) If $\xi(\mathscr{A}) = \{A_1, \ldots, A_k\}$ where $m(A_i) = 1/k$ $\forall i$ then

$$H(\mathscr{A}) = -\sum_{i=1}^{k} \frac{1}{k} \log \frac{1}{k} = \log k.$$

We shall show later (Corollary 4.2.1) that $\log k$ is the maximum value for the entropy of a partition with k sets. The greatest uncertainty about the outcome should occur when the outcomes are equally likely.

(3) $H(\mathscr{A}) \geq 0$.
(4) If $\mathscr{A} \doteq \mathscr{C}$ then $H(\mathscr{A}) = H(\mathscr{C})$.
(5) If $T: X \to X$ is measure-preserving then $H(T^{-1} \mathscr{A}) = H(\mathscr{A})$.

Several properties of entropy are implied by the following elementary result.

Theorem 4.2. *The function $\phi:[0, \infty) \to R$ defined by*

$$\phi(x) = \begin{cases} 0 & \text{if } x = 0 \\ x \cdot \log x & \text{if } x \neq 0 \end{cases}$$

is strictly convex, i.e., $\phi(\alpha x + \beta y) \leq \alpha\phi(x) + \beta\phi(y)$ if $x, y \in [0, \infty)$, $\alpha, \beta \geq 0$, $\alpha + \beta = 1$; with equality only when $x = y$ or $\alpha = 0$ or $\beta = 0$.

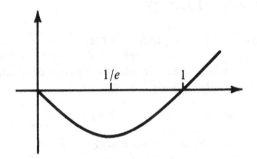

By induction we get

$$\phi\left(\sum_{i=1}^{k} \alpha_i x_i\right) \leq \sum_{i=1}^{k} \alpha_i \phi(x_i)$$

if $x_i \in [0, \infty)$, $\alpha_i \geq 0$, $\sum_{i=1}^{k} \alpha_i = 1$; and equality holds only when all the x_i, corresponding to non-zero α_i, are equal.

PROOF. We have

$$\phi'(x) = 1 + \log x$$

$$\phi''(x) = \frac{1}{x} > 0 \quad \text{on } (0, \infty).$$

Fix α, β with $\alpha > 0$, $\beta > 0$. Suppose $y > x$. By the mean value theorem

$$\phi(y) - \phi(\alpha x + \beta y) = \phi'(z)\alpha(y - x) \quad \text{for some } z \text{ with } \alpha y + \beta y < z < y$$

and

$$\phi(\alpha x + \beta y) - \phi(x) = \phi'(w)\beta(y - x) \quad \text{for some } w \text{ with } x < w < \alpha x + \beta y.$$

Since $\phi'' > 0$, we have $\phi'(z) > \phi'(w)$ and hence

$$\beta(\phi(y) - \phi(\alpha x + \beta y)) = \phi'(z)\alpha\beta(y - x) > \phi'(w)\alpha\beta(y - x)$$
$$= \alpha(\phi(x + \beta y) - \phi(x)).$$

Therefore $\phi(\alpha x + \beta y) < \alpha\phi(x) + \beta\phi(y)$ if $x, y > 0$. It clearly holds also if $x, y \geq 0$ and $x \neq y$. $\square$

Corollary 4.2.1. *If $\xi = \{A_1, \ldots, A_k\}$ then $H(\xi) \le \log k$, and $H(\xi) = \log k$ only when $m(A_i) = 1/k$ for all i.*

PROOF. Put $\alpha_i = 1/k$ and $x_i = m(A_i)$, $1 \le i \le k$.   □

§4.3 Conditional Entropy

Conditional entropy is not required in order to give the definition of the entropy of a transformation. It is useful in deriving properties of entropy, and we discuss it now before we consider the entropy of a transformation.

Let $\mathscr{A}, \mathscr{C}$ be finite sub-σ-algebras of $\mathscr{B}$ and

$$\xi(\mathscr{A}) = \{A_1, \ldots, A_k\}, \qquad \xi(\mathscr{C}) = \{C_1, \ldots, C_p\}.$$

The discussion in §4.2 suggests the following definition.

Definition 4.7. The *entropy of $\mathscr{A}$ given $\mathscr{C}$* is the number

$$H(\xi(\mathscr{A})/\xi(\mathscr{C})) = H(\mathscr{A}/\mathscr{C}) = -\sum_{j=1}^{p} m(C_j) \sum_{i=1}^{k} \frac{m(A_i \cap C_j)}{m(C_j)} \log \frac{m(A_i \cap C_j)}{m(C_j)}$$

$$= -\sum_{i,j} m(A_i \cap C_j) \log \frac{m(A_i \cap C_j)}{m(C_j)},$$

omitting the j-terms when $m(C_j) = 0$.

So to get $H(\mathscr{A}/\mathscr{C})$ one considers C_j as a measure space with normalized measure $m(\cdot)/m(C_j)$ and calculates the entropy of the partition of the set C_j induced by $\xi(\mathscr{A})$ (this gives

$$-\sum_{i=1}^{k} \frac{m(A_i \cap C_j)}{m(C_j)} \log \frac{m(A_i \cap C_j)}{m(C_j)})$$

and then averages the answer taking into account the size of C_j. ($H(\mathscr{A}/\mathscr{C})$ measures the uncertainty about the outcome of $\mathscr{A}$ given that we will be told the outcome of $\mathscr{C}$.)

Let $\mathscr{N}$ denote the σ-field $\{\phi, X\}$. Then $H(\mathscr{A}/\mathscr{N}) = H(\mathscr{A})$. (Since $\mathscr{N}$ represents the outcome of the trivial experiment one gains nothing from knowledge of it.)

Remarks

(1) $H(\mathscr{A}/\mathscr{C}) \ge 0$.
(2) If $\mathscr{A} \doteq \mathscr{D}$ then $H(\mathscr{A}/\mathscr{C}) = H(\mathscr{D}/\mathscr{C})$.
(3) If $\mathscr{C} \doteq \mathscr{D}$ then $H(\mathscr{A}/\mathscr{C}) = H(\mathscr{A}/\mathscr{D})$.

Theorem 4.3. *Let* $(X, \mathscr{B}, m)$ *be a probability space. If* $\mathscr{A}$, $\mathscr{C}$, $\mathscr{D}$ *are finite subalgebras of* $\mathscr{B}$ *then*:

(i) $H(\mathscr{A} \vee \mathscr{C}/\mathscr{D}) = H(\mathscr{A}/\mathscr{D}) + H(\mathscr{C}/\mathscr{A} \vee \mathscr{D})$.
(ii) $H(\mathscr{A} \vee \mathscr{C}) = H(\mathscr{A}) + H(\mathscr{C}/\mathscr{A})$.
(iii) $\mathscr{A} \subseteq \mathscr{C} \Rightarrow H(\mathscr{A}/\mathscr{D}) \leq H(\mathscr{C}/\mathscr{D})$.
(iv) $\mathscr{A} \subseteq \mathscr{C} \Rightarrow H(\mathscr{A}) \leq H(\mathscr{C})$.
(v) $\mathscr{C} \subseteq \mathscr{D} \Rightarrow H(\mathscr{A}/\mathscr{C}) \geq H(\mathscr{A}/\mathscr{D})$.
(vi) $H(\mathscr{A}) \geq H(\mathscr{A}/\mathscr{D})$.
(vii) $H(\mathscr{A} \vee \mathscr{C}/\mathscr{D}) \leq H(\mathscr{A}/\mathscr{D}) + H(\mathscr{C}/\mathscr{D})$.
(viii) $H(\mathscr{A} \vee \mathscr{C}) \leq H(\mathscr{A}) + H(\mathscr{C})$.
(ix) *If* T *is measure-preserving then*:

$$H(T^{-1}\mathscr{A}/T^{-1}\mathscr{C}) = H(\mathscr{A}/\mathscr{C}), \text{ and}$$

(x) $H(T^{-1}\mathscr{A}) = H(\mathscr{A})$.

(The reader should think of the intuitive meaning of each statement. This enables one to remember these results easily.)

PROOF. Let $\xi(\mathscr{A}) = \{A_i\}$, $\xi(\mathscr{C}) = \{C_j\}$, $\xi(\mathscr{D}) = \{D_k\}$ and assume, without loss of generality, that all sets have strictly positive measure (since if $\xi(\mathscr{A}) = \{A_1, \ldots, A_k\}$ with $m(A_i) > 0$, $1 \leq i \leq r$ and $m(A_i) = 0$, $r < i \leq k$ we can replace $\xi(\mathscr{A})$ by $\{A_1, \ldots, A_{r-1}, A_r \cup A_{r+1} \cup \cdots \cup A_k\}$ (see remarks (2), (3) above).

(i) $H(\mathscr{A} \vee \mathscr{C}/\mathscr{D}) = -\sum_{i,j,k} m(A_i \cap C_j \cap D_k) \log \dfrac{m(A_i \cap C_j \cap D_k)}{m(D_k)}$.

But

$$\frac{m(A_i \cap C_j \cap D_k)}{m(D_k)} = \frac{m(A_i \cap C_j \cap D_k)}{m(A_i \cap D_k)} \frac{m(A_i \cap D_k)}{m(D_k)},$$

unless $m(A_i \cap D_k) = 0$ and then the left hand side is zero and we need not consider it; and therefore

$$H(\mathscr{A} \vee \mathscr{C}/\mathscr{D}) = -\sum_{i,j,k} m(A_i \cap C_j \cap D_k) \log \frac{m(A_i \cap D_k)}{m(D_k)}$$

$$- \sum_{i,j,k} m(A_i \cap C_j \cap D_k) \log \frac{m(A_i \cap C_j \cap D_k)}{m(A_i \cap D_k)}$$

$$= -\sum_{i,k} m(A_i \cap D_k) \log \frac{m(A_i \cap D_k)}{m(D_k)} + H(\mathscr{C}/\mathscr{A} \vee \mathscr{D})$$

$$= H(\mathscr{A}/\mathscr{D}) + H(\mathscr{C}/\mathscr{A} \vee \mathscr{D}).$$

(ii) Put $\mathscr{D} = \mathscr{N} = \{\phi, X\}$ in (i).

(iii) By (i)

$$H(\mathscr{C}/\mathscr{D}) = H(\mathscr{A} \vee \mathscr{C}/\mathscr{D}) = H(\mathscr{A}/\mathscr{D}) + H(\mathscr{C}/\mathscr{A} \vee \mathscr{D}) \geq H(\mathscr{A}/\mathscr{D}).$$

(iv) Put $\mathscr{D} = \mathscr{N}$ in (iii).

(v) Fix i, j and let

$$\alpha_k = \frac{m(D_k \cap C_j)}{m(C_j)}, \qquad x_k = \frac{m(A_i \cap D_k)}{m(D_k)}.$$

Then by Theorem 4.2

$$\phi\left(\sum_k \frac{m(D_k \cap C_j)}{m(C_j)} \frac{m(A_i \cap D_k)}{m(D_k)} \right) \leq \sum_k \frac{m(D_k \cap C_j)}{m(C_j)} \phi\left(\frac{m(A_i \cap D_k)}{m(D_k)} \right),$$

but since $\mathscr{C} \subseteq \mathscr{D}$ the left hand side equals

$$\phi\left(\frac{m(A_i \cap C_j)}{m(C_j)} \right) = \frac{m(A_i \cap C_j)}{m(C_j)} \log \frac{m(A_i \cap C_j)}{m(C_j)}.$$

Multiply both sides by $m(C_j)$ and sum over i and j to give

$$\sum_{i,j} m(A_i \cap C_j) \log \frac{m(A_i \cap C_j)}{m(C_j)} \leq \sum_{i,j,k} m(D_k \cap C_j) \frac{m(A_i \cap D_k)}{m(D_k)} \log \frac{m(A_i \cap D_k)}{m(D_k)}$$

$$= \sum_{i,k} m(D_k) \frac{m(A_i \cap D_k)}{m(D_k)} \log \frac{m(A_i \cap D_k)}{m(D_k)}$$

or $-H(\mathscr{A}/\mathscr{C}) \leq -H(\mathscr{A}/\mathscr{D})$. Therefore, $H(\mathscr{A}/\mathscr{D}) \leq H(\mathscr{A}/\mathscr{C})$.

(vi) Put $\mathscr{C} = \mathscr{N}$ in (v).

(vii) Use (i) and (v).

(viii) Set $\mathscr{D} = \mathscr{N}$ in (vii).

(ix), (x) Clear from definitions. $\square$

The following also fits in with our intuitive ideas

Theorem 4.4. Let $\mathscr{A}, \mathscr{C}$ be finite sub-algebras of $\mathscr{B}$. Then

(i) $H(\mathscr{A}/\mathscr{C}) = 0$ (i.e. $H(\mathscr{A} \vee \mathscr{C}) = H(\mathscr{C})$) iff $\mathscr{A} \subseteq \mathscr{C}$.

(ii) $H(\mathscr{A}/\mathscr{C}) = H(\mathscr{A})$ (i.e. $H(\mathscr{A} \vee \mathscr{C}) = H(\mathscr{A}) + H(\mathscr{C})$) iff $\mathscr{A}$ and $\mathscr{C}$ are independent (i.e. $m(A \cap C) = m(A) \cdot m(C)$ whenever $A \in \mathscr{A}, C \in \mathscr{C}$).

PROOF. Let $\xi(\mathscr{A}) = \{A_1, \ldots, A_k\}$, $\xi(\mathscr{C}) = \{C_1, \ldots, C_p\}$. Without loss of generality we can assume all these sets have non-zero measure.

(i) $\mathscr{A} \subseteq \mathscr{C}$ means for each i and each j either $m(A_i \cap C_j) = m(C_j)$ or $m(A_i \cap C_j) = 0$. Clearly this implies $H(\mathscr{A}/\mathscr{C}) = 0$. Suppose $H(\mathscr{A}/\mathscr{C}) = 0$. Then

$$0 = -\sum_{i=1}^{k} \sum_{j=1}^{p} m(A_i \cap C_j) \log \frac{m(A_i \cap C_j)}{m(C_j)}$$

and since

$$-m(A_i \cap C_j) \log \frac{m(A_i \cap C_j)}{m(C_j)} \geq 0$$

we must have

$$m(A_i \cap C_j) \log \frac{m(A_i \cap C_j)}{m(C_j)} = 0$$

for each i, j. Hence either $m(A_i \cap C_j) = m(C_j)$ or $m(A_i \cap C_j) = 0$. Therefore $\mathscr{A} \doteq \mathscr{C}$.

(ii) If $\mathscr{A}$ and $\mathscr{C}$ are independent we quickly see from the definition of $H(\mathscr{A}/\mathscr{C})$ that $H(\mathscr{A}/\mathscr{C}) = H(\mathscr{A})$. To prove the converse suppose $H(\mathscr{A}/\mathscr{C}) = H(\mathscr{A})$. Then

$$-\sum_{i=1}^{k} \sum_{j=1}^{p} m(A_i \cap C_j) \log \frac{m(A_i \cap C_j)}{m(C_j)} = -\sum_{i=1}^{k} m(A_i) \log m(A_i). \qquad (*)$$

If we fix i and apply Theorem 4.2. with $\alpha_j = m(C_j)$ and $x_j = m(A_i \cap C_j)/m(C_j)$ we get

$$-\sum_{j=1}^{p} m(A_i \cap C_j) \log \frac{m(A_i \cap C_j)}{m(C_j)} \leq -m(A_i) \log m(A_i) \qquad (**)$$

with equality only when $m(A_i \cap C_j)/m(C_j)$ does not depend on j. If a_i denotes this constant value then by summing the equations $m(A_i \cap C_j) = a_i m(C_j)$ over j we have $a_i = m(A_i)$. Hence equality holds only when $m(A_i \cap C_j) = m(C_j)m(A_i)$. However equation $(*)$ says equality holds in $(**)$ for each i and so $m(A_i \cap C_j) = m(C_j)m(A_i)$ for all i, j. Therefore $m(A \cap C) = m(C)m(A)$ whenever $A \in \mathscr{A}, C \in \mathscr{C}$. $\qquad \square$

Theorem 4.5. *Let V denote the space of all finite sub-algebras of $\mathscr{B}$ where two finite algebras $\mathscr{A}, \mathscr{C}$ are identified if $\mathscr{A} \doteq \mathscr{C}$. Then $d(\mathscr{A}, \mathscr{C}) = H(\mathscr{A}/\mathscr{C}) + H(\mathscr{C}/\mathscr{A})$ is a metric on V.*

(A corresponding statement about the space of finite partitions can be made).

PROOF. We have $d(\mathscr{A}, \mathscr{C}) \geq 0$ and equality holds iff $\mathscr{A} \doteq \mathscr{C}$ (Theorem 4.4). Also $H(\mathscr{A}/\mathscr{D}) \leq H(\mathscr{A} \vee \mathscr{C}/\mathscr{D}) = H(\mathscr{C}/\mathscr{D}) + H(\mathscr{A}/\mathscr{C} \vee \mathscr{D}) \leq H(\mathscr{C}/\mathscr{D}) + H(\mathscr{A}/\mathscr{C})$ and similarly $H(\mathscr{D}/\mathscr{A}) \leq H(\mathscr{C}/\mathscr{A}) + H(\mathscr{D}/\mathscr{C})$. Therefore $d(\mathscr{A}, \mathscr{D}) \leq d(\mathscr{A}, \mathscr{C}) + d(\mathscr{C}, \mathscr{D})$. $\qquad \square$

We can also define conditional entropy $H(\mathscr{A}/\mathscr{F})$ when $\mathscr{A}$ is a finite sub-σ-algebra of $\mathscr{B}$ and $\mathscr{F}$ is an arbitrary sub-σ-algebra of $\mathscr{B}$. To do this we use the conditional expectation map $E(\cdot/\mathscr{F}): L^1(X, \mathscr{B}, m) \to L^1(X, \mathscr{F}, m)$. If $\mathscr{C}$ is a finite sub-σ-algebra of $\mathscr{B}$ with $\xi(\mathscr{C}) = \{C_1, \ldots, C_p\}$ then $E(f/\mathscr{C})(x) = \sum_{j=1}^{p} \chi_{C_j}(x)(1/m(C_j)) \int_{C_j} f \, dm$. If $\mathscr{A}$ is also finite and $\xi(\mathscr{A}) = \{A_1, \ldots, A_k\}$

then

$$H(\mathscr{A}/\mathscr{C}) = -\sum_{i,j} m(A_i \cap C_j) \log \frac{m(A_i \cap C_j)}{m(C_j)}$$

$$= -\sum_{i=1}^{k} \int \chi_{A_i} \log E(\chi_{A_i}/\mathscr{C}) \, dm$$

$$= -\int \sum_{i=1}^{k} E(\chi_{A_i}/\mathscr{C}) \log E(\chi_{A_i}/\mathscr{C}) \, dm.$$

This leads to the following definition.

Definition 4.8. Let $(X, \mathscr{B}, m)$ be a probability space. If $\mathscr{A}$ is a finite sub-σ-algebra of $\mathscr{B}$ with $\zeta(\mathscr{A}) = \{A_1, \ldots, A_k\}$, and $\mathscr{F}$ is an arbitrary sub-σ-algebra of $\mathscr{B}$ the entropy of $\mathscr{A}$ given $\mathscr{F}$ is the number

$$H(\mathscr{A}/\mathscr{F}) = -\int \sum_{i=1}^{k} E(\chi_{A_i}/\mathscr{F}) \log E(\chi_{A_i}/\mathscr{F}) \, dm.$$

Remark. Since $E(\ /\mathscr{F})$ is a positive linear operator and $\sum_{i=1}^{k} \chi_{A_i} = 1$ we have $0 \leq E(\chi_{A_i}/\mathscr{F})(x) \leq 1$ a.e., and therefore

$$-\sum_{i=1}^{k} E(\chi_{A_i}/\mathscr{F})(x) \log E(\chi_{A_i}/\mathscr{F})(x) \leq k \max_{t \in [0,1]} (-t \log t) = ke.$$

Hence $H(\mathscr{A}/\mathscr{F})$ is finite.

One can show that the properties listed in Theorem 4.3 are satisfied by this more general conditional entropy. However, they can also be deduced from Theorem 4.3 (in the case when $(X, \mathscr{B}, m)$ has a countable basis) by using a limit theorem that we shall use for another purpose. To prepare for the proof we give the following lemma. If $\{\mathscr{F}_n\}_1^{\infty}$ is a family of sub-σ-algebras of $\mathscr{B}$ we let $\bigvee_{n=1}^{\infty} \mathscr{F}_n$ denote the smallest sub-σ-algebra containing all the $\mathscr{F}_n$.

Lemma 4.6. Let $(X, \mathscr{B}, m)$ be a probability space and let $\{\mathscr{F}_n\}_1^{\infty}$ be an increasing sequence of sub-σ-algebras of $\mathscr{B}$. Denote $\bigvee_{n=1}^{\infty} \mathscr{F}_n$ by $\mathscr{F}$. For each $f \in L^2(X, \mathscr{B}, m)$ we have

$$\|E(f/\mathscr{F}_n) - E(f/\mathscr{F})\|_2 \to 0.$$

PROOF. Recall that $E(\cdot/\mathscr{F}_n)$ is the orthogonal projection of $L^2(X, \mathscr{B}, m)$ onto $L^2(X, \mathscr{F}_n, m)$. Let $B \in \mathscr{F}$. Choose $B_n \in \mathscr{F}_n$ with $m(B_n \triangle B) \to 0$. Since $E(\chi_B/\mathscr{F}_n)$ is that member of $L^2(X, \mathscr{F}_n, m)$ closest to χ_B we have

$$\|E(\chi_B/\mathscr{F}_n) - \chi_B\|_2^2 \leq \|\chi_{B_n} - \chi_B\|_2^2 = m(B_n \triangle B) \to 0.$$

Since finite linear combinations of characteristic functions are dense in $L^2(X, \mathscr{F}, m)$ we have $\|E(h/\mathscr{F}_n) - h\|_2 \to 0$ for all $h \in L^2(X, \mathscr{F}, m)$. Hence if $f \in L^2(X, \mathscr{B}, m)$,

$$\|E(f/\mathscr{F}_n) - E(f/\mathscr{F})\|_2 \to 0 \quad \text{because } E(E(f/\mathscr{F})/\mathscr{F}_n) = E(f/\mathscr{F}_n). \qquad \square$$

Remarks

(1) The same result holds for a decreasing sequence of $\{\mathscr{F}_n\}_1^\infty$ sub-σ-algebras with $\bigcap_{n=1}^\infty \mathscr{F}_n = \mathscr{F}$.

(2) If $f \in L^1(X, \mathscr{B}, m)$ and $\{\mathscr{F}_n\}_1^\infty$ increase to $\mathscr{F}$ then Doob's martingale theorem implies $E(f/\mathscr{F}_n) \to E(f/\mathscr{F})$ a.e. and in $L^1(X, \mathscr{B}, m)$. The corresponding statement holds for a decreasing sequence of σ-algebras (see Parthasarathy [2], p. 230).

Theorem 4.7. *Let $(X, \mathscr{B}, m)$ be a probability space. Let $\mathscr{A}$ be a finite subalgebra of $\mathscr{B}$ and let $\{\mathscr{F}_n\}_1^\infty$ be an increasing sequence of sub-σ-algebras of $\mathscr{B}$ with $\bigvee_{n=1}^\infty \mathscr{F}_n = \mathscr{F}$. Then $H(\mathscr{A}/\mathscr{F}_n) \to H(\mathscr{A}/\mathscr{F})$.*

PROOF. Let $\xi(\mathscr{A}) = \{A_1, \ldots, A_k\}$. From Lemma 4.6. we know that for each i

$$\|E(\chi_{A_i}/\mathscr{F}_n) - E(\chi_{A_i}/\mathscr{F})\|_2 \to 0.$$

Therefore $E(\chi_{A_i}/\mathscr{F}_n)$ converges in measure to $E(\chi_{A_i}/\mathscr{F})$ and hence

$$-\sum_{i=1}^k E(\chi_{A_i}/\mathscr{F}_n) \log E(\chi_{A_i}/\mathscr{F}_n)$$

converges in measure to $-\sum_{i=1}^k E(\chi_{A_i}/\mathscr{F}) \log E(\chi_{A_i}/\mathscr{F})$. Since all these functions are bounded by ke we know we have convergence in $L^1(m)$ too. Therefore $H(\mathscr{A}/\mathscr{F}_n) \to H(\mathscr{A}/\mathscr{F})$. $\qquad \square$

Remarks

(1) The same result holds for a decreasing sequence $\{\mathscr{F}_n\}_1^\infty$ of sub-σ-algebras with $\bigcap_{n=1}^\infty \mathscr{F}_n = \mathscr{F}$.

(2) When $(X, \mathscr{B}, m)$ has a countable basis the statements of Theorem 4.3 (where $\mathscr{D}$ is now an arbitrary sub-σ-algebra of $\mathscr{B}$) hold by choosing an increasing sequence $\{\mathscr{D}_n\}_1^\infty$ of finite sub-algebras with $\mathscr{D}_n \nearrow \mathscr{D}$ and using Theorem 4.7.

We have the following extension of Theorem 4.4.

Theorem 4.8. *Let $(X, \mathscr{B}, m)$ be a probability space and let $\mathscr{A}$, $\mathscr{F}$ be sub-σ-algebras of $\mathscr{B}$ with $\mathscr{A}$ finite. Then*

(i) $H(\mathscr{A}/\mathscr{F}) = 0$ *iff* $\mathscr{A} \subset\!\!\!\!\!= \mathscr{F}$.
(ii) $H(\mathscr{A}/\mathscr{F}) = H(\mathscr{A})$ *iff* $\mathscr{A}$ *and* $\mathscr{F}$ *are independent.*

PROOF. Let $\xi(\mathscr{A}) = \{A_1, \ldots, A_k\}$.

(i) If $\mathscr{A} \subset\!\!\!\!\!= \mathscr{F}$ then $E(\chi_{A_i}/\mathscr{F})(x)$ takes only the values 0, 1 so $H(\mathscr{A}/\mathscr{F}) = 0$. Conversely, if $0 = H(\mathscr{A}/\mathscr{F}) = \int -\sum_{i=1}^k E(\chi_{A_i}/\mathscr{F}) \log E(\chi_{A_i}/\mathscr{F}) \, dm$ then since $-E(\chi_{A_i}/\mathscr{F})(x)\log E(\chi_{A_i}/\mathscr{F})(x) \geq 0$ we have that for each i, $E(\chi_{A_i}/\mathscr{F})$ takes only the values 0, 1. Therefore $\mathscr{A} \stackrel{\subset}{=} \mathscr{F}$.

(ii) Suppose $H(\mathscr{A}/\mathscr{F}) = H(\mathscr{A})$. Let $B \in \mathscr{F}$. Let $\mathscr{D}$ be the finite sub-algebra consisting of the sets $\{\phi, B, X\backslash B, X\}$. Then $\mathscr{D} \subset \mathscr{F}$ and $H(\mathscr{A}) \geq H(\mathscr{A}/\mathscr{D}) \geq H(\mathscr{A}/\mathscr{F}) = H(\mathscr{A})$. Hence $H(\mathscr{A}) = H(\mathscr{A}/\mathscr{D})$ so $m(A \cap B) = m(A)m(B)$ for all $A \in \mathscr{A}$, by Theorem 4.4. Therefore $\mathscr{A}$ and $\mathscr{F}$ are independent.

If $\mathscr{A}$ and $\mathscr{F}$ are independent then for each $A \in \mathscr{A}$ $E(\chi_A/\mathscr{F}) = m(A)$ (because $\int_F E(\chi_A/\mathscr{F})\,dm = \int_F \chi_A\,dm = m(A)m(F)$ for all $F \in \mathscr{F}$ and $E(\chi_A/\mathscr{F})$ is the only $\mathscr{F}$-measurable function with this property). Therefore $H(\mathscr{A}/\mathscr{F}) = H(\mathscr{A})$. $\square$

§4.4 Entropy of a Measure-Preserving Transformation

Recall that if $\xi(\mathscr{A}) = \{A_1, \ldots, A_k\}$ then

$$H(\xi(\mathscr{A})) = H(\mathscr{A}) = -\sum_{i=1}^{k} m(A_i)\log m(A_i).$$

The second stage of the definition of the entropy of a measure-preserving transformation T is given in the next definition. Recall that the elements of the partition $\xi(\bigvee_{i=0}^{n-1} T^{-i}\mathscr{A}) = \bigvee_{i=0}^{n-1} T^{-i}\xi(\mathscr{A})$ are all sets of the form $\bigcap_{i=0}^{n-1} T^{-i}A_{j_i}$.

Definition 4.9. Suppose $T: X \to X$ is a measure-preserving transformation of the probability space $(X, \mathscr{B}, m)$. If $\mathscr{A}$ is a finite-sub-σ-algebra of $\mathscr{B}$ then

$$h(T, \xi(\mathscr{A})) = h(T, \mathscr{A}) = \lim_{n\to\infty} \frac{1}{n} H\left(\bigvee_{i=0}^{n-1} T^{-i}\mathscr{A}\right)$$

is called the *entropy of T with respect to $\mathscr{A}$.* (Later (in Corollary 4.9.1) we will show that the above limit always exists. In fact $(1/n)H(\bigvee_{i=0}^{n-1} T^{-i}\mathscr{A})$ decreases to $h(T, \mathscr{A})$.)

This means that if we think of an application of T as a passage of one day of time, then $\bigvee_{i=0}^{n-1} T^{-i}\mathscr{A}$ represents the combined experiment of performing the original experiment, represented by $\mathscr{A}$, on n consecutive days. Then $h(T, \mathscr{A})$ is the average information per day that one gets from performing the original experiment daily forever.

Remark. $h(T, \mathscr{A}) \geq 0$.

We can now give the final stage of the definition of the entropy of a measure-preserving transformation.

Definition 4.10. If $T:X \to X$ is a measure-preserving transformation of the probability space $(X, \mathcal{B}, m)$ then $h(T) = \sup h(T, \mathcal{A})$, where the supremum is taken over all finite sub-algebras $\mathcal{A}$ of $\mathcal{B}$, is called the *entropy* of T. Equivalently $h(T) = \sup h(T, \xi)$ where the supremum is taken over all finite partitions of $(X, \mathcal{B}, m)$.

If, as above, we think of an application of T as a passage of one day of time then $h(T)$ is the maximum average information per day obtainable by performing the same experiment daily.

Remarks

(1) $h(T) \geq 0$. $h(T)$ could be $+\infty$.
(2) $h(id_X) = 0$. If $h(T) = 0$ then $h(T, \mathcal{A}) = 0$ for every finite $\mathcal{A}$, which implies that $\bigvee_{i=0}^{n-1} T^{-i}\mathcal{A}$ does not change much as $n \to \infty$.
(3) If logarithms of some other base are used then the entropy of a transformation is changed by a multiplicative constant that depends only on the base. Some authors use logarithms of base 2.

We shall now show the existence of the limit in Definition 4.9. We shall do this in two ways. The first method uses a simple result on sequences of real numbers and can also be applied to prove the corresponding result for topological entropy, while the second method uses properties of conditional entropy but gives a stronger result.

Theorem 4.9. *If* $\{a_n\}_{n \geq 1}$ *is a sequence of real numbers such that* $a_{n+p} \leq a_n + a_p$ *$\forall n, p$ then* $\lim_{n \to \infty} a_n/n$ *exists and equals* $\inf_n a_n/n$. *(The limit could be* $-\infty$ *but if the* a_n *are bounded below then the limit will be non-negative.)*

PROOF. Fix $p > 0$. Each $n > 0$ can be written $n = kp + i$ with $0 \leq i < p$. Then

$$\frac{a_n}{n} = \frac{a_{i+kp}}{i+kp} \leq \frac{a_i}{kp} + \frac{a_{kp}}{kp} \leq \frac{a_i}{kp} + \frac{ka_p}{kp} = \frac{a_i}{kp} + \frac{a_p}{p}.$$

As $n \to \infty$ then $k \to \infty$ so

$$\overline{\lim} \frac{a_n}{n} \leq \frac{a_p}{p}$$

and therefore

$$\overline{\lim} \frac{a_n}{n} \leq \inf \frac{a_p}{p}.$$

But

$$\inf \frac{a_p}{p} \leq \underline{\lim} \frac{a_n}{n}$$

so that $\lim a_n/n$ exists and equals $\inf a_n/n$. $\qquad \square$

Corollary 4.9.1. *If* $T: X \to X$ *is measure-preserving and* $\mathscr{A}$ *is a finite sub-algebra of* $\mathscr{B}$ *then* $\lim_{n \to \infty} (1/n)H(\bigvee_{i=0}^{n-1} T^{-i}\mathscr{A})$ *exists.*

PROOF. Let $a_n = H(\bigvee_{i=0}^{n-1} T^{-i}A) \geq 0$. Then

$$a_{n+p} = H\left(\bigvee_{i=0}^{n+p-1} T^{-i}\mathscr{A}\right)$$

$$\leq H\left(\bigvee_{i=0}^{n-1} T^{-i}\mathscr{A}\right) + H\left(\bigvee_{i=n}^{n+p-1} T^{-i}\mathscr{A}\right) \quad \text{by Theorem 4.3(viii).}$$

$$= a_n + H\left(\bigvee_{i=0}^{p-1} T^{-i}\mathscr{A}\right) \quad \text{by Theorem 4.3(x)}$$

$$= a_n + a_p.$$

We then apply Theorem 4.9. □

Theorem 4.10. *If* $T: X \to X$ *is measure-preserving and* $\mathscr{A}$ *is a finite sub-algebra of* $\mathscr{B}$ *then* $(1/n)H(\bigvee_{i=0}^{n-1} T^{-i}\mathscr{A})$ *decreases to* $h(T, \mathscr{A})$.

PROOF. We first show, by induction, that

$$H\left(\bigvee_{i=0}^{n-1} T^{-i}\mathscr{A}\right) = H(\mathscr{A}) + \sum_{j=1}^{n-1} H\left(\mathscr{A} \middle/ \bigvee_{i=1}^{j} T^{-i}\mathscr{A}\right).$$

For $n = 1$ it is clear, and if we assume it true for $n = p$ then it also holds for $n = p + 1$ because

$$H\left(\bigvee_{i=0}^{p} T^{-i}\mathscr{A}\right) = H\left(\bigvee_{i=1}^{p} T^{-i}\mathscr{A} \vee \mathscr{A}\right)$$

$$= H\left(\bigvee_{i=1}^{p} T^{-i}\mathscr{A}\right) + H\left(\mathscr{A} \middle/ \bigvee_{i=1}^{p} T^{-i}\mathscr{A}\right) \quad \text{by Theorem 4.3(ii)}$$

$$= H\left(\bigvee_{i=0}^{p-1} T^{-i}\mathscr{A}\right) + H\left(\mathscr{A} \middle/ \bigvee_{i=1}^{p} T^{-i}\mathscr{A}\right) \quad \text{by Theorem 4.3(x)}$$

$$= H(\mathscr{A}) + \sum_{j=1}^{p} H\left(\mathscr{A} \middle/ \bigvee_{i=1}^{j} T^{-i}\mathscr{A}\right) \quad \begin{array}{l}\text{by the induction} \\ \text{assumption.}\end{array}$$

Thus the claimed formula holds for all n. From this formula and Theorem 4.3(iii) we have $H(\bigvee_{i=0}^{n-1} T^{-i}\mathscr{A}) \geq nH(\mathscr{A}/\bigvee_{i=1}^{n} T^{-i}\mathscr{A})$ so that

$$nH\left(\bigvee_{i=0}^{n} T^{-i}\mathscr{A}\right) = n\left[H\left(\bigvee_{i=0}^{n-1} T^{-i}\mathscr{A}\right) + H\left(\mathscr{A} \middle/ \bigvee_{i=1}^{n} T^{-i}\mathscr{A}\right)\right]$$

$$\leq (n+1)H\left(\bigvee_{i=0}^{n-1} T^{-i}\mathscr{A}\right).$$

Hence

$$\frac{1}{n+1} H\left(\bigvee_{i=0}^{n} T^{-i}\mathcal{A}\right) \leq \frac{1}{n} H\left(\bigvee_{i=0}^{n-1} T^{-i}\mathcal{A}\right). \qquad \square$$

We now show $h(T)$ is a conjugacy invariant.

Theorem 4.11. *Entropy is a conjugacy invariant and hence an isomorphism invariant.*

PROOF. Let $T_1: X_1 \to X_1$, $T_2: X_2 \to X_2$ be measure-preserving and let $\Phi: (\tilde{\mathcal{B}}_2, \tilde{m}_2) \to (\tilde{\mathcal{B}}_1, \tilde{m}_1)$ be an isomorphism of measure algebras such that $\Phi \tilde{T}_2^{-1} = \tilde{T}_1^{-1}\Phi$. Let $\mathcal{A}_2$ be finite, $\mathcal{A}_2 \subset \mathcal{B}_2$, and $\xi(\mathcal{A}_2) = \{A_1, \ldots, A_r\}$. Choose $B_i \in \mathcal{B}_1$ such that $\tilde{B}_i = \Phi(\tilde{A}_i)$ and so that $\eta = \{B_1, \ldots, B_r\}$ forms a partition of $(X_1, \mathcal{B}_1, m_1)$. Let $\mathcal{A}_1 = \mathcal{A}(\eta)$.

Now $\bigcap_{i=0}^{n-1} T_1^{-i} B_{q_i}$ (where $q_i \in \{1, \ldots, r\}$) has the same measure as $\bigcap_{i=0}^{n-1} T_2^{-i} A_{q_i}$ since

$$\Phi\left(\bigcap_{i=0}^{n-1} (T_2^{-i}A_{q_i})^{\sim}\right) = \Phi\left(\bigcap_{i=0}^{n-1} \tilde{T}_2^{-i}\tilde{A}_{q_i}\right) = \bigcap_{i=0}^{n-1} \tilde{T}_1^{-i}\Phi(\tilde{A}_{q_i})$$

$$= \bigcap_{i=0}^{n-1} \tilde{T}_1^{-i}\tilde{B}_{q_i} = \bigcap_{i=0}^{n-1} (T_1^{-i}B_{q_i})^{\sim}.$$

Thus $H(\bigvee_{i=0}^{n-1} T_1^{-i}\mathcal{A}_1) = H(\bigvee_{i=0}^{n-1} T_2^{-i}\mathcal{A}_2)$ which implies that $h(T_1, \mathcal{A}_1) = h(T_2, \mathcal{A}_2)$ which in turn implies $h(T_1) \geq h(T_2)$. By symmetry we then get that $h(T_1) = h(T_2)$. $\qquad \square$

The proof of Theorem 4.11 also shows that if T_2 is a factor of T_1 (or a semi-conjugate image) then $h(T_2) \leq h(T_1)$.

After we have developed some properties of $h(T, \mathcal{A})$ and $h(T)$ we shall consider the problem of how to calculate $h(T)$. These calculations and Theorem 4.11 will allow us to give examples of non-conjugate measure-preserving transformations.

§4.5 Properties of $h(T, \mathcal{A})$ and $h(T)$

Recall that

$$h(T, \mathcal{A}) = \lim_{n \to \infty} \frac{1}{n} H\left(\bigvee_{i=0}^{n-1} T^{-i}\mathcal{A}\right).$$

Theorem 4.12. *Suppose $\mathcal{A}, \mathcal{C}$ are finite subalgebras of $\mathcal{B}$ and T is a measure-preserving transformation of the probability space $(X, \mathcal{B}, m)$. Then*

(i) $h(T, \mathcal{A}) \leq H(\mathcal{A})$.
(ii) $h(T, \mathcal{A} \vee \mathcal{C}) \leq h(T, \mathcal{A}) + h(T, \mathcal{C})$.

(iii) $\mathscr{A} \subseteq \mathscr{C} \Rightarrow h(T, \mathscr{A}) \leq h(T, \mathscr{C})$.

(iv) $h(T, \mathscr{A}) \leq h(T, \mathscr{C}) + H(\mathscr{A}/\mathscr{C})$.

(v) $h(T, T^{-1}\mathscr{A}) = h(T, \mathscr{A})$.

(vi) If $k \geq 1$, $h(T, \mathscr{A}) = h(T, \bigvee_{i=0}^{k-1} T^{-i}\mathscr{A})$.

(vii) If T is invertible and $k \geq 1$ then

$$h(T, \mathscr{A}) = h\left(T, \bigvee_{i=-k}^{k} T^i \mathscr{A}\right).$$

PROOF

(i) $\dfrac{1}{n} H\left(\bigvee_{i=0}^{n-1} T^{-i}\mathscr{A}\right) \leq \dfrac{1}{n} \sum_{i=0}^{n-1} H(T^{-i}\mathscr{A})$ by Theorem 4.3

$$= \frac{1}{n} \sum_{i=0}^{n-1} H(\mathscr{A}) \quad \text{by Theorem 4.3(x)}$$

$$= H(\mathscr{A}).$$

(ii) $H\left(\bigvee_{i=0}^{n-1} T^{-i}(\mathscr{A} \vee \mathscr{C})\right) = H\left(\bigvee_{i=0}^{n-1} T^{-i}\mathscr{A} \vee \bigvee_{i=0}^{n-1} T^{-i}\mathscr{C}\right)$

$$\leq H\left(\bigvee_{i=0}^{n-1} T^{-i}\mathscr{A}\right) + H\left(\bigvee_{i=0}^{n-1} T^{-i}\mathscr{C}\right)$$

$$\text{by Theorem 4.3(viii)}.$$

(iii) If $\mathscr{A} \subseteq \mathscr{C}$ then

$$\bigvee_{i=0}^{n-1} T^{-i}\mathscr{A} \subseteq \bigvee_{i=0}^{n-1} T^{-i}\mathscr{C}, \qquad n \geq 1$$

so one uses Theorem 4.3(iv).

(iv) $H\left(\bigvee_{i=0}^{n-1} T^{-i}\mathscr{A}\right) \leq H\left(\left(\bigvee_{i=0}^{n-1} T^{-i}\mathscr{A}\right) \vee \left(\bigvee_{i=0}^{n-1} T^{-i}\mathscr{C}\right)\right)$

$$\text{by Theorem 4.3(iv)}$$

$$= H\left(\bigvee_{i=0}^{n-1} T^{-i}\mathscr{C}\right) + H\left(\left(\bigvee_{i=0}^{n-1} T^{-i}\mathscr{A}\right)\Big/\left(\bigvee_{i=0}^{n-1} T^{-i}\mathscr{C}\right)\right)$$

$$\text{by Theorem 4.3(ii)}.$$

But by Theorem 4.3(vii),

$$H\left(\left(\bigvee_{i=0}^{n-1} T^{-i}\mathscr{A}\right)\Big/\left(\bigvee_{i=0}^{n-1} T^{-i}\mathscr{C}\right)\right) \leq \sum_{i=0}^{n-1} H\left(T^{-i}\mathscr{A}\Big/\bigvee_{j=0}^{n-1} T^{-j}\mathscr{C}\right)$$

$$\leq \sum_{i=0}^{n-1} H(T^{-i}\mathscr{A}/T^{-i}\mathscr{C}) \quad \text{by Theorem 4.3(v)}$$

$$= nH(\mathscr{A}/\mathscr{C}) \quad \text{by Theorem 4.3(ix)}.$$

Thus,

$$H\left(\bigvee_{i=0}^{n-1} T^{-i}\mathscr{A}\right) \le H\left(\bigvee_{i=0}^{n-1} T^{-i}\mathscr{C}\right) + nH(\mathscr{A}/\mathscr{C}).$$

(v) $H\left(\bigvee_{i=1}^{n} T^{-i}\mathscr{A}\right) = H\left(\bigvee_{i=0}^{n-1} T^{-i}\mathscr{A}\right)$ by Theorem 4.3(x), so

$$h(T, T^{-1}\mathscr{A}) = h(T, \mathscr{A}).$$

(vi) $h\left(T, \bigvee_{0}^{k} T^{-i}\mathscr{A}\right) = \lim_{n\to\infty} \frac{1}{n} H\left(\bigvee_{j=0}^{n-1} T^{-j}\left(\bigvee_{i=0}^{k} T^{-i}\mathscr{A}\right)\right)$

$$= \lim_{n\to\infty} \frac{1}{n} H\left(\bigvee_{i=0}^{k+n-1} T^{-i}\mathscr{A}\right)$$

$$= \lim_{n\to\infty} \left(\frac{k+n-1}{n}\right)\frac{1}{k+n-1} H\left(\bigvee_{i=0}^{k+n-1} T^{-i}\mathscr{A}\right)$$

$$= h(T, \mathscr{A}).$$

(vii) $h\left(T, \bigvee_{i=-k}^{k} T^{-i}\mathscr{A}\right) = h\left(T, \bigvee_{0}^{2k} T^{-i}\mathscr{A}\right)$ by (v)

$$= h(T, \mathscr{A}) \quad \text{by (vi).} \qquad \square$$

Corollary 4.12.1. *If $\mathscr{A}, \mathscr{C}$ are finite sub-algebras of $\mathscr{B}$ we have $|h(T, \mathscr{A}) - h(T, \mathscr{C})| \le d(\mathscr{A}, \mathscr{C})$, so that $h(T, \cdot)$ is a continuous real-valued function on the metric space (V, d) introduced in Theorem 4.5.*

PROOF. By (iii)

$$|h(T, \mathscr{A}) - h(T, \mathscr{C})| \le \max(H(\mathscr{A}/\mathscr{C}), H(\mathscr{C}/\mathscr{A}))$$
$$\le d(\mathscr{A}, \mathscr{C}). \qquad \square$$

We can deduce from Theorem 4.12 some simple properties of $h(T)$.

Theorem 4.13. *Let T be a measure-preserving transformation of the probability space $(X, \mathscr{B}, m)$.*

(i) *For $k > 0$, $h(T^k) = kh(T)$.*
(ii) *If T is invertible then $h(T^k) = |k|h(T) \; \forall k \in Z$.*

PROOF

(i) We first show that

$$h\left(T^k, \bigvee_{i=0}^{k-1} T^{-i}\mathscr{A}\right) = kh(T, \mathscr{A}) \quad \text{if } k > 0.$$

This follows since

$$\lim_{n \to \infty} \frac{1}{n} H\left(\bigvee_{j=0}^{k-1} T^{-kj} \left(\bigvee_{i=0}^{k-1} T^{-i}\mathscr{A} \right) \right) = \lim_{n \to \infty} \frac{k}{nk} H\left(\bigvee_{i=0}^{nk-1} T^{-i}\mathscr{A} \right)$$

$$= kh(T, \mathscr{A}).$$

Thus,

$$kh(T) = k \cdot \sup_{\mathscr{A} \text{ finite}} h(T, \mathscr{A}) = \sup_{\mathscr{A}} h\left(T^k, \bigvee_{i=0}^{k-1} T^{-i}\mathscr{A} \right)$$

$$\leq \sup_{\mathscr{C}} h(T^k, \mathscr{C}) = h(T^k).$$

Also, $h(T^k, \mathscr{A}) \leq h(T^k, \bigvee_{i=0}^{k-1} T^{-i}\mathscr{A}) = kh(T, \mathscr{A})$ by Theorem 4.12(iii) and so, $h(T^k) \leq kh(T)$. The result follows from these two inequalities.

   (ii) It suffices to show that $h(T^{-1}) = h(T)$ and all we need to show is that $h(T^{-1}, \mathscr{A}) = h(T, \mathscr{A})$ for all finite $\mathscr{A}$. But

$$H\left(\bigvee_{i=0}^{n-1} T^i\mathscr{A} \right) = H\left(T^{-(n-1)} \bigvee_{i=0}^{n-1} T^i\mathscr{A} \right) \quad \text{by Theorem 4.3(x)}$$

$$= H\left(\bigvee_{j=0}^{n-1} T^{-j}\mathscr{A} \right). \qquad \square$$

   We shall obtain more information on how $h(T)$ behaves relative to natural operations on transformations when we have proved some results that make these calculations simpler. The following result allows us to understand when $h(T, \mathscr{A})$ is zero (Corollary 4.14.1) and allows us to conclude that a non-invertible measure-preserving transformation T which is not mod 0 invertible (i.e. $T^{-1}\mathscr{B} \neq \mathscr{B}$) must have (strictly) positive entropy (Corollary 4.14.3).

Theorem 4.14. *If $\mathscr{A}$ is a finite sub-algebra of $\mathscr{B}$ and T is a measure-preserving transformation of $(X, \mathscr{B}, m)$ then*

$$h(T, \mathscr{A}) = \lim_{n \to \infty} H\left(\mathscr{A} \,\Big/\, \left(\bigvee_{i=1}^{n} T^{-i}\mathscr{A} \right) \right) = H\left(\mathscr{A} \,\Big/\, \bigvee_{i=1}^{\infty} T^{-i}\mathscr{A} \right).$$

PROOF. The limit exists since the right hand side is non-increasing in n by virtue of Theorem 4.3(v). We know from the proof of Theorem 4.10 that for $n \geq 1$

$$H\left(\bigvee_{i=0}^{n-1} T^{-i}\mathscr{A} \right) = H(\mathscr{A}) + \sum_{j=1}^{n-1} H\left(\mathscr{A} \,\Big/\, \left(\bigvee_{i=1}^{j} T^{-i}\mathscr{A} \right) \right).$$

The desired result follows from dividing by n and taking the limit, since the Cesaro limit of a convergent sequence of real numbers equals the ordinary limit. The last equality is by Theorem 4.7. $\qquad \square$

   We can now deduce which finite sub-algebras have $h(T, \mathscr{A}) = 0$.

Corollary 4.14.1. *Let T be a measure-preserving transformation of the probability space $(X, \mathscr{B}, m)$. Let $\mathscr{A}$ be a finite subalgebra of $\mathscr{B}$. Then $h(T, \mathscr{A}) = 0$ iff $\mathscr{A} \overset{e}{\subset} \bigvee_{i=1}^{\infty} T^{-i}\mathscr{A}$.*

PROOF. By Theorems 4.14 and 4.8. $\qquad\qquad\qquad\qquad\qquad\qquad\qquad\Box$

Intuitively, this result says that the average information per day from performing the experiment, represented by $\mathscr{A}$, is zero exactly when the outcome on the first day can be determined from combined knowledge of the outcomes on all subsequent days (the future determines the present).

Corollary 4.14.2. *Let T be a measure-preserving transformation of the probability space $(X, \mathscr{B}, m)$. Then $h(T) = 0$ iff for every finite sub-algebra $\mathscr{A}$ of $\mathscr{B}$ we have $\mathscr{A} \overset{e}{\subset} \bigvee_{i=1}^{\infty} T^{-i}\mathscr{A}$.*

We can now conclude that a genuinely non-invertible measure-preserving transformation must have non-zero entropy.

Corollary 4.14.3. *Let T be a measure-preserving transformation of the probability space $(X, \mathscr{B}, m)$. If $h(T) = 0$ then $T^{-1}\mathscr{B} \overset{e}{=} \mathscr{B}$ (so T is invertible mod 0 if $(X, \mathscr{B}, m)$ is a Lebesgue space or a complete separable metric space).*

PROOF. Let $B \in \mathscr{B}$ and let $\mathscr{A}$ be the finite algebra $\{\phi, B, X \backslash B, X\}$. By Corollary 4.14.2 we have $\mathscr{A} \overset{e}{\subset} \bigvee_{i=1}^{\infty} T^{-i}\mathscr{A} \subset T^{-1}\mathscr{B}$. Since B is an arbitrary element of $\mathscr{B}$ we have $\mathscr{B} \overset{e}{=} T^{-1}\mathscr{B}$. $\qquad\qquad\qquad\qquad\Box$

We can strengthen this to the following.

Corollary 4.14.4. *Let T be a measure-preserving transformation of the probability space $(X, \mathscr{B}, m)$. Suppose $h(T) = 0$. If $\mathscr{F}$ is a sub-σ-algebra of $\mathscr{B}$ with $T^{-1}\mathscr{F} \overset{e}{\subset} \mathscr{F}$ then $T^{-1}\mathscr{F} \overset{e}{=} \mathscr{F}$.*

PROOF. The transformation T induces a measure-preserving transformation $T|_{(X, \mathscr{F}, m)}$ of $(X, \mathscr{F}, m)$ and it clearly has zero entropy. Apply Corollary 4.14.3 to this transformation. $\qquad\qquad\qquad\qquad\qquad\qquad\qquad\qquad\Box$

Remark. There is the following important result known as the Shannon–McMillan–Brieman theorem: Let T be an ergodic measure-preserving transformation of the probability space $(X, \mathscr{B}, m)$ and let ξ be a finite partition of $(X, \mathscr{B}, m)$. Let $B_n(x)$ denote the member of the partition $\bigvee_{i=0}^{n-1} T^{-i}\xi$ to which x belongs. Then $-(1/n) \log m(B_n(x)) \to h(T, \xi)$ a.e. and in $L^1(X, \mathscr{B}, m)$. (For a proof see Parry [2].) Hence if $h(T, \xi) > 0$ one can say that $m(B_n(x))$ goes to zero with exponential rate $e^{-h(T, \xi)}$ for a.e. $x \in X$. One can deduce that if $\varepsilon \in (0, 1)$ is given and $g_n(\varepsilon)$ denotes the smallest number of elements of

$\bigvee_{i=0}^{n-1} T^{-i}\xi$ required to give a set of total measure at least $1 - \varepsilon$, then $(1/n)$ $\log g_n(\varepsilon) \to h(T, \xi)$.

One can obtain Corollary 4.9.1 from the Shannon–McMillan–Brieman theorem by integration.

§4.6 Some Methods for Calculating $h(T)$

It is difficult to calculate $h(T)$ from its definition because one would need to calculate $h(T, \mathscr{A})$ for every finite sub-algebra $\mathscr{A}$. We consider what conditions on a finite sub-algebra $\mathscr{A}$ are needed to ensure $h(T) = h(T, \mathscr{A})$ (and hence simplify the calculation of $h(T)$), and also consider what conditions on a sequence $\{\mathscr{A}_n\}$ of sub-algebras would imply $h(T) = \lim_{n \to \infty} h(T, \mathscr{A}_n)$. These results lead to methods of calculating $h(T)$ for specific examples of measure-preserving transformations and they also lead to proofs of further properties of $h(T)$.

The main ingredient in the proofs of the above results is Theorem 4.16. We shall prove it using the following.

Lemma 4.15. *Let $r \geq 1$ be a fixed integer. For each $\varepsilon > 0$ there exists $\delta > 0$ such that if $\xi = \{A_1, \ldots, A_r\}$, $\eta = \{C_1, \ldots, C_r\}$ are any two partitions of $(X, \mathscr{B}, m)$ into r sets with $\sum_{i=1}^{r} m(A_i \triangle C_i) < \delta$ then $H(\xi/\eta) + H(\eta/\xi) < \varepsilon$.*

PROOF. Let $\varepsilon > 0$ be given. Choose $\delta > 0$ so that $\delta < \frac{1}{4}$ and $-r(r-1)\delta \log$ $\delta - (1 - \delta)\log(1 - \delta) < \varepsilon/2$. Let ζ be the partition into the sets $A_i \cap C_j$ $(i \neq j)$, and $\bigcup_{i=1}^{r} (A_i \cap C_i)$. Then $\xi \vee \eta = \eta \vee \zeta$ and since $A_i \cap C_j \subset \bigcup_{n=1}^{r} (A_n \triangle C_n)$ $(i \neq j)$ we have

$$m(A_i \cap C_j) < \delta \ (i \neq j) \quad \text{and} \quad m\left(\bigcup_{i=1}^{r} (A_i \cap C_i)\right) > 1 - \delta.$$

Hence $H(\zeta) < r(r-1)\delta \log \delta - (1 - \delta)\log(1 - \delta) < \varepsilon/2$. Therefore $H(\eta) + H(\xi/\eta) = H(\xi \vee \eta) = H(\eta \vee \zeta) \leq H(\eta) + H(\zeta) < H(\eta) + \varepsilon/2$, and so $H(\xi/\eta) < \varepsilon/2$. By symmetry (since $\xi \vee \eta = \xi \vee \zeta$) we have $H(\eta/\xi) < \varepsilon/2$. $\quad\square$

Remark. For $r \geq 1$ let $\tilde{V}_r$ denote the space of all ordered finite partitions of $(X, \mathscr{B}, m)$ into r sets where two partitions $\xi = \{A_1, \ldots, A_r\}$ $\eta = \{C_1, \ldots, C_r\}$ are identified if $m(A_i \triangle C_i) = 0$ for all i. A metric on $\tilde{V}_r$ is given by $\rho(\xi, \eta) = \sum_{i=1}^{r} m(A_i \triangle C_i)$ if $\xi = \{A_1, \ldots, A_r\}$ $\eta = \{C_1, \ldots, C_r\}$. If (V, d) denotes the metric space introduced in Theorem 4.5 then Lemma 4.15 says that the inclusion map from $(\tilde{V}_r, \rho)$ to (V, d) is uniformly continuous.

Theorem 4.16. *Let $(X, \mathscr{B}, m)$ be a probability space and $\mathscr{B}_0$ be an algebra such that the σ-algebra generated by $\mathscr{B}_0$ (denoted by $\mathscr{B}(\mathscr{B}_0)$) satisfies $\mathscr{B}(\mathscr{B}_0) \doteq \mathscr{B}$. Let $\mathscr{C}$ be a finite sub-algebra of $\mathscr{B}$. Then for every $\varepsilon > 0$, there exists a finite algebra $\mathscr{D}, \mathscr{D} \subseteq \mathscr{B}_0$ such that $H(\mathscr{D}/\mathscr{C}) + H(\mathscr{C}/\mathscr{D}) < \varepsilon$.*

PROOF. Let $\xi(\mathcal{C}) = \{C_1, \ldots, C_r\}$. Let $\varepsilon > 0$ and choose δ to correspond to r and ε in Lemma 4.15. It suffices to show that for each $\sigma > 0$ we can find a partition $\{D_1, \ldots, D_r\}$ with $D_i \in \mathcal{B}_0$ and $m(C_i \triangle D_i) < \sigma$ for each i. To do this choose $\lambda > 0$ such that $\lambda(r - 1)[1 + r(r - 1)] < \sigma$, and for each i choose $B_i \in \mathcal{B}_0$ such that $m(C_i \triangle B_i) < \lambda$. If $i \neq j$ then $B_i \cap B_j \subset (B_i \triangle C_i) \cup (B_j \triangle C_j)$ so that $m(B_i \cap B_j) < 2\lambda$. Let $N = \bigcup_{i \neq j}(B_i \cap B_j)$. We have $m(N) < r(r - 1)\lambda$. Set $D_i = B_i \backslash N$ for $1 \leq i < r$ and $D_r = X \backslash \bigcup_{i=1}^{r-1} D_i$. $\{D_1, \ldots, D_r\}$ is a partition of X and each $D_i \in \mathcal{B}_0$. If $i < r$ then $D_i \triangle C_i \subset (B_i \triangle C_i) \cup N$ and so,

$$m(D_i \triangle C_i) < \lambda[1 + r(r - 1)] < \sigma.$$

However, $D_r \triangle C_r \subset \bigcup_{i=1}^{r-1}(D_i \triangle C_i)$ and therefore

$$m(D_r \triangle C_r) < (r - 1)\lambda[1 + r(r - 1)] < \sigma.$$

So the theorem is proved. $\square$

If $\{\mathscr{A}_n\}$ is a sequence of sub-σ-algebras of $\mathscr{B}$ then $\bigvee_n \mathscr{A}_n$ denotes the sub-σ-algebra of $\mathscr{B}$ generated by $\{\mathscr{A}_n\}$ i.e. $\bigvee_n \mathscr{A}_n$ is the intersection of all those sub-σ-algebras of $\mathscr{B}$ that contain every $\mathscr{A}_n$. We shall use $\bigcup_n \mathscr{A}_n$ to denote the collection of sets that belong to some $\mathscr{A}_n$. If $\{\mathscr{A}_n\}_{n=1}^\infty$ is an increasing sequence of sub-σ-algebras then $\bigcup_n \mathscr{A}_n$ is an algebra, but not necessarily a σ-algebra.

Corollary 4.16.1. *If $\{\mathscr{A}_n\}$ is an increasing sequence of finite sub-algebras of $\mathscr{B}$ and $\mathscr{C}$ is a finite sub-algebra with $\mathscr{C} \overset{\circ}{\subset} \bigvee_n \mathscr{A}_n$, then $H(\mathscr{C}/\mathscr{A}_n) \to 0$ as $n \to \infty$.*

PROOF. If $\mathscr{B}_0 = \bigcup_{j=1}^\infty \mathscr{A}_j$ then $\mathscr{B}_0$ is an algebra and $\mathscr{C} \overset{\circ}{\subset} \mathscr{B}(\mathscr{B}_0)$ by hypothesis. Let $\varepsilon > 0$. By Theorem 4.16 there exists a finite sub-algebra $\mathscr{D}_\varepsilon$ of $\mathscr{B}_0$ such that $H(\mathscr{C}/\mathscr{D}_\varepsilon) < \varepsilon$. But $\mathscr{D}_\varepsilon \subseteq \mathscr{A}_{j_0}$ for some j_0 since $\mathscr{D}_\varepsilon$ is finite. If $j \geq j_0$ Theorem 4.3(v) gives $H(\mathscr{C}/\mathscr{A}_j) \leq H(\mathscr{C}/\mathscr{A}_{j_0}) \leq H(\mathscr{C}/\mathscr{D}_\varepsilon) < \varepsilon$.
    Thus $H(\mathscr{C}/\mathscr{A}_n) \to 0$. $\square$

The main methods for calculating $h(T)$ are supplied by the next two theorems.

Theorem 4.17 (Kolmogorov–Sinai Theorem). *Let T be an invertible measure-preserving transformation of the probability space $(X \cdot \mathscr{B}, m)$ and let $\mathscr{A}$ be a finite sub-algebra of $\mathscr{B}$ such that $\bigvee_{n=-\infty}^\infty T^n \mathscr{A} \overset{\circ}{=} \mathscr{B}$. Then $h(T) = h(T, \mathscr{A})$.*

PROOF. Let $\mathscr{C} \subseteq \mathscr{B}$ be finite. We want to show that $h(T, \mathscr{C}) \leq h(T, \mathscr{A})$. For $n \geq 1$,

$$h(T, \mathscr{C}) \leq h\left(T, \bigvee_{i=-n}^n T^i \mathscr{A}\right) + H\left(\mathscr{C} \bigg/ \bigvee_{i=-n}^n T^i \mathscr{A}\right) \quad \text{by Theorem 4.12(iv)}$$

$$= h(T, \mathscr{A}) + H\left(\mathscr{C} \bigg/ \bigvee_{i=-n}^n T^{-i} \mathscr{A}\right) \quad \text{by Theorem 4.12(vii)}.$$

Let $\mathscr{A}_n = \bigvee_{i=-n}^n T^i \mathscr{A}$. It suffices to show that $H(\mathscr{C}/\mathscr{A}_n)$ goes to zero as $n \to \infty$. This follows by Corollary 4.16.   □

A similar result holds when T is not necessarily invertible:

Theorem 4.18. *If T is a measure-preserving transformation (but not necessarily invertible) of the probability space $(X, \mathscr{B}, m)$ and if $\mathscr{A}$ is a finite sub-algebra of $\mathscr{B}$ with $\bigvee_{i=0}^\infty T^{-i} \mathscr{A} \doteq \mathscr{B}$ then $h(T) = h(T, \mathscr{A})$.*

PROOF. This is similar to the proof of the previous theorem; use $\bigvee_{i=0}^{n-1} T^{-i} \mathscr{A}$ in the place of $\bigvee_{i=-n}^n T^i \mathscr{A}$, and Theorem 4.12(vi).   □

The following is sometimes useful in showing transformations have zero entropy. We shall use it later to show a rotation of the unit circle has zero entropy.

Corollary 4.18.1. *If T is an invertible measure-preserving transformation of the probability space $(X, \mathscr{B}, m)$ and $\bigvee_{i=0}^\infty T^{-i} \mathscr{A} \doteq \mathscr{B}$ for some finite sub-algebra $\mathscr{A}$ then $h(T) = 0$.*

PROOF. By Theorem 4.18

$$h(T) = h(T, \mathscr{A})$$

$$= \lim_{n \to \infty} H\left(\mathscr{A} \Big/ \bigvee_{i=1}^n T^{-i} \mathscr{A}\right) \quad \text{by Theorem 4.14.}$$

But $\bigvee_{i=1}^\infty T^{-i} \mathscr{A} \doteq T^{-1} \mathscr{B} \doteq \mathscr{B}$. Let $\mathscr{A}_n = \bigvee_{i=1}^n T^{-i} \mathscr{A}$. Then $\mathscr{A}_1 \subset \mathscr{A}_2 \subset \cdots$ and $\bigvee_{n=1}^\infty \mathscr{A}_n \doteq \mathscr{B}$. By Corollary 4.16.1 we have $H(\mathscr{A}/\mathscr{A}_n) \to 0$, so that $h(T) = 0$.   □

Remarks

(1) Entropy can be defined for any countable partition of $(X, \mathscr{B}, m)$ as follows: If $\xi = \{A_1, A_2, \ldots\}$ then

$$H(\xi) = -\sum_i m(A_i) \log m(A_i)$$

(which may be infinite). One can show $h(T) = \sup h(T, \xi)$ where the supremum is then over all countable partitions ξ with $H(\xi) < \infty$.

A countable partition ξ of X is called a *generator* for an invertible measure-preserving transformation T if

$$\bigvee_{n=-\infty}^\infty T^n \mathscr{A}(\xi) \doteq \mathscr{B}.$$

As in Theorem 4.16 one can prove that if ξ is a generator and $H(\xi) < \infty$ then $h(T) = h(T, \xi)$.

The basic theorem on existence of generators ξ with $H(\xi) < \infty$ was given by Rohlin in 1963.

Let us suppose $(X, \mathcal{B}, m)$ is a Lebesgue space and T is an invertible measure-preserving transformation of $(X, \mathcal{B}, m)$. We say that T is *aperiodic* if

$$m \left(\bigcup_{\substack{n \in Z \\ n \neq 0}} \{x \in X : T^n(x) = x\} \right) = 0.$$

Note that if T is ergodic then T is aperiodic unless X differs from a finite set by a set a measure 0.

Theorem 4.19. (See Rohlin [3].) *Suppose $(X, \mathcal{B}, m)$ is a Lebesgue space (which we assume is not isomorphic to a finite set) and T is an invertible measure-preserving transformation of $(X, \mathcal{B}, m)$. Then T has a generator ξ with $H(\xi) < \infty$ iff $h(T) < \infty$ and T is aperiodic.*

Thus, if T is ergodic and $h(T) < \infty$ then T has a generator ξ with $H(\xi) < \infty$. In 1970 Krieger [1] proved:

Theorem 4.20. *If $(X, \mathcal{B}, m)$ is a Lebesgue space and T is an ergodic invertible measure-preserving transformation of $(X, \mathcal{B}, m)$ with $h(T) < \infty$ then T has a finite generator*

$$\xi = \{A_1, \ldots, A_n\}.$$

In fact ξ may be taken so that $e^{h(T)} \leq n \leq e^{h(T)} + 1$.

Hence finite generators exist in the most interesting cases, although they may be difficult to find.

(2) Another important use of partitions is in relating measure-preserving transformations to shift mappings. Suppose T is an invertible measure-preserving transformation of the probability space $(X, \mathcal{B}, m)$ and suppose $\xi = \{A_1, \ldots, A_k\}$ is a finite partition of $(X, \mathcal{B}, m)$. Let Y denote the product space $\{1, 2, \ldots, k\}^Z$ so that each point of Y is a bisequence $y = (y_n)_{-\infty}^{\infty}$ where each $y_n \in \{1, 2, \ldots, k\}$. Let $S : Y \to Y$ be the transformation defined by $S(y_n) = (z_n)$ where $z_n = y_{n+1}$. In other words S maps the point $(\ldots, y_{-1} \overset{*}{y_0} y_1 y_2, \ldots)$ to $(\ldots, y_{-1} y_0 \overset{*}{y_1} y_2, \ldots)$. If we put the discrete topology on $\{1, 2, \ldots, k\}$ and the product topology on Y then Y is a compact metrisable space and S is a homeomorphism of Y. We shall study this shift homeomorphism in Chapter 5. We shall now show that there is a probability measure μ on the σ-algebra $\mathcal{B}(Y)$ of Borel subsets of Y which is preserved by S and a measure-preserving map $\phi : (X, \mathcal{B}, m) \to (Y, \mathcal{B}(Y), \mu)$ such that $\phi T = S \phi$. For $x \in X$ let $\phi(x) = (y_n)_{-\infty}^{\infty}$ if $T^n x \in A_{y_n}$. This defines a transformation $\phi : X \to Y$. Also $S \phi(x) = \phi(Tx)$, $x \in X$. To see that ϕ is measurable notice that if for

$s < r$ $_s[i_s, \ldots, i_r]_r = \{y = (y_n) | y_j = i_j s \leq j \leq r\}$ then

$$\phi^{-1}(_s[i_s, \ldots, i_r]_r) = \bigcap_{j=s}^{r} T^{-j} A_{i_j}$$

and the sets of the form $_s[i_s, \ldots, i_r]_r$ generate $\mathscr{B}(Y)$. Since

$$\{B \in \mathscr{B}(Y) | \phi^{-1}(B) \in \mathscr{B}\}$$

is a σ-algebra we must have $\phi^{-1} \mathscr{B}(Y) \subset \mathscr{B}$. The measure μ on $(Y, \mathscr{B}(Y))$ is defined by $\mu = m \circ \phi^{-1}$. Clearly ϕ is measure-preserving and from $S\phi = \phi T$ we see S preserves μ. (The measure μ is not in general a product measure or a Markov measure.) The map $\phi : X \to Y$ can be very far from invertible: if T is the identity map then $\phi(x)$ is constant on each set A_i. In this example the map $\tilde{\phi}^{-1} : (\tilde{\mathscr{B}}(Y), \tilde{\mu}) \to (\tilde{\mathscr{B}}, \tilde{m})$ is not onto since only the members of $\tilde{\mathscr{B}}$ corresponding to the sets A_i are in the image of $\tilde{\phi}^{-1}$. If we want ϕ to be a conjugacy (i.e. $\tilde{\phi}^{-1} : (\tilde{\mathscr{B}}(Y), \tilde{\mu}) \to (\tilde{\mathscr{B}}, \tilde{m})$ is onto) then we need $\bigvee_{n=-\infty}^{\infty} T^n \mathscr{A}(\xi) \doteq \mathscr{B}$, in other words we need ξ to be a generator. So a generator for T gives a natural conjugacy between T and the shift homeomorphism on a product space equipped with some shift invariant probability measure defined on the Borel subsets of the product space. By Krieger's theorem we can suppose that each ergodic invertible measure-preserving transformation is represented in this way, in this case the Borel measure on the shift space is an ergodic invariant measure for the shift. (This implies that there are many different ergodic shift invariant measures on the Borel subsets of a given product space.)

For non-invertible transformations similar results are true when the one-sided shift map $S(y_0, y_1, \ldots) = (y_1, y_2, \ldots)$ is used on the one-sided shift space $\{1, 2, \ldots, k\}^{Z^+}$. Then one considers one-sided generators (i.e. $\bigvee_{n=0}^{\infty} T^{-n} \mathscr{A}(\xi) \doteq \mathscr{B}$). $\quad\square$

We now prove some more results that are useful for the computation of entropy.

Theorem 4.21. *Let $(X, \mathscr{B}, m)$ be a probability space. If $\mathscr{B}_0$ is a sub-algebra of $\mathscr{B}$ with $\mathscr{B}(\mathscr{B}_0) \doteq \mathscr{B}$ then for each measure-preserving transformation $T : X \to X$ we have*

$$h(T) = \sup h(T, \mathscr{A})$$

where the supremum is taken over all finite sub-algebras $\mathscr{A}$ of $\mathscr{B}_0$.

PROOF. Let $\varepsilon > 0$. Let $\mathscr{C} \subseteq \mathscr{B}$ be finite. By Theorem 4.16 there exists a finite $\mathscr{D}_\varepsilon \subseteq \mathscr{B}_0$ such that

$$H(\mathscr{C}/\mathscr{D}_\varepsilon) < \varepsilon.$$

Thus

$$h(T, \mathscr{C}) \leq h(T, \mathscr{D}_\varepsilon) + H(\mathscr{C}/\mathscr{D}_\varepsilon) \quad \text{by Theorem 4.12(iv)}$$
$$\leq h(T, \mathscr{D}_\varepsilon) + \varepsilon.$$

Therefore $h(T, \mathscr{C}) \leq \varepsilon + \sup\{h(T, \mathscr{D}): \mathscr{D} \subset \mathscr{B}_0, \mathscr{D} \text{ finite}\}$ and thus $h(T) \leq \sup\{h(T, \mathscr{D}): \mathscr{D} \subset \mathscr{B}_0, \mathscr{D} \text{ finite}\}$. The opposite inequality is obvious. □

When it is difficult to find a generator for a given measure-preserving transformation the following result is often useful to calculate entropy.

Theorem 4.22. *Let $(X, \mathscr{B}, m)$ be a probability space and let $\{\mathscr{A}_n\}_1^\infty$ be finite sub-algebras of $\mathscr{B}$ such that $\mathscr{A}_1 \subseteq \mathscr{A}_2 \subseteq \cdots$ and $\bigvee_{n=1}^\infty \mathscr{A}_n \doteq \mathscr{B}$. If $T: X \to X$ is measure-preserving then $h(T) = \lim_{n \to \infty} h(T, \mathscr{A}_n)$.*

PROOF. We note that $h(T, \mathscr{A}_n)$ is an increasing sequence by Theorem 4.12(iii). Also $\mathscr{B}_0 = \bigcup_{n=1}^\infty \mathscr{A}_n$ is an algebra and $\mathscr{B}(\mathscr{B}_0) \doteq \mathscr{B}$. By Theorem 4.21 $h(T) = \sup\{h(T, \mathscr{C}): \mathscr{C} \subset \mathscr{B}_0, \mathscr{C} \text{ finite}\}$. If $\mathscr{C} \subseteq \mathscr{B}_0$ is finite then $\mathscr{C} \subseteq \mathscr{A}_{n_0}$ for some n_0. Thus

$$h(T, \mathscr{C}) \leq h(T, \mathscr{A}_{n_0}),$$

which implies $h(T) \leq \lim_{n \to \infty} h(T, \mathscr{A}_n)$ and hence $h(T) = \lim_{n \to \infty} h(T, \mathscr{A}_n)$.
□

Theorem 4.21 allows us to prove the following formula for the entropy of a direct product of measure-preserving transformations.

Theorem 4.23. *Let $(X_1, \mathscr{B}_1, m_1)$, $(X_2, \mathscr{B}_2, m_2)$ be probability spaces and let $T_1: X_1 \to X_1$, $T_2: X_2 \to X_2$ be measure-preserving. Then*

$$h(T_1 \times T_2) = h(T_1) + h(T_2).$$

PROOF. If $\mathscr{A}_1 \subseteq \mathscr{B}_1$, $\mathscr{A}_2 \subseteq \mathscr{B}_2$ are finite then $\mathscr{A}_1 \times \mathscr{A}_2$ is finite, where

$$\xi(\mathscr{A}_1 \times \mathscr{A}_2) = \{A_1 \times A_2 : A_1 \in \xi(\mathscr{A}_1), A_2 \in \xi(\mathscr{A}_2)\}.$$

Let $\mathscr{F}_0$ denote the algebra of finite unions of measurable rectangles. Then $\mathscr{B}(\mathscr{F}_0) = \mathscr{B}_1 \times \mathscr{B}_2$ by definition of $\mathscr{B}_1 \times \mathscr{B}_2$, and by Theorem 4.21,

$$h(T_1 \times T_2) = \sup\{h(T_1 \times T_2, \mathscr{C}): \mathscr{C} \subset \mathscr{F}_0, \mathscr{C} \text{ finite}\}.$$

But if $\mathscr{C}$ is finite and $\mathscr{C} \subseteq \mathscr{F}_0$ then $\mathscr{C} \subseteq \mathscr{A}_1 \times \mathscr{A}_2$ for some finite $\mathscr{A}_1 \subseteq \mathscr{B}_1$, $\mathscr{A}_2 \subseteq \mathscr{B}_2$. Hence, by Theorem 4.12(v),

$$h(T_1 \times T_2) = \sup\{h(T_1 \times T_2, \mathscr{A}_1 \times \mathscr{A}_2): \mathscr{A}_1 \subset \mathscr{B}_1, \mathscr{A}_2 \subset \mathscr{B}_2; \mathscr{A}_1, \mathscr{A}_2 \text{ finite}\}.$$

We have

$$h\left(\bigvee_{i=0}^{n-1} (T_1 \times T_2)^{-i}(\mathscr{A}_1 \times \mathscr{A}_2)\right)$$

$$= H\left(\left(\bigvee_{i=0}^{n-1} T_1^{-i} \mathscr{A}_1\right) \times \left(\bigvee_{i=0}^{n-1} T_2^{-i} \mathscr{A}_2\right)\right)$$

$$= -\sum (m_1 \times m_2)(C_k \times D_j) \cdot \log(m_1 \times m_2)(C_k \times D_j)$$

where $\{C_k\}$ are the members of $\xi(\bigvee_{i=0}^{n-1} T_1^{-i}\mathscr{A}_1)$, and $\{D_j\}$ are the members of $\xi(\bigvee_{i=0}^{n-1} T_2^{-i}\mathscr{A}_2)$

$$= -\sum m_1(C_k)m_2(D_j) \cdot \log(m_1(C_k)m_2(D_j))$$
$$= -\sum m_1(C_k)m_2(D_j) \cdot [\log m_1(C_k) + \log m_2(D_j)]$$
$$= -\sum m_1(C_k) \cdot \log m_1(C_k) - \sum m_2(D_j) \cdot \log m_2(D_j)$$
$$= H\left(\bigvee_{i=0}^{n-1} T_1^{-i}\mathscr{A}_1\right) + H\left(\bigvee_{i=0}^{n-1} T_2^{-i}\mathscr{A}_2\right).$$

Thus $h(T_1 \times T_2, \mathscr{A}_1 \times \mathscr{A}_2) = h(T_1, \mathscr{A}_1) + h(T_2, \mathscr{A}_2)$ so that $h(T_1 \times T_2) = h(T_1) + h(T_2)$.   □

Remark. Theorem 4.23 readily extends to the direct product of any finite number of measure-preserving transformations.

§4.7 Examples

We shall now calculate the entropy of our examples.

(1) If $I:(X, \mathscr{B}, m) \to (X, \mathscr{B}, m)$ is the identity, then $h(I) = 0$. This is because $h(I, \mathscr{A}) = \lim(1/n)H(\mathscr{A}) = 0$. Also, if $T^p = I$ for some $p \neq 0$ then $h(T) = 0$. This follows since $0 = h(T^p) = |p| \cdot h(T)$ by Theorem 4.13. In particular any measure-preserving transformation of a finite space has zero entropy.

(2) **Theorem 4.24.** *Any rotation, $T(z) = az$, of the unit circle K has zero entropy.*

Case 1: Suppose $\{a^n : n \in Z\}$ is not dense, i.e., a is a root of unity. Thus $a^p = 1$ for some $p \neq 0$; and $T^p(z) = a^p z = z$ so $h(T) = 0$ by example (1).

Case 2: Suppose $\{a^n : n \in Z\}$ is dense in K. Then $\{a^n : n < 0\}$ is dense in K. Let $\xi = \{A_1, A_2\}$ where A_1 is the upper half circle $[1, -1)$, and A_2 is the

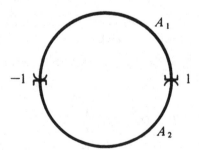

lower half circle $[-1, 1)$. For $n > 0$, $T^{-n}\xi$ consists of semi-circles beginning at a^{-n} and $-a^{-n}$. Since $\{a^{-n} : n > 0\}$ is dense any semi-circle belongs to

$\bigvee_{n=0}^{\infty} T^{-n}\mathscr{A}(\xi)$. Hence any arc belongs to $\bigvee_{n=0}^{\infty} T^{-n}\mathscr{A}(\xi)$. Thus, $\mathscr{B} = \bigvee_{n=0}^{\infty} T^{-n}\mathscr{A}(\xi)$ and so, $h(T) = 0$ by Corollary 4.18.1.

(3) **Theorem 4.25.** *Any rotation of a compact metric abelian group has entropy zero.*

PROOF

(a) Suppose $X = K^n$, the n-torus, and $T(z_1, \ldots, z_n) = (a_1 z_1, \ldots, a_n z_n)$. Then $T = T_1 \times T_2 \times \cdots \times T_n$ where $T_i : K \to K$ is defined by $T_i(z) = a_i z$. By example (2) $h(T_i) = 0$ for all i so by Theorem 4.23

$$h(T) = \sum_{i=1}^{n} h(T_i) = 0.$$

(b) *General Case.* Let $T : G \to G$ be $T(x) = ax$. Let $\hat{G} = \{\gamma_1, \gamma_2, \ldots\}$. Let $H_n = \operatorname{Ker} \gamma_1 \cap \cdots \cap \operatorname{Ker} \gamma_n$. Then H_n is a closed subgroup of G and $\widehat{(G/H_n)}$ is the group generated by $\{\gamma_1, \ldots, \gamma_n\}$ (by 6 of §0.7.) Thus

$$\widehat{(G/H_n)} = \text{finite group} \times Z^{i_n},$$

so

$$G/H_n = F_n \times K^{i_n}$$

where F_n is a finite group and K^{i_n} is a finite-dimensional torus.

The rotation T induces a map $T_n : G/H_n \to G/H_n$ by $T_n(gH_n) = agH_n$. The map T_n is a rotation on G/H_n, so that it can be written $T_n = T_{n,1} \times T_{n,2}$ where $T_{n,1}$ is a rotation of F_n and $T_{n,2}$ is a rotation of K^{i_n}. Thus

$$h(T_n) = h(T_{n,1}) + h(T_{n,2}) = 0$$

by example (1) and case (a) of this proof.

Note that $\bigvee_n \mathscr{A}(G/H_n) \doteq \mathscr{B}$, where $\mathscr{A}(G/H_n)$ denotes the σ-algebra consisting of those elements of $\mathscr{B}$ that are unions of cosets of H_n, because γ_n is measurable relative to $\mathscr{A}(G/H_n)$ and so every member of $L^2(m)$ is measurable relative to $\bigvee_n \mathscr{A}(G/H_n)$.

Therefore if $\mathscr{B}_0 = \bigcup_{n=1}^{\infty} \mathscr{A}(G/H_n)$ then by Theorem 4.21

$$h(T) = \sup_{\substack{\mathscr{C} \subseteq \mathscr{B}_0 \\ \mathscr{C} \text{ finite}}} h(T, \mathscr{C}).$$

However, if $\mathscr{C} \subseteq \mathscr{B}_0$ is finite then $\mathscr{C} \subseteq \mathscr{A}(G/H_n)$ for some n and so $h(T, \mathscr{C}) \le h(T_n) = 0$. Thus $h(T) = 0$. $\qquad \square$

Corollary 4.25.1. *Any ergodic transformation with discrete spectrum has zero entropy.*

This follows from Theorem 3.6. (Actually we have shown the result only when $(X, \mathscr{B}, m)$ has a countable basis since the above calculation was for a metric group G.)

(4) *Endomorphisms of Compact Groups.* If A is an endomorphism of the n-torus K^n onto K^n we shall show in Chapter 8 that $h(A) = \sum \log|\lambda_i|$ where the summation is over all eigenvalues of the matrix $[A]$ with absolute value greater than one.

One can write down a complicated formula for the entropy of an endomorphism of a general compact metric abelian group. See Yuzvinskii [1].

(5) *Affine Transformations.* We shall show in Theorem 8.10 that when $T = a \cdot A$ is an affine transformation of K^n then $h(T) = h(A)$.

(6) **Theorem 4.26.** *The two-sided $(p_0, \ldots, p_{k-1})$-shift has entropy $-\sum_{i=0}^{k-1} p_i \cdot \log p_i$.*

PROOF. Let $Y = \{0, 1, \ldots, k-1\}$, $X = \prod_{-\infty}^{\infty} Y$, and let T be the shift. Let $A_i = \{\{x_k\} : x_0 = i\}$, $0 \le i \le k-1$. Then $\xi = \{A_0, \ldots, A_{k-1}\}$ is a partition of X. For ease of notation let $\mathscr{A}$ denote $\mathscr{A}(\xi)$. By the definition of the product σ-algebra, $\mathscr{B}$, we have

$$\bigvee_{i=-\infty}^{\infty} T^i \mathscr{A} = \mathscr{B}.$$

By the Kolmogorov–Sinai Theorem (4.17),

$$h(T) = \lim_{n \to \infty} \frac{1}{n} H(\mathscr{A} \vee T^{-1}\mathscr{A} \vee \cdots \vee T^{-(n-1)}\mathscr{A}).$$

A typical element of $\xi(\mathscr{A} \vee T^{-1}\mathscr{A} \vee \cdots \vee T^{-(n-1)}\mathscr{A})$ is

$$A_{i_0} \cap T^{-1}A_{i_1} \cap \cdots \cap T^{-(n-1)}A_{i_{n-1}}$$
$$= \{\{x_n\} : x_0 = i_0, x_1 = i_1, \ldots, x_{n-1} = i_{n-1}\}$$

which has measure $p_{i_0} \cdot p_{i_1} \cdot \ldots \cdot p_{i_{n-1}}$. Thus,

$$H(\mathscr{A} \vee T^{-1}\mathscr{A} \vee \cdots \vee T^{-(n-1)}\mathscr{A})$$
$$= -\sum (p_{i_0} \cdot \ldots \cdot p_{i_{n-1}}) \cdot \log(p_{i_0} \cdot \ldots \cdot p_{i_{n-1}})$$
$$= -\sum_{i_0,\ldots,i_{n-1}=0}^{k-1} (p_{i_0} \cdot \ldots \cdot p_{i_{n-1}})[\log p_{i_0} + \cdots + \log p_{i_{n-1}}]$$
$$= -n \sum_{i=0}^{k-1} p_i \cdot \log p_i.$$

Therefore, $h(T) = h(T, \mathscr{A}) = -\sum_{i=0}^{k-1} p_i \cdot \log p_i.$ $\square$

Remark. The 2-sided $(\frac{1}{2}, \frac{1}{2})$-shift has entropy $\log 2$; the 2-sided $(\frac{1}{3}, \frac{1}{3}, \frac{1}{3})$-shift has entropy $\log 3$. Thus these transformations cannot be conjugate.

(7) The 1-sided $(p_0, \ldots, p_{k-1})$-shift has entropy $-\sum_{i=0}^{k-1} p_i \cdot \log p_i$. The proof is very similar to the one in example (6) but Theorem 4.18 is used instead of Theorem 4.17.

(8) The following is an example of a transformation with infinite entropy. Let $I = (0, 1]$ with Borel sets and Lebesgue measure. Let $X = \prod_{-\infty}^{\infty} I$ with product measure and let T be the shift on X. Then $h(T) = \infty$.

To see this let

$$A_{n,i} = \left\{ \{x_j\} : \frac{i-1}{n} < x_0 \le \frac{i}{n}, n > 0, 1 \le i \le n \right\}.$$

Then $m(A_{n,i}) = 1/n$ and $\xi_n = \{A_{n,1}, \ldots, A_{n,n}\}$ is a partition of X. Hence $h(T, \xi_n) = \log n$ by the same argument as used in example (6) (using the independence of $\xi_n, T^{-1}\xi_n, \ldots, T^{-k}\xi_n$). Therefore, $h(T) \ge \log n$ for each n, and so $h(T) = \infty$.

(9) **Theorem 4.27.** *The two-sided* $(\mathbf{p}, P)$ *Markov shift has entropy* $-\sum_{i,j} p_i p_{ij} \log p_{ij}$.

PROOF. We shall use the notation of example (6). We have $\sum_{i=0}^{k-1} p_i p_{ij} = p_j$. As in (6) $h(T) = \lim_{n \to \infty} (1/n) H(\bigvee_0^{n-1} T^{-i} \mathscr{A})$. The element $A_{i_0} \cap T^{-i} A_{i_1} \cap \cdots \cap T^{-(n-1)} A_{i_{n-1}}$ has measure $p_{i_0} p_{i_0 i_1} \cdots p_{i_{n-2} i_{n-1}}$ so

$$
\begin{aligned}
H\left(\bigvee_{i=0}^{n-1} T^{-i} \mathscr{A} \right) \\
= -\sum_{i_0, i_1, \ldots, i_{n-1}=0}^{k-1} p_{i_0} p_{i_0 i_1} \cdots p_{i_{n-2} i_{n-1}} \log(p_{i_0} p_{i_0 i_1} \cdots p_{i_{n-2} i_{n-1}}) \\
= -\sum p_{i_0} p_{i_0 i_1} \cdots p_{i_{n-2} i_{n-1}} [\log p_{i_0} + \log p_{i_0 i_1} + \cdots + \log p_{i_{n-2} i_{n-1}}] \\
= -\sum_{i_0=0}^{k-1} p_{i_0} \log p_{i_0} - (n-1) \sum_{i,j=0}^{k-1} p_i p_{ij} \log p_{ij},
\end{aligned}
$$

where we have used the relationships $\sum_{i=0}^{k-1} p_i p_{ij} = p_j$ and $\sum_{j=0}^{k-1} p_{ij} = 1$. Hence $h(T) = -\sum_{i,j} p_i p_{ij} \log p_{ij}$. $\square$

The same formula holds for the one-sided $(\mathbf{p}, P)$ Markov shift.

§4.8 How Good an Invariant is Entropy?

An invariant P for an equivalence relation is a complete invariant if whenever T and S both have the property P then T and S are equivalent.

Entropy is, in general, far from complete for the equivalence relation of conjugacy on the set of all measure-preserving transformations. The following is an example of two ergodic measure-preserving transformations with equal entropy which are not conjugate.

Let $T:K \to K$ be defined by $T(z) = az$, $a \in K$, where $\{a^n\}_{-\infty}^{\infty}$ is dense in K, and let $S:K \to K$ be defined by $S(z) = bz$, $b \in K$, where $\{b^n\}_{-\infty}^{\infty}$ is dense in K. We know T and S are ergodic and $h(T) = 0 = h(S)$ by example (2) of the previous section. If we choose a, b so that $\{a^n\}_{-\infty}^{\infty} \neq \{b^n\}_{-\infty}^{\infty}$ then T and S are not conjugate (in fact, they are not even spectrally equivalent) by Theorem 3.4.

The following is an example (due to Anzai) of two ergodic and spectrally equivalent measure-preserving transformations with equal entropy which are not conjugate.

Let $T:K^2 \to K^2$ and $S:K^2 \to K^2$ be defined by

$$T(z, w) = (az, z^p w) \qquad S(z, w) = (az, z^q w),$$

where $\{a^n\}_{-\infty}^{\infty}$ is dense in K and p, q are non-zero integers. Observe that T and S are affine transformations, T and S are ergodic (see the Remarks after Theorem 1.11) and $h(T) = h(S) = 0$ by examples (5) and (4) of §4.7.

For $n \in Z$ define $g_n:K^2 \to K$ by $g_n(z, w) = z^n$. Then $g_n \circ T = a^n g_n$ and $g_n \circ S = a^n g_n$. By considering the other characters of K^2 one can readily show that there are functions $\{f_i : i \geq 0\}$ such that $L^2(m)$ has a basis consisting of the functions $\{g_n : n \in Z\}$ together with $\{U_T^j f_i : j \in Z, i \geq 0\}$. Similarly there are functions $\{h_i : i \geq 0\}$ such that functions $\{g_n \mid n \in Z\}$ together with $\{U_S^j h_i : j \in Z, i \geq 0\}$ form a basis of $L^2(m)$. One can then define a unitary operator $W:L^2(m) \to L^2(m)$ by $W(g_n) = g_n$ and $W(U_T^j f_q) = U_S^j h_q$ and extending. Clearly $WU_T = U_S W$ showing T and S are spectrally isomorphic.

However, if $p \neq \pm q$, T and S are not conjugate. As mentioned before (Theorem 2.6) conjugacy and isomorphism coincide for measure-preserving transformations of K^2 equipped with Haar measure m. We shall show T and S are not isomorphic. Suppose $\phi T = S\phi$ and $\phi(z, w) = (f(z, w), g(z, w))$. The maps f and g are only defined almost everywhere but this will not affect our argument as we shall consider them as members of $L^2(m)$. We have $f(T) = af$ and $g(T) = f^q g$. Since f is an eigenfunction with eigenvalue a, Theorem 3.1(iii) implies $f(z, w) = c \cdot z$ for some $c \in K$. The second equation then becomes $g(T(z, w)) = c^q z^q g(z, w)$. If one now expresses g as a Fourier series, then it is straightforward to show that $g(z, w) = dz^n w^t$ where $d \in K$, $pt = q$ for some $t \in Z$ and $a^n = c^q$. So $\phi(z, w) = (cz, dz^n w^t)$ is an affine transformation and for ϕ to be an invertible measure-preserving transformation one needs $t = \pm 1$, i.e., $p = \pm q$.

Hence entropy is not a good invariant for the relation of conjugacy on the class of transformations with zero entropy. It is desirable to find a class of measure-preserving transformations for which entropy is a good invariant. It turns out that such a class is provided by transformations that are "the opposites" of transformations with zero entropy. We consider these transformations in the next section.

§4.9 Bernoulli Automorphisms and Kolmogorov Automorphisms

Definition 4.11. Let $(Y, \mathcal{F}, \mu)$ be a probability space. Let $(X, \mathcal{B}, m) = \prod_{-\infty}^{\infty}$ $(Y, \mathcal{F}, \mu)$ and let $T: X \to X$ be the shift $T(\{y_n\}) = \{x_n\}$ where $x_n = y_{n+1} n \in Z$. Then T is an invertible measure-preserving transformation and is called the *Bernoulli shift* with *state space* $(Y, \mathcal{F}, \mu)$.

EXAMPLES OF BERNOULLI SHIFTS

(1) The two sided $(p_0, \ldots, p_{k-1})$-shift. Here $Y = \{0, 1, \ldots, k-1\}$.
(2) The example (8) of §4.7. Here $Y = (0, 1]$.
(3) If T is a Bernoulli shift so is T^2. The state space for T^2 is $(Y \times Y, \mathcal{F} \times \mathcal{F}, \mu \times \mu)$.
(4) If T_1 and T_2 are Bernoulli shifts so is $T_1 \times T_2$. The state space for $T_1 \times T_2$ is the direct product of the state spaces for T_1 and T_2.

Remark. The method used to calculate the entropy of examples 6 and 8 of §4.7 can be used to show that if T is a Bernoulli shift then $h(T) < \infty$ iff there exists a countable partition η of $(Y, \mathcal{F}, \mu)$ such that $H(\eta) < \infty$ and $\mathcal{A}(\eta) \doteq \mathcal{F}$. In this case $h(T) = H(\mathcal{F})$.

In 1958 Kolmogorov asked if entropy is a complete isomorphism invariant on the collection of all Bernoulli shifts. This was answered in 1969 by Ornstein (see Ornstein [1]).

Theorem 4.28 (Ornstein). *Let T_1, T_2 be Bernoulli shifts whose state spaces are Lebesgue spaces. If $h(T_1) = h(T_2)$ then T_1 is conjugate to T_2, and hence isomorphic by the assumption on the state spaces (a countable direct product of Lebesgue spaces is a Lebesgue space).*

The proof of this deep theorem is presented in Ornstein [1], Shields [1], and Moser et al [1]. Certain special cases had been worked out earlier by Meshalkin [1] and by Blum and Hanson [1]. This result reduces the conjugacy problem for Bernoulli shifts to their state spaces, since the entropy depends only on the state space. It is possible, for example, for a Bernoulli shift with a state space of two points to be conjugate to a Bernoulli shift with a countably infinite state space.

Remark. Since the map $(p_1, \ldots, p_n) \to -\sum p_i \log p_i$, defined on $\{(p_1, \ldots, p_n) | p_i \geq 0 \sum_1^n p_i = 1\}$, has image $(0, \log n]$, for each $x > 0$ there is a Bernoulli shift with entropy x.

Since we are interested in measure-preserving transformations up to conjugacy we make the following definition.

Definition 4.12. An invertible measure-preserving transformation of a Lebesgue space $(X, \mathscr{B}, m)$ is called a *Bernoulli automorphism* if it is conjugate to a Bernoulli shift.

(The word "automorphism" is used because an invertible measure-preserving transformation is a bijective structure preserving map of a measure space.)

The above remarks apply to Bernoulli automorphisms and Ornstein's theorem says that two Bernoulli automorphisms with the same entropy are conjugate. This implies the following.

Corollary 4.28.1

(i) *Every Bernoulli automorphism has an n-th root.* (S *is an n-th root of* T *if* $S^n = T$.)

(ii) *Every Bernoulli automorphism is conjugate to a direct product of two Bernoulli automorphisms.*

(iii) *Every Bernoulli automorphism* T *is conjugate to its inverse.*

PROOF

(i) Let T be a Bernoulli automorphism and $n > 0$. Let S be a Bernoulli automorphism with $h(S) = (1/n)h(T)$. Then S^n is a Bernoulli automorphism with entropy $h(T)$, and therefore S^n and T are conjugate.

(ii) Let T be a Bernoulli automorphism. Let S be Bernoulli with $h(S) = \frac{1}{2} \cdot h(T)$. Then $h(S \times S) = h(T)$ and, since $S \times S$ is Bernoulli, $S \times S$ is conjugate to T.

(iii) T, T^{-1} are Bernoulli automorphisms with the same entropy.   □

The following theorem is a summary of results about the structure of the space of Bernoulli automorphisms, and shows this space is closed under some natural operations on measure-preserving transformations. The proofs are given in Ornstein [1]. All the results are due to Ornstein.

Theorem 4.29

(i) *Every root of a Bernoulli automorphism is a Bernoulli automorphism.*

(ii) *Every factor of a Bernoulli automorphism is a Bernoulli automorphism.*

(iii) *If* $\{\mathscr{F}_n\}_1^\infty$ *is a sequence of sub-σ-algebras of* $\mathscr{B}$ *with* $T\mathscr{F}_n = \mathscr{F}_n$, $\mathscr{F}_1 \subset \mathscr{F}_2 \subset \cdots$, $\bigvee_{n=1}^\infty \mathscr{F}_n \doteq \mathscr{B}$, *and if the factor transformation (see §2.3) associated with each* $\mathscr{F}_n$ *is a Bernoulli automorphism then* T *is a Bernoulli automorphism (i.e. an inverse limit of Bernoulli automorphisms is a Bernoulli automorphism).*

Ornstein has given a criterion for a measure-preserving transformation to be a Bernoulli automorphism. This criterion is used in some of the proofs of the results in Theorem 4.29 and is also useful when checking if concrete examples are Bernoulli automorphisms. At the end of this section we list some examples of Bernoulli automorphisms and Ornstein's criterion (or some variant of it) is the usual method used to show these examples are Bernoulli automorphisms.

Since a Bernoulli shift is really an independent identically distributed stochastic process indexed by the integers we can think of a Bernoulli automorphism as an abstraction of such a stochastic process. In 1958 Kolmogorov introduced the following class of measure-preserving transformations as abstractions of regular, identically distributed stochastic processes.

Definition 4.13. An invertible measure-preserving transformation T of a probability space $(X, \mathscr{B}, m)$ is a *Kolmogorov automorphism* (K-automorphism) if there exists a sub-σ-algebra $\mathscr{K}$ of $\mathscr{B}$ such that:

(i) $\mathscr{K} \subset T\mathscr{K}$.

(ii) $\bigvee_{n=0}^{\infty} T^n\mathscr{K} \doteq \mathscr{B}$.

(iii) $\bigcap_{n=0}^{\infty} T^{-n}\mathscr{K} \doteq \mathscr{N} = \{X, \phi\}$.

We always assume $\mathscr{N} \neq \mathscr{B}$ (since if not the identity is the only measure-algebra automorphism). Hence $\mathscr{K} \neq T\mathscr{K}$. In fact the space $(X, \mathscr{B}, m)$ is usually taken to be a Lebesgue space.

Theorem 4.30. *Every Bernoulli automorphism is a Kolmogorov automorphism.*

PROOF. Let the state space for T be $(Y, \mathscr{F}, \mu)$. If $F \in \mathscr{F}$, let $\tilde{F} = \{\{x_n\} \in X : x_0 \in F\} \in \mathscr{B}$. Let $\mathscr{G} = \{\tilde{F} : F \in \mathscr{F}\}$, which is called the time-0 σ-algebra. Let $\mathscr{K} = \bigvee_{i=-\infty}^{0} T^i\mathscr{G}$. We now verify that $\mathscr{K}$ satisfies the conditions for a Kolmogorov automorphism.

(i) $\mathscr{K} = \bigvee_{i=-\infty}^{0} T^i\mathscr{G} \subset \bigvee_{i=-\infty}^{1} T^i\mathscr{G} = T\mathscr{K}$.

(ii) $\bigvee_{n=0}^{\infty} T^n\mathscr{K} = \bigvee_{n=0}^{\infty} \bigvee_{i=-\infty}^{n} T^i\mathscr{G} = \bigvee_{-\infty}^{\infty} T^i\mathscr{G} = \mathscr{B}$ by definition of $\mathscr{B}$.

(iii) We have to show $\bigcap_{0}^{\infty} T^{-n}\mathscr{K} \doteq \mathscr{N} = \{X, \phi\}$. Fix $A \in \bigcap_{0}^{\infty} T^{-n}\mathscr{K} = \bigcap_{n=0}^{\infty} \bigvee_{-\infty}^{-n} T^i\mathscr{G}$. Let $B \in \bigvee_{k=j}^{\infty} T^k\mathscr{G}$, for some fixed $j \in Z$. Since $A \in \bigvee_{i<j} T^i\mathscr{G}$, A and B are independent, and therefore $m(A \cap B) = m(A)m(B)$. The collection of all sets B for which $m(A \cap B) = m(A)m(B)$ is a monotone class, and, by the above, contains $\bigcup_{j=-\infty}^{\infty} \bigvee_{k=j}^{\infty} T^k\mathscr{G}$. Therefore $\forall B \in \mathscr{B}$, $m(A \cap B) = m(A)m(B)$. Put $B = A$, then $m(A) = m(A)^2$ which implies $m(A) = 0$ or 1. Hence

$$\bigcap_{n=0}^{\infty} T^{-n}\mathscr{K} \doteq \mathscr{N}. \qquad \square$$

It was an open problem from 1958 to 1969 as to whether the converse of Theorem 4.30 was true, i.e., whether a Kolmogorov automorphism acting on a Lebesgue space is a Bernoulli automorphism. This was shown to be false by Ornstein.

Theorem 4.31 (Ornstein). *There is an example of a Kolmogorov automorphism T which is not a Bernoulli automorphism.*

Corollary 4.31.1. *Entropy is not a complete invariant for the class of Kolmogorov automorphisms.*

PROOF. Let T be the example of Ornstein. By Corollary 4.14.4 $h(T) > 0$. Choose a Bernoulli automorphism S with $h(S) = h(T)$. S and T are not isomorphic.   □

The following results show that the class of Kolmogorov automorphisms does not share all the properties the class of Bernoulli automorphisms enjoys. The proofs are given in the references cited.

Theorem 4.32

(i) *There are uncountably many non-conjugate Kolmogorov automorphisms with the same entropy (Ornstein and Shields [1]).*

(ii) *There is a Kolmogorov automorphism T not conjugate to its inverse T^{-1} (Ornstein and Shields [1]).*

(iii) *There is a Kolmogorov automorphism which has no n-th roots for any $n \geq 2$ (Clark [1]).*

(iv) *There are non-conjugate Kolmogorov automorphisms T, S with $T^2 = S^2$ (Rudolf [1]).*

(v) *There are two non-conjugate Kolmogorov automorphisms each of which is a factor of the other (Polit [1] and Rudolf [2]).*

Remarks

(1) Statement (ii) of Theorem 4.32 contrasts with the behaviour of ergodic transformations with pure point spectrum (see Corollary 3.4.1).

(2) Ornstein's example for Theorem 4.31 is defined by induction and so is fairly complicated to describe. It is therefore important to check whether the more "natural" examples of Kolmogorov automorphisms are Bernoulli automorphisms or not. We consider some of these at the end of this section and give an (easy to describe) example of a Kolmogorov automorphism that was recently shown not to be a Bernoulli automorphism.

(3) Sinai has proved that if T is an ergodic invertible measure-preserving transformation of a Lebesgue space $(X, \mathscr{B}, m)$ with $h(T) > 0$ and if S is a Bernoulli automorphism with $h(S) \leq h(T)$ then there exists a measure-preserving transformation ϕ such that $\phi T = S\phi$, i.e., S is a factor of T (see Rohlin [3], p. 45).

The next theorem shows that all Kolmogorov automorphisms are spectrally the same.

Theorem 4.33 (Rohlin). *If $(X, \mathscr{B}, m)$ is a probability space with a countable basis then any Kolmogorov automorphism $T: X \to X$ has countable Lebesgue spectrum.*

PROOF. Recall that we are assuming $\mathscr{B} \neq \{X, \phi\} = \mathscr{N}$. We have (i) $\mathscr{K} \subset T\mathscr{K}$, (ii) $\bigvee T^n \mathscr{K} \doteq \mathscr{B}$, (iii) $\bigcap T^{-n} \mathscr{K} \doteq \mathscr{N}$. We split the proof into three parts:

(a) We first show that $\mathscr{K}$ has no atoms, i.e., if $C \in \mathscr{K}$ and $m(C) > 0$ then $\exists D \in \mathscr{K}$ with $D \subset C$ and $m(D) < m(C)$.

Suppose C is an atom of $\mathscr{K}$ with $m(C) > 0$. Then TC is an atom of $T\mathscr{K}$ and since $\mathscr{K} \subset T\mathscr{K}$ either $TC \subset C$ or $m(C \cap TC) = 0$. If $TC \subset C$ then $TC \doteq C$ since both sets have the same measure so that $C \overset{\circ}{\in} \bigcap_{n=0}^{\infty} T^{-n} \mathscr{K}$ and therefore $m(C) = 1$. Hence $\mathscr{K} \doteq \mathscr{N}$ so $\mathscr{B} \doteq \mathscr{N}$, a contradiction. On the other hand, suppose $m(TC \cap C) = 0$. Then either for some $k > 0$ $T^k C \overset{\circ}{\subset} C$ (and we use the above proof to get a contradiction) or $m(T^k C \cap C) = 0$ $\forall k > 0$ and then $C \cup TC \cup T^2 C \cup \cdots$ has infinite measure, a contradiction.

(b) Let $\mathscr{H} = \{f \in L^2(m): f \text{ is } \mathscr{K}\text{-measurable}\}$. Then $U_T \mathscr{H} \subset \mathscr{H}$. Let $\mathscr{H} = V \oplus U_T \mathscr{H}$. From $U_T^{-n} \mathscr{H} = \bigoplus_{-n}^{m} U_T^i V \oplus U_T^{m+1} \mathscr{H} (n, m > 0)$ it follows that $L^2(m) = \bigoplus_{-\infty}^{\infty} U_T^n V \oplus C$ where C is the subspace of constants. It suffices to show V is infinite-dimensional since if $\{f_1, f_2, f_3, \ldots\}$ is a basis for V, then $\{f_0 \equiv 1, U_T^n f_j : n \in Z, j > 0\}$ is a basis for $L^2(m)$.

(c) We now show V is infinite-dimensional. Since $T\mathscr{K} \not\doteq \mathscr{K}$ (we are assuming $\mathscr{B} \not\doteq \mathscr{N}$) we know $V \neq \{0\}$. Let $g \in V$, $g \neq 0$ and then $G = \{x : g(x) \neq 0\}$ satisfies $m(G) > 0$. Since g is $\mathscr{K}$-measurable we have $G \in \mathscr{K}$ and using (a) we know $\chi_G \mathscr{H} = \{\chi_G f : f \in \mathscr{H}\}$ is infinite-dimensional. Also $\chi_G \mathscr{H} = V' \oplus \chi_G U_T \mathscr{H}$ where $V' \subset V$ so either V' is infinite-dimensional (and hence V is) or $\chi_G U_T \mathscr{H}$ is infinite-dimensional. In this second case there is a linearly independent sequence of functions $\{\chi_G U_T f_n\}$ where the f_n are bounded functions in $\mathscr{H}$. Then $\{g U_T f_n\}$ are linearly independent in $\mathscr{H}$. It suffices to show these functions are in V. But if $f \in \mathscr{H}$ then

$$(g U_T f_n, U_T f) = (g, U_T(f \bar{f}_n)) = 0$$

so $g U_T f_n \in V$. $\square$

Corollary 4.33.1 *A Kolmogorov automorphism is strong-mixing.*

PROOF. By Theorem 2.12. $\square$

Kolmogorov automorphisms are connected to entropy theory by the following result (half of which was proved by Pinsker).

Theorem 4.34 (Rohlin and Sinai, see Rohlin [3]). *Let $(X, \mathscr{B}, m)$ be a Lebesgue space and let $T: X \to X$ be an invertible measure-preserving transformation. Then T is a Kolmogorov automorphism iff $h(T, \mathscr{A}) > 0$ for all finite $\mathscr{A} \not\doteq \mathscr{N}$.*

Remarks

(1) One says that T has completely positive entropy when the latter condition holds. Hence T has completely positive entropy iff it is a Kolmogorov automorphism.

(2) This shows that K-automorphisms are "the opposites" of transformations with zero entropy (since $h(T, \mathscr{A}) = 0 \; \forall \mathscr{A}$ in the zero entropy case).

(3) We already know from Corollary 4.14.4 that a Kolmogorov automorphism has positive entropy (since $K \neq T\mathscr{K}$).

(4) It follows from Theorem 4.34 that if T is a Kolmogorov automorphism then so is T^{-1}. We know from Theorem 4.32(ii) that T^{-1} need not be a conjugate to T.

(5) The results of Rohlin and Sinai (Rohlin [3], page 37) show that T is a Kolmogorov automorphism iff whenever $\mathscr{A}$ is a finite sub-algebra of $\mathscr{B}$ then $\bigcap_{n=0}^{\infty} \bigvee_{j=n}^{\infty} T^{-j}\mathscr{A} \doteq \mathscr{N}$. One can see from this how much stronger the Kolmogorov property is than strong-mixing because Sucheston has shown that T is strong-mixing iff for every subsequence Γ of positive integers and every finite sub-algebra $\mathscr{A}$ of $\mathscr{B}$ there exists a subsequence $\{b_j\}$ of Γ with $\bigcap_{n=1}^{\infty} \bigvee_{j=n}^{\infty} T^{-b_j}\mathscr{A} \doteq \mathscr{N}$.

(6) There is another result that shows the Kolmogorov property is a uniform strong-mixing condition. Let T be an invertible measure-preserving transformation of a Lebesgue space $(X, \mathscr{B}, m)$. For $B \in \mathscr{B}$ and $k > 0$ let $\mathscr{B}(B, k)$ denote the smallest σ-algebra containing all the sets $T^{-n}B$ for $n \geq k$. Then T is a Kolmogorov automorphism iff for all $A, B \in \mathscr{B}$

$$\lim_{k \to \infty} \sup_{C \in \mathscr{B}(B,k)} \left| m(A \cap C) - m(A)m(C) \right| = 0.$$

In the following examples the word "automorphism" is used in two senses. When we say "group automorphism of a compact group" we mean automorphism in the sense of topological groups, whereas in the terminology "Kolmogorov automorphism" the word automorphism refers to an invertible structure preserving map of a probability space.

Examples

(1) *Group Automorphisms.* Rohlin proved that any ergodic automorphism of a compact abelian metric group is a Kolmogorov automorphism [4] and later Yusinskii proved the theorem in the non-abelian case. [2]. Katznelson [1] has shown that ergodic automorphisms of finite-dimensional tori are Bernoulli automorphisms. Lind [1] and Miles and Thomas [1] have proved, using different methods that any ergodic automorphism of a compact metric group is a Bernoulli automorphism.

(2) *Markov Shifts.* Let T be the two-sided $(\mathbf{p}, P)$ Markov-shift. We have seen that T is ergodic iff P is irreducible (i.e., $\forall$ pairs of states $i, j \; \exists n > 0$ with $p_{ij}^{(n)} > 0$) and T is strong mixing iff P is irreducible and aperiodic (i.e., $\exists N > 0$ with $p_{ij}^{(N)} > 0$ for all states i, j). Friedman and Ornstein [1] have shown that

such a Markov shift is a Bernoulli automorphism. Therefore from the point of view of ergodic theory, mixing Markov chains are the same as Bernoulli automorphisms. One can easily deduce that an ergodic Markov shift is the direct product of a Bernoulli automorphism and a rotation on a finite group.

The proof of Friedman and Ornstein consists of showing that a transformation with a certain property is isomorphic to a Bernoulli shift. This is a generalisation of the deep result of Ornstein (Theorem 4.28). It is however easy to show that the $(\mathbf{p}, P)$ Markov shift is a Kolmogorov automorphism iff P is aperiodic and irreducible. This generalises Theorem 4.30.

Theorem 4.35. *The two-sided $(\mathbf{p}, P)$ Markov shift is a Kolmogorov automorphism iff P is irreducible and aperiodic.*

PROOF. Let T denote the two sided $(\mathbf{p}, P)$ Markov shift and let $(X, \mathscr{B}, m)$ be the space on which it acts. If T is a Kolmogorov automorphism then T is strong-mixing (Corollary 4.33.1) and therefore P is irreducible and aperiodic (Theorem 1.27). Now suppose P is irreducible and aperiodic. Let $\xi = \{A_0, A_1, \ldots, A_{k-1}\}$ be the natural partition into states at time zero i.e. $A_i = \{\{x_n\}_{-\infty}^{\infty} \in X \,|\, x_0 = i\}$. Let $\mathscr{K}$ be the smallest σ-algebra containing all sets in the partitions $T^{-n}\xi$, $n \geq 0$. i.e. $\mathscr{K} = \bigvee_{n=0}^{\infty} T^{-n}\mathscr{A}(\xi)$. Then $\mathscr{K} \subset T\mathscr{K}$ and $\bigvee_{n=0}^{\infty} T^n\mathscr{K} = \mathscr{B}$ so it remains to show $\bigcap_{r=0}^{\infty} T^{-r}\mathscr{K} \doteq \{\phi, X\}$. We shall do this by generalising the argument used in the proof of Theorem 4.30.

Recall from Theorem 1.27 that since P is irreducible and aperiodic we have $\lim_{n \to \infty} p_{ij}^{(n)} = p_j$ for all states i, j. Suppose A is a cylinder block $_a[i_0, \ldots, i_r]_{a+r}$ and B is a cylinder block $_b[j_0, \ldots, j_s]_{b+s}$ with $b + s < a$. Then $m(A \cap B) = p_{j_0} p_{j_0 j_1} \cdots p_{j_{s-1} j_s} p_{j_s i_0}^{(a-b-s)} p_{i_0 i_1} \cdots p_{i_{r-1} i_r}$ so that

$$m(A)m(B) \min_{i,j} \left(\frac{p_{ij}^{(a-b-s)}}{p_j} \right) \leq m(A \cap B) \leq m(A)m(B) \max_{i,j} \left(\frac{p_{ij}^{(a-b-s)}}{p_j} \right).$$

The same inequality is true if $A \in \bigvee_{n=a}^{a+r} T^{-n}\mathscr{A}(\xi)$ and $B \in \bigvee_{n=b}^{b+s} T^{-n}\mathscr{A}(\mathscr{B})$. Fix $B \in \bigvee_{n=b}^{b+s} T^{-n}\mathscr{A}(\xi)$ and fix $N > b + s$. Consider the collection $\mathscr{M}$ of all measurable sets A with

$$m(A)m(B) \inf_{\substack{i,j \\ a \geq N}} \left(\frac{p_{ij}^{(a-b-s)}}{p_j} \right) \leq m(A \cap B) \leq m(A)m(B) \sup_{\substack{i,j \\ a \geq N}} \left(\frac{p_{ij}^{(a-b-s)}}{p_j} \right).$$

The collection $\mathscr{M}$ is a monotone class and contains the algebra

$$\bigcup_{a=N}^{\infty} \bigcup_{r=0}^{\infty} \bigvee_{n=a}^{a+r} T^{-n}\mathscr{A}(\xi)$$

and hence contains the σ-algebra $\bigvee_{n=N}^{\infty} T^{-n}\mathscr{A}(\xi) = T^{-N}\mathscr{K}$. Let $\varepsilon > 0$ be given and choose t_0 so that $t \geq t_0$ implies

$$\left| \frac{p_{ij}^{(t)}}{p_j} - 1 \right| < \varepsilon.$$

for all i, j. Then if $B \in \bigvee_{n=b}^{b+s} T^{-n}\mathscr{A}(\xi)$ we have $|m(A \cap B) - m(A)m(B)| < \varepsilon$ whenever $A \in T^{-N}\mathscr{K}$ and N is large. In particular this is true for $A \in \bigcap_{N=0}^{\infty} T^{-N}\mathscr{K}$. Now fix $A \in \bigcap_{N=0}^{\infty} T^{-N}\mathscr{K}$ and consider the collection $\mathscr{R}$ of all measurable sets B with $|m(A \cap B) - m(A)m(B)| < \varepsilon$. The collection $\mathscr{R}$ is a monotone class and contains $\bigvee_{n=b}^{b+s} T^{-n}\mathscr{A}(\xi)$ for all b and all $s \geq 0$. Therefore $\mathscr{R} = \mathscr{B}$ so that $|m(A \cap B) - m(A)m(B)| < \varepsilon$ for all $B \in \mathscr{B}$ and all $A \in \bigcap_{n=0}^{\infty} T^{-n}\mathscr{K}$. We then get $m(A \cap B) = m(A)m(B)$ for all $B \in \mathscr{B}$ and all $A \in \bigcap_{n=0}^{\infty} T^{-n}\mathscr{K}$, so by putting $B = A$ we have $m(A) = 0$ or 1 whenever $A \in \bigcap_{n=0}^{\infty} T^{-n}\mathscr{K}$. □

(3) One can generalize the notion of a finite-dimensional torus to obtain another kind of homogeneous space called a nilmanifold. Let N be a connected, simple connected, nilpotent Lie group and D a discrete subgroup of N so that the quotient space N/D is compact. N/D is called a *nilmanifold*. When $N = R^n$ and $D = Z^n$ we get an n-torus. The Haar measure on N determines a normalized Borel measure on N/D. If $\bar{A}:N \to N$ is a (continuous) automorphism with $\bar{A}D = D$ then this induces a map $A:N/D \to N/D$, which is called an *automorphism* of N/D. The automorphism A always preserves the measure m. Parry has investigated the ergodic theory of such maps and has shown that if A is ergodic then A is a K-automorphism. A subclass of the ergodic automorphisms of N/D are known to be Bernoulli automorphisms (see Marcuard [1]).

The simplest examples are as follows: Let

$$N = \left\{ \begin{pmatrix} 1 & x & z \\ 0 & 1 & y \\ 0 & 0 & 1 \end{pmatrix} : \quad x, y, z \in R \right\}.$$

N satisfies the above conditions with the operation of matrix multiplication and the natural topology from R^3. Let

$$D = \left\{ \begin{pmatrix} 1 & m & p \\ 0 & 1 & n \\ 0 & 0 & 1 \end{pmatrix} : \quad m, n, p \in Z \right\}.$$

Then N/D is a nilmanifold. The automorphism

$$\begin{pmatrix} 1 & x & z \\ 0 & 1 & y \\ 0 & 0 & 1 \end{pmatrix} \to \begin{pmatrix} 1 & 2x+y & z+x^2+y+\dfrac{y^2}{2} \\ 0 & 1 & x+y \\ 0 & 0 & 1 \end{pmatrix}$$

of N induces an ergodic automorphism of N/D.

(4) The following is an example of a Kolmogorov automorphism that is not a Bernoulli automorphism. The proof of this is due to S. Kalikow [1].

Let $T:X \to X$ denote the two-sided $(\frac{1}{2}, \frac{1}{2})$-shift. We shall define a transformation of the direct product measure space $X \times X$. If $x = (x_n)_{-\infty}^{\infty}$ $y = (y_n)_{-\infty}^{\infty}$, where $x_n, y \in \{0, 1\}$, put $S(x, y) = (Tx, T^{\varepsilon(x)}y)$ where $\varepsilon(x) = -1$ if $x_0 = 0$ and $\varepsilon(x) = 1$ if $x_0 = 1$. Then S is a Kolmogorov automorphism but not a Bernoulli automorphism.

§4.10 The Pinsker σ-Algebra of a Measure-Preserving Transformation

Let T be a measure-preserving transformation of a Lebesgue space. Let

$$\mathscr{P}(T) = \bigvee \{ \mathscr{A} : \mathscr{A} \subset \mathscr{B}, \ \mathscr{A} \text{ finite}, \ h(T, \mathscr{A}) = 0 \}.$$

This is called the *Pinsker σ-algebra of T*.

One can show that $T^{-1}\mathscr{P}(T) = \mathscr{P}(T)$. One can also prove that if $\mathscr{A}$ is finite then $A \subset \mathscr{P}(T)$ iff $h(T, \mathscr{A}) = 0$. Thus $\mathscr{P}(T)$ is the maximum σ-algebra such that T restricted to $(X, \mathscr{P}(T), m|_{\mathscr{P}(T)})$ has zero entropy. Note that $\mathscr{P}(T) = \mathscr{B}$ iff $h(T) = 0$ and $\mathscr{P}(T) = \mathscr{N}$ iff T is a Kolmogorov automorphism (by Theorem 4.34). See Rohlin [1] or Parry [2] for a full account of these results.

Theorem 4.36 (Rohlin). *If T is an invertible measure-preserving transformation of a Lebesgue space with $h(T) > 0$ then U_T has countable Lebesgue spectrum in the orthogonal complement of $L^2(\mathscr{P}(T))$ in $L^2(\mathscr{B})$.*

This reduces the study of the spectrum of invertible measure-preserving transformations to those with zero entropy.

The types of spectrum that occur for zero entropy transformations are unknown. There are examples of zero entropy transformations with countable Lebesgue spectrum (from Gaussian processes and horocycle flows).

Another important result is

Theorem 4.37. *Let $T: X \to X$ be an invertible measure-preserving transformation of a Lebesgue space $(X, \mathscr{B}, m)$. Suppose $\mathscr{F}$ is a sub-σ-algebra of $\mathscr{B}$ with $T\mathscr{F} = \mathscr{F}$ and such that T has completely positive entropy on $(X, \mathscr{F}, m)$ (i.e., if $\mathscr{A}$ is finite $\mathscr{A} \neq \mathscr{N}$ and $\mathscr{A} \subset \mathscr{F}$ then $h(T, \mathscr{A}) > 0$). Then $\mathscr{F}$ and $\mathscr{P}(T)$ are independent i.e. if $F \in \mathscr{F}$ and $A \in \mathscr{P}(T)$ then $m(F \cap A) = m(F)m(A)$.*

For a proof see Parry ([2], Chapter 6).

Because of this theorem Pinsker conjectured that any ergodic measure-preserving transformation could be written as a direct product of one with zero entropy and one with completely positive entropy. However, Theorem 4.32(ii) shows this conjecture is false because if $T: X \to X$ is the example of Ornstein with no square root then the transformation S of the direct product measure space $\{0, 1\} \times X$ (where the measure on $\{0, 1\}$ gives measure $\frac{1}{2}$ to each point) defined by $S(0, x) = (1, x)$, $S(1, x) = (0, Tx)$ provides a counter-example to the Pinsker conjecture. (It is not difficult to show that $\mathscr{P}(S)$ consists of the four sets ϕ, $\{0\} \times X$, $\{1\} \times X$, $\{0, 1\} \times X$, and then one shows that if there is a sub-σ-algebra $\mathscr{G}$ with $S\mathscr{G} = \mathscr{G}$ and $\mathscr{G}$ being an independent complement for $\mathscr{P}(S)$ then T must have a square root.) This example is not strong-mixing (since S^2 is not ergodic) but Ornstein has also constructed a strong mixing transformation that violates Pinsker's conjecture.

§4.11 Sequence Entropy

Let $(X, \mathscr{B}, m)$ be a probability space and let $T: X \to X$ be an invertible measure-preserving transformation. Let $\Gamma = \{t_1, t_2, \dots\}$ be a sequence of integers. Let $\mathscr{A}$ be a finite sub-algebra of $\mathscr{B}$. Define

$$h_\Gamma(T, \mathscr{A}) = \limsup_{n \to \infty} \frac{1}{n} H(T^{t_1}\mathscr{A} \vee \cdots \vee T^{t_n}\mathscr{A})$$

and define

$$h_\Gamma(T) = \sup_{\mathscr{A} \text{ finite}} h_\Gamma(T, \mathscr{A}).$$

It is easily shown that $h_\Gamma(T)$ is a conjugacy invariant for each Γ. Entropy and spectral properties are connected by the following:

Theorem 4.38 (Kushnirenko [1]). *Let T be an ergodic measure-preserving transformation. Then T has discrete spectrum iff $h_\Gamma(T) = 0$ for every sequence Γ.*

One can also show that if T is ergodic either $\sup_\Gamma h_\Gamma(T) = \infty$ or $\log k$, for some positive integer. Moreover, those T with $\sup_\Gamma h_\Gamma(T) = \log k$ are those ergodic measure-preserving transformations for which there exists an ergodic measure-preserving transformation S with discrete spectrum and a measure-preserving transformation ϕ with $\phi T = S\phi$ and for almost all y the set $\phi^{-1}(y)$ consists of k points.

D. Newton has given a formula of the form $h_\Gamma(T) = a(\Gamma)h(T)$ except in the cases when T has zero entropy and Γ has large gaps (i.e. $a(\Gamma) = \infty$) (Krug and Newton [1]). The number $a(\Gamma)$ depends only on the sequence Γ and not on T. Therefore only when $h(T) = 0$ can sequence entropy provide new conjugacy invariants. Kushnirenko used it to find two non-conjugate measure-preserving transformations with zero entropy and countable Lebesgue spectrum. In fact he showed that if T is the time-one map of the horocycle flow on a two dimensional manifold of constant negative curvature then T and $T \times T$ have different values of sequence entropy for the sequence $\Gamma = \{2^n \mid n \geq 1\}$.

P. Hulse [1] has given some information on which sequences Γ give $h_\Gamma(T) > 0$ when T has quasi-discrete spectrum.

One can also formulate the mixing concepts in terms of sequence entropy. The following is due to A. Saleski [1].

Theorem 4.39. *Let T be an invertible measure-preserving transformation of the Lebesgue space $(X, \mathcal{B}, m)$. Then*

(i) *T is weak-mixing iff $\sup_\Gamma h_\Gamma(T, \mathcal{A}) = H(\mathcal{A})$ for all finite sub-algebras $\mathcal{A}$ of $\mathcal{B}$, where the supremum is taken over all subsequences.*

(ii) *T is strong-mixing iff for every increasing Γ of positive integers and every finite sub-algebra $\mathcal{A}$ of $\mathcal{B}$ we have $\sup\{h_\Lambda(T, \mathcal{A}) : \Lambda \subset \Gamma\} = H(\mathcal{A})$.*

§4.12 Non-invertible Transformations

Suppose $(X, \mathcal{B}, m)$ is a probability space and $T : X \to X$ is measure-preserving. We have $\mathcal{B} \supset T^{-1}\mathcal{B} \supset T^{-2}\mathcal{B} \supset \cdots$. We know that if $\mathcal{B} \doteq T^{-1}\mathcal{B}$ (i.e. $\tilde{\mathcal{B}} = \tilde{T}^{-1}\tilde{\mathcal{B}}$) then $\tilde{T}^{-1}$ is an automorphism of the measure algebra $(\tilde{\mathcal{B}}, \tilde{m})$ and hence T is invertible mod 0 if $(X, \mathcal{B}, m)$ is a Lebesgue space or if X is a complete separable metric space and $\mathcal{B}$ is its σ-algebra of Borel subsets. Let $\mathcal{B}_\infty = \bigcap_{n=0}^\infty T^{-n}\mathcal{B}$. Then $T^{-1}\mathcal{B}_\infty = \mathcal{B}_\infty$ and $\mathcal{B}_0$ is the largest σ-algebra with this property. One can show that U_T has one-sided countable Lebesgue spectrum on the subspace $L^2(X, \mathcal{B}, m) \ominus L^2(X, \mathcal{B}_\infty, m)$ (i.e. there is an ortho-normal basis for this subspace of the form $\{U_T^n f_k : n \geq 0, k \geq 1\}$). It is essentially the same as the proof of Theorem 4.33 since one only has to show $L^2(X, \mathcal{B}, m) \ominus L^2(X, T^{-1}\mathcal{B}, m)$ is infinite dimensional.) This reduces the study of spectral properties of measure-preserving transformations to that of invertible ones and hence, by Theorem 4.36, to those of zero entropy.

We have $\mathcal{P}(T) \doteq \mathcal{B}_\infty$, because if $\mathcal{A}$ is a finite sub-algebra of $\mathcal{B}$ with $h(T, \mathcal{A}) = 0$ then $\mathcal{A} \doteq \bigvee_{i=1}^\infty T^{-i}\mathcal{A}$ (by Corollary 4.14.1) and hence $\mathcal{A} \doteq \bigcap_{n=0}^\infty \bigvee_{i=n}^\infty T^{-i}\mathcal{A}$ $(n \geq 1)$ so $\mathcal{A} \doteq \mathcal{B}_\infty$.

The analogous concept to a Kolmogorov automorphism is an exact endomorphism. It is the abstraction of a regular identically distributed stochastic process indexed by the non-negative integers.

Definition 4.14. A measure-preserving transformation T, of the probability space $(X, \mathcal{B}, m)$, is an *exact endomorphism* if

$$\bigcap_{n=0}^\infty T^{-n}\mathcal{B} \doteq \mathcal{N}; \quad \text{i.e., } \mathcal{B}_\infty \doteq \mathcal{N}.$$

So exact endomorphisms are as far from being invertible as possible. Examples of exact endomorphisms are the one-sided Bernoulli shifts. Exact endomorphisms have one-sided countable Lebesgue spectrum and hence are strong-mixing (by a proof like that of Theorem 2.12).

It was conjectured that every ergodic measure-preserving transformation is a product of an exact endomorphism and an invertible measure-preserving transformation. This is not so (Parry and Walters [1], Walters [1]).

Also, one-sided Bernoulli shifts with the same entropy are not necessarily conjugate since an m-to-1 map cannot be conjugate to an n-to-1 map if $m \neq n$. So entropy is far from complete for one-sided Bernoulli shifts. In fact, the one-sided $(p_0, \ldots, p_{k-1})$-shift is conjugate to the one-sided $(q_0, \ldots, q_{l-1})$-shift iff $k = l$ and $(p_0, \ldots, p_{k-1})$ is a permutation of $(q_0, \ldots, q_{l-1})$. (We have assumed that no p_i or q_j is zero, which is no loss of generality since the $(p_0, \ldots, p_{k-1})$-shift is conjugate to the $(0, p_0, \ldots, p_{k-1})$-shift) (Parry and Walters [1], Walters [1]). The proof of this uses an invariant which is not present in the invertible case, namely the Jacobian which was introduced by Parry [2].

Another invariant of conjugacy for non-invertible transformations is the decreasing sequence of σ-algebras $\{T^{-n}\mathcal{B}\}_{n=0}^{\infty}$. However the three invariants of entropy, Jacobian and the sequence $\{T^{-n}\mathcal{B}\}_{n=0}^{\infty}$ are not complete for the relation of conjugacy on the class of exact endomorphisms because there are two exact endomorphisms S, T with $S^{-n}\mathcal{B} = T^{-n}\mathcal{B}$ $n \geq 0$, $S^2 = T^2$ ($\Rightarrow h(S) = h(T)$), S and T having equal Jacobians but with S and T not conjugate.

Also a one-sided Markov chain which is exact need not be conjugate to a one-sided Bernoulli shift (Parry and Walters [1]).

§4.13 Comments

Entropy was introduced as a conjugacy invariant for measure-preserving transformations. It was soon realized that entropy theory was more than just an assignment of a number to each transformation. Kolmogorov automorphisms and transformations with zero entropy have received the most treatment. They are "opposites" from the point of view of entropy. Kolmogorov automorphisms are important for applications as it seems that the most interesting smooth systems are Kolmogorov and even Bernoulli.

By Theorem 4.36 we know that the spectral theory of invertible measure-preserving transformations reduces to that for the zero entropy case. The following is still an open problem: If $h(T) = 0$ what kind of spectrum can U_T have?

For transformations with zero entropy the isomorphism problem is only solved for ergodic transformations with discrete spectrum, totally ergodic transformations with quasi-discrete spectrum and some other special cases. Sequence entropy may play a role in the isomorphism problem for zero entropy transformations.

In the weak topology on the set of all invertible measure-preserving transformations on a given space $(X, \mathcal{B}, m)$, the set of transformations with zero entropy is a dense G_δ (countable intersection of open sets) (Rohlin [5]). Since the set of weak mixing transformations is also a dense G_δ and the set of strong mixing transformations is a set of first category it follows that "most" transformations are weak mixing, have zero entropy, but are not strong mixing. However the ones of interest for applications are often not in this class.

The main problem to consider for Kolmogorov automorphisms seems to be to find more examples of Kolmogorov automorphisms that are not Bernoulli automorphisms. One should first check whether the usual ways of constructing new transformations from old ones transform a Bernoulli automorphism to a Bernoulli automorphism (e.g., is a weak mixing group extension of a Bernoulli automorphism a Bernoulli automorphism?). Several results in this direction are known. If one of these constructions leads to a Kolmogorov automorphism which is not a Bernoulli automorphism then this may lead to a new invariant that may be complete for Kolmogorov automorphisms.

CHAPTER 5
Topological Dynamics

In measure theoretic ergodic theory one studies the asymptotic properties of measure-preserving transformations. In topological dynamics one studies the asymptotic properties of continuous maps. We shall study continuous transformations of compact metric spaces. The compactness assumption is a "finiteness" assumption which is similar to the assumption of a finite measure in the measure-theoretic work. The assumption of metrisability is not needed for some of the results but it often shortens proofs and most applications are for metric spaces. The notations we shall use are given in §0.10.

If X is compact metric and $T: X \to X$ is continuous one has an induced map $U_T: C(X) \to C(X)$ given by $U_T f = f \circ T$. The map U_T is clearly linear and multiplicative (i.e. $U_T(f \cdot g) = (U_T f)(U_T g)$). If T maps X onto X then U_T is an isometry and if T is a homeomorphism then U_T is an isometric automorphism in the sense of Banach algebras (i.e. a multiplicative linear isometry of $C(X)$ onto $C(X)$).

In the first section we give a list of examples and in subsequent sections we discuss dynamical properties. We shall connect these properties with measures in Chapter 6, when we study the family of invariant probability measures for a given continuous transformation.

§5.1 Examples

(1) The identity, I, on any X.

(2) A rotation $Tx = ax$ on a compact metric group (recall from §0.6 that such a group has a rotation invariant metric).

(3) A surjective endomorphism of a compact metric group; in particular, of a torus.

(4) An affine transformation $Tx = a \cdot A(x)$ where A is a surjective endomorphism of a compact group G and $a \in G$. This example includes examples 2 and 3.

(5) Let $Y = \{0, 1, \ldots, k - 1\}$ with the discrete topology. Let $X = \prod_{-\infty}^{\infty} Y$ with the product topology. A neighbourhood basis of a point $\{x_n\}$ consists of the sets $U_N = \{\{y_n\} \mid y_n = x_n \text{ for } |n| \leq N\}$, $N \geq 1$. A metric on X is given by

$$d(\{x_n\}, \{y_n\}) = \sum_{n=-\infty}^{\infty} \frac{|x_n - y_n|}{2^{|n|}}.$$

The two-sided shift T, defined by $T\{x_n\} = \{y_n\}$ with $y_n = x_{n+1}$, is a homeomorphism of X. We sometimes write this $T(\ldots, x_{-1}\overset{*}{x}_0 x_1, \ldots) = (\ldots, x_{-1} x_0 \overset{*}{x}_1 x_2, \ldots)$ where the symbol $*$ occurs over the 0-th coordinate of each point. Note that here we have a special case of (iii) since X is a compact group under the operation

$$\{x_n\} + \{y_n\} = \{(x_n + y_n)\bmod(k)\},$$

and T is an automorphism of X.

(6) There is a one-sided shift map corresponding to the two-sided shift in (5). If Y is as in example (5) then let $X = \prod_0^{\infty} Y$ be equipped with the product topology. The one-sided shift $T: X \to X$ is defined by $T\{x_n\} = \{y_n\}$ where $y_n = x_{n+1}$ i.e., $T(x_0, x_1, x_2, \ldots)) = (x_1, x_2, \ldots)$. The one-sided shift is a continuous transformation. The preimage under T of any point consists of k points.

(If we replaced the special space Y by any compact metric space then we can clearly define the two-sided and one-sided shift maps with "state space" Y.)

If one has a continuous map $T: X \to X$ of a compact space and a closed subset Y of X with $TY \subset Y$ then $T|_Y$ is a continuous map of the compact space Y. The map $T|_Y$ is sometimes called a subsystem of T. There are many interesting subsystems of the shift maps of example (5). The following is one of them and is the topological analogue of a Markov chain.

(7) Let $T: X \to X$ be the two-sided shift as in example 5. Let $A = (a_{ij})_{i,j=0}^{k-1}$ be a $k \times k$ matrix with $a_{ij} \in \{0, 1\}$ for all i, j. Let $X_A = \{(x_n)_{-\infty}^{\infty} \mid a_{x_n x_{n+1}} = 1$ $\forall n \in Z\}$. In other words X_A consists of all the bisequences $(x_n)_{-\infty}^{\infty}$ whose neighbouring pairs are allowed by the matrix A. The complement of X_A is clearly open so X_A is a closed subset of X. Also $TX_A = X_A$ so that $T|_{X_A}$ is a homeomorphism of X_A and is called the two-sided topological Markov chain (or subshift of finite type) determined by the matrix A. For simplicity we shall write $T: X_A \to X_A$ rather than $T|_{X_A}$. If $(a_{ij}) = 1$ all i, j then $X_A = X$. If $A = I$, the identity matrix, then X_A consists of only k points. Sometimes X_A is empty; for example when $A = \left(\begin{smallmatrix} 0 & 0 \\ 1 & 0 \end{smallmatrix}\right)$ and $X = \prod_{-\infty}^{\infty} \{0, 1\}$. Two matrices can define the same topological Markov chain; for example $A_1 = \left(\begin{smallmatrix} 0 & 1 \\ 0 & 1 \end{smallmatrix}\right)$, $A_2 = \left(\begin{smallmatrix} 0 & 0 \\ 0 & 1 \end{smallmatrix}\right)$ when $X = \prod_{-\infty}^{\infty} \{0, 1\}$.

Topological Markov chains are very important as a source of examples and as models for important diffeomorphisms [see Bowen [2]]. One can also define one-sided topological Markov chains using example 6 rather than 5.

(8) This example is one of the simplest used in the qualitative study of diffeomorphisms of compact manifolds. It is called the north-south map. Consider the unit circle K and suppose it is positioned so that it is tangent to the real line, R, at $0 \in R$. Consider the map $x \to x/2$ on R and let $T : K \to K$

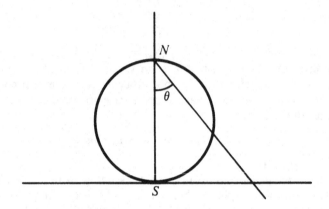

be the map derived from this using stereographic projection. In other words $T(N) = N$, $T(S) = S$ and if $\theta \in (-\pi/2, \pi/2)$ is the angle shown in the diagram then T maps the point of K cutting the line with angle θ to the point of K cutting the line with angle $\tan^{-1}(\tan(\theta)/2)$. So if $x \notin \{N, S\}$ then $T(x)$ is closer to S than x was and $T^n(x) \to S$ as $n \to \infty$. Also $T^{-n}x \to N$ as $n \to \infty$ if $x \neq S$.

§5.2 Minimality

In this section X will denote a compact metric space and $T : X \to X$ a homeomorphism. We shall study homeomorphisms in this section and later we will consider continuous transformations. We would like to find a concept of "irreducibility" to play the rôle ergodicity played for measure-preserving transformations.

Definition 5.1. A homeomorphism $T : X \to X$ is *minimal* if $\forall x \in X$ the set $\{T^N x : n \in Z\}$ is dense in X. The set $0_T(x) = \{T^n x : n \in Z\}$ is called the *T-orbit* of x.

Theorem 5.1. *The following are equivalent for a homeomorphism* $T : X \to X$ *of a compact metric space.*

    (i) *T is minimal.*
    (ii) *The only closed subsets E of X with TE = E are $\emptyset$ and X.*
    (iii) *For every non-empty open subset U of X we have $\bigcup_{-\infty}^{\infty} T^n U = X$.*

PROOF

    (i) $\Rightarrow$ (ii). Suppose T is minimal and let E be closed, $E \neq \emptyset$ and $TE = E$. If $x \in E$ then $0_T(x) \subset E$ so $X = \overline{0_T(x)} \subset E$. Hence $X = E$.
    (ii) $\Rightarrow$ (iii) If U is non-empty and open then $E = X \backslash \bigcup_{-\infty}^{\infty} T^n U$ is closed and $TE = E$. Since $E \neq X$ we have $E = \emptyset$.
    (iii) $\Rightarrow$ (i). Let $x \in X$ and let U be any non-empty open subset of X. By (iii) $x \in T^n U$ for some $n \in Z$ so $T^{-n} x \in U$ and $0_T(x)$ is dense in X. $\qquad \square$

A subset E of X is called T-invariant if $TE = E$. If E is closed and T-invariant then $T|_E$ is a homeomorphism of the compact metric space E.

Definition 5.2. Let $T : X \to X$ be a homeomorphism. A closed subset E of X which is T-invariant is called a *minimal set* with respect to T if $T|_E$ is minimal.

Theorem 5.2. *Any homeomorphism $T : X \to X$ has a minimal set.*

PROOF. Let $\mathscr{E}$ denote the collection of all closed non-empty T-invariant subsets of X. Clearly $\mathscr{E} \neq \emptyset$ since X belongs to $\mathscr{E}$. The set $\mathscr{E}$ is a partially ordered set under inclusion. Every linearly ordered subset of $\mathscr{E}$ has a least element (the intersection of the elements of the chain. The least element is non-empty by Cantor's intersection property.) Thus, by Zorn's lemma, $\mathscr{E}$ has a minimum element. This element is a minimal set for T. $\qquad \square$

Remark. Ergodicity has the properties:

    (i) An ergodic transformation is "indecomposable" in the sense that it cannot be decomposed into two transformations (see §1.5).
    (ii) Every measure-preserving transformation on a decent measure space can be decomposed into ergodic pieces in a nice way.
By its definition, a minimal transformation is "indecomposable". We know that each homeomorphism $T : X \to X$ has a minimal set. However, in general, one cannot partition X into T-invariant closed sets E_α such that $X = \bigcup_\alpha E_\alpha$, $TE_\alpha = E_\alpha \forall \alpha$, and $T|_{E_\alpha}$ is minimal (although we can in some important cases). If T has such a decomposition it is sometimes called *semi-simple*. An example of a transformation not admitting such a decomposition is an ergodic automorphism of a compact metric group. This is because there are some points x with $0_T(x)$ dense and some points where $0_T(x)$ is not dense. We shall see this in §5.4.

As one might expect a minimal transformation can have no invariant non-constant continuous functions.

Theorem 5.3. *If* $T: X \to X$ *is minimal homeomorphism and* $f \in C(X)$ *then* $f \circ T = f$ *implies* f *is a constant.*

PROOF. Since $f \circ T = f$ we have $f \circ T^n = f \ \forall n \in Z$, so if we pick some $x \in X$ we know f is constant on the dense set $0_T(x)$. Since f is continuous it must be constant.   □

Remark. It is clear from the above proof that the conclusion of the theorem is true if there is some point with a dense orbit (rather than all points having a dense orbit), so the conclusion of Theorem 5.3 does not characterise minimality.

Since a minimal transformation cannot have non-trivial closed invariant sets it will not have any finite invariant sets unless X is finite. The points of finite invariant sets are called periodic points:

Definition 5.3. If $T: X \to X$ is a homeomorphism then x is a *periodic point* of T if $T^n x = x$ for some $n \geq 1$. The smallest such n is the *period* of x. A periodic point of period one is called a *fixed point*.

As we mentioned above, if X is infinite no minimal homeomorphism of X can have any periodic points. However we shall see that there are homeomorphisms of X with a dense set of periodic points and with $\{x \in X \,|\, 0_T(x) \text{ is dense}\}$ also being dense. An ergodic automorphism of a finite-dimensional torus will have this property.

We now check whether the examples mentioned in §5.1 are minimal or not.

(1) The identity map of X is minimal iff X consists of a single point.

(2) **Theorem 5.4.** *Let* G *be a compact metric group and* $T(x) = ax$. *Then* T *is minimal iff* $\{a^n : n \in Z\}$ *is dense in* X.

PROOF. Let e denote the identity element of G. Since $0_T(e) = \{a^n : n \in Z\}$, the minimality of T implies $\{a^n : n \in Z\}$ is dense. Now suppose the powers of a are dense. Let $x \in X$. We must show that $\overline{0_T(x)} = X$. Let $y \in X$. There exists n_i such that $a^{n_i} \to yx^{-1}$, so that

$$T^{n_i}(x) = a^{n_i} \cdot x \to y.$$

Therefore, $0_T(x)$ is dense in X.   □

(3) An automorphism A of a compact metric group G is minimal iff $G = \{e\}$. This is because $A(e) = e$.

(4) For affine transformations of compact metric groups necessary and sufficient conditions for minimality are known. For example, if G is also abelian and connected then $T = a \cdot A$ is minimal iff

$$\bigcap_{n=0}^{\infty} B^n G = \{e\} \quad \text{and} \quad [a, BG] = G$$

where B is the endomorphism of G defined by $B(x) = x^{-1} \cdot A(x)$ and $[a, BG]$ denotes the smallest closed subgroup of G containing a and BG. This was proved by Hoare and Parry [1].

(5) The shift on k symbols is minimal iff $k = 0$. This is seen from (3) above.

(6) The north–south map of K is not minimal because the orbit of the point N is not dense.

§5.3 The Non-wandering Set

One basic difference between an ergodic rotation of K and the north-south map of K is that the points of K have a recurrence property for ergodic rotations (if $x \in K$ then points in the orbit of x come very close to x, since the orbit is dense) whereas the points of K, except for N and S, do not have a recurrence property for the north-south map (if $x \neq N$ then $T^n(x) \to S$). This difference motivates the definitions of this section.

Definition 5.4. Let $T : X \to X$ be a continuous transformation of a compact metric space and let $x \in X$. The ω-*limit set of* x consists of all the limit points of $\{T^n x \mid n \geq 0\}$ i.e. $\omega(x) = \{y \in X \mid \exists n_i \nearrow \infty \text{ with } T^{n_i}(x) \to y\}$.

Theorem 5.5. *Let* $T : X \to X$ *be a continuous transformation of a compact metric space and* $x \in X$. *Then*

(i) $\omega(x) \neq \varnothing$.
(ii) $\omega(x)$ *is a closed subset of* X.
(iii) $T\omega(x) = \omega(x)$.

PROOF. (i) is clear.

(ii) Let $y_k \in \omega(x)$ for $k \geq 1$ and $y_k \to y \in X$. We want to show $y \in \omega(x)$. For each $j \geq 1$ choose k_j with $d(y_{k_j}, y) < \frac{1}{2j}$. Now choose n_j with $d(T^{n_j}x, y_{k_j}) < \frac{1}{2j}$ and so that $n_j < n_{j+1}$ for all j. Then $d(T^{n_j}x, y) < 1/j$ so $y \in \omega(x)$.

(iii) It is clear that $T\omega(x) \subset \omega(x)$. Let $y \in \omega(x)$ and suppose $T^{n_i}x \to y$. Then $\{T^{n_i-1}(x)\}$ has a convergent subsequence so $T^{n_{i_j}-1}(x) \to z$ for some $z \in X$. Then $T^{n_{i_j}}(x) \to T(z)$ so that $T(z) = y$. Since $z \in \omega(x)$ we have $T\omega(x) = \omega(x)$. □

Remarks

(1) If T is not a homeomorphism then $T^{-1}\omega(x)$ can be larger than $\omega(x)$. We can see this for the one-sided shift by choosing x to be the point $\{x_n\}_{-\infty}^{\infty}$ with $x_n = 0$ for all n. Then $\omega(x) = x$ but $T^{-1}x$ has k points.

(2) If T is a homeomorphism then we can define the ω-limit sets for T^{-1} and these are called the α-limit for sets for T. Hence $\alpha(x) = \{z \in X \mid \exists n_i \nearrow \infty \text{ with } T^{-n_i}(x) \to z\}$.

(3) If T is a homeomorphism and E is a minimal set for T then $\omega(x) = E$ for all $x \in E$, by Theorem 5.5.

Definition 5.5. Let $T: X \to X$ be continuous. A point x is called *wandering* for T if there is an open neighbourhood U of x such that the sets $T^{-n}U$, $n \geq 0$, are mutually disjoint. The *non-wandering set for* T, $\Omega(T)$, consists of all the points that are not wandering for T, hence

$$\Omega(T) = \{x \in X \,|\, \text{for every neighbourhood } U \text{ of } x \,\exists n \geq 1$$

$$\text{with } T^{-n}U \cap U \neq \varnothing\}.$$

Remarks

(1) The set $\Omega(T)$ consists of those points with a weak recurrence property. The interesting action of T takes place in $\Omega(T)$. We shall see later that any probability measure on the Borel subsets of X which is preserved by T gives zero measure to $X \backslash \Omega(T)$.

(2) If T is a homeomorphism then $T^{-n}U \cap U = T^{-n}(U \cap T^n U)$ so $\Omega(T^{-1}) = \Omega(T)$ and $\Omega(T) = \{x \in X \,|\, \text{for every neighbourhood } U \text{ of } x \,\exists n \neq 0$ with $T^{-n}U \cap U \neq \phi\}$.

Theorem 5.6. *Let* $T: X \to X$ *be continuous. Then*

    (i) $\Omega(T)$ *is closed.*
    (ii) $\bigcup_{x \in X} \omega(x) \subset \Omega(T)$ *(in particular* $\Omega(T) \neq \varnothing$*).*
    (iii) *All periodic points belong to* $\Omega(T)$.
    (iv) $T\Omega(T) \subset \Omega(T)$, *and if* T *is a homeomorphism then* $T\Omega(T) = \Omega(T)$.

PROOF

(i) From the definition of $\Omega(T)$ it is clear that $X \backslash \Omega(T)$ is open.

(ii) Let $x \in X$ and $y \in \omega(x)$. We want to show $y \in \Omega(T)$. Let V be a neighbourhood of y. We want to find $n \geq 1$ with $T^{-n}V \cap V \neq \varnothing$; so we seek $n \geq 1$ and some $z \in V$ with $T^n z \in V$. We know $T^{n_i}(x) \to y$ for some subsequence $\{n_i\}$ of the natural numbers so choose $n_{i_0} < n_{i_1}$ with $T^{n_{i_0}}(x) \in V$ and $T^{n_{i_1}}(x) \in V$. Then take $n = n_{i_1} - n_{i_0}$ and $z = T^{n_{i_0}}(x)$.

(iii) If $T^n x = x, n > 0$, and U is a neighbourhood of x then $x \in T^{-n}U \cap U$.

(iv) Let $x \in \Omega(T)$ and let V be a neighbourhood of $T(x)$. Then $T^{-1}V$ is a neighbourhood of x so there is some $n > 0$ with $T^{-(n+1)}V \cap T^{-1}V \neq \varnothing$. Therefore $T^{-n}V \cap V \neq \varnothing$ so $T(x) \in \Omega(T)$.

If T is a homeomorphism we know $\Omega(T) = \Omega(T^{-1})$ and therefore $T^{-1}\Omega(T) \subset \Omega(T)$. $\qquad\qquad\qquad\qquad\qquad\qquad\qquad\qquad\qquad\qquad\square$

Remark. If E is a minimal set for T then $E \subset \Omega(T)$ because if U is open, $\{T^n U \,|\, n \in Z\}$ are pairwise disjoint, and $U \cap E \neq \varnothing$ then $E \subset \bigcup_{-\infty}^{\infty} T^n U$ and this contradicts compactness.

Another way to state the definition of $\Omega(T)$ is given in the following.

Theorem 5.7. *If $T:X \to X$ is a continuous transformation of a compact metric space then $\Omega(T) = \{x \in X \mid$ for every neighbourhood U of x and every $N \geq 1$ there exists $n \geq N$ with $T^{-n}U \cap U \neq \varnothing\}$.*

PROOF. Clearly the stated set is a subset of $\Omega(T)$. Suppose $x \in \Omega(T)$ and U is a neighbourhood of x and $N \geq 1$. If x is a periodic point then clearly $T^{-n}U \cap U \neq \varnothing$ for some $n \geq N$. Suppose x is not a periodic point. Choose $r > 0$ so that $B(x; r) \subset U$. We now show we can choose $\delta < r$ so that $B(x; \delta) \cap T^{-i}B(x; \delta) = \varnothing$ for all $1 \leq i \leq N - 1$. If no such δ exists then if $1/n < r$ there exists $x_n \in B(x; 1/n) \cap T^{-i_n}B(x; 1/n)$ for some $1 \leq i_n \leq N - 1$. Choose a subsequence of natural numbers $\{n_j\}$ such that i_{n_j} is independent of j, say $i_{n_j} = k$ for all j. Then $x_{n_j} \in B(x; 1/n_j)$ so $x_{n_j} \to x$, and $T^k(x_{n_j}) \in B(x; 1/n_j)$ so $T^k(x_{n_j}) \to x$. Since $T^k(x_{n_j})$ converges to both $T^k(x)$ and x we must have $T^k(x) = x$ contradicting the fact that x is not a periodic point. Therefore there exists $\delta < r$ with $B(x; \delta) \cap T^{-i}B(x; \delta) = \varnothing$ if $1 \leq i \leq N - 1$. However $x \in \Omega(T)$ so there is some $n_1 \geq 1$ with $B(x; \delta) \cap T^{-n_1}B(x, \delta) \neq \varnothing$ and hence $n_1 \geq N$. Therefore $U \cap T^{-n_1}U \neq \varnothing$. $\qquad\square$

If $T:X \to X$ is a homeomorphism then $T|_{\Omega(T)}$ is a homeomorphism of the compact set $\Omega(T)$ and we can consider its non-wandering set $\Omega(T|_{\Omega(T)})$ which we shall denote by $\Omega_2(T)$. The following is an example where $\Omega_2(T) \neq \Omega(T)$.

Let X be the closed unit disc in the plane represented in polar coordinates, so that $X = \{(r, 2\pi\theta) \mid 0 \leq r \leq 1, \theta \in [0, 1)\}$. Define $T:X \to X$ by $T(r, 2\pi\theta) = (r^{1/2}, 2\pi(\theta^2 + 1 - r) \bmod 2\pi)$. Then T is a homeomorphism which fixes the origin. On the boundary, ∂X, of X, T is the homeomorphism $2\pi\theta \to 2\pi\theta^2$ of the unit circle and hence has one fixed point at $\theta = 0$ and the other points move clockwise around the circle towards $\theta = 0$. Hence $\Omega(T|_{\partial X})$ consists of the point $(1, 0)$. All other points of the disc spiral out towards ∂X under action of T. We claim $\Omega(T) = \{(0,0)\} \cup \partial X$. If $(r, 2\pi\theta) \notin \{(0,0)\} \cup \partial X$ then it is wandering. Clearly $(0,0) \in \Omega(T)$. Now let $(1, 2\pi\theta) \in \partial X$, and let U be a connected neighbourhood of $(1, 2\pi\theta)$ (say, an intersection of an open ball in

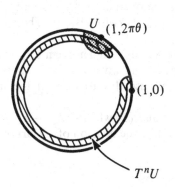

R^2 with X). Then T^nU will also be a connected set, and for large n will have the shape shown in the diagram. Hence $T^nU \cap U \neq \varnothing$ and hence $\Omega(T) = \{(0,0)\} \cup \partial X$. Hence $\Omega_2(T) = \{(0,0) \cup (1,0)\}$ so $\Omega_2(T) \neq \Omega(T)$.

If we put $\Omega_1(T) = \Omega(T)$ and define $\Omega_2(T)$ as above then we can define $\Omega_n(T)$ inductively by $\Omega_n(T) = \Omega(T|_{\Omega_{n-1}(T)})$. So $\Omega_1(T) \supset \Omega_2(T) \supset \cdots$ is a decreasing sequence of closed subsets of X and the intersection $\bigcap_{n=1}^{\infty} \Omega_n(T)$ is denoted by $\Omega_\infty(T)$ and called the centre of T.

In §6.4 we shall state formally the connections between $\Omega(T)$ and probability measures on the Borel subsets of X which are invariant for T. We shall mention these results briefly now because they help to determine $\Omega(T)$ from knowledge of invariant measures. If μ is a probability measure on the Borel subsets of X and is invariant for T and if $\mu(U) > 0$ whenever U is a non-empty open set, then $\Omega(T) = X$. (This is part of Theorem 6.15 and the proof is easy because if $x \notin \Omega(T)$ there is an open set U with the sets $\{T^{-n}U\}_{n=0}^{\infty}$ pairwise disjoint and this cannot happen when $\mu(U) > 0$). Because of this any affine transformation T of a compact group G has $\Omega(T) = G$ because T preserves Haar measure which is strictly positive on non-empty open sets.

Consider now the north-south map of K. For this map $\Omega(T) = \{N, S\}$. Clearly $\{N, S\} \subset \Omega(T)$ since N, S are both fixed points. Let $x \notin \{N, S\}$ and we show x is wandering. Choose y between $T^{-1}x$ and x. Then Ty lies between

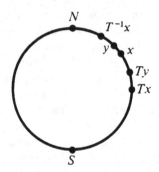

x and Tx. Let U be the open arc between y and Ty. Then U is a neighbourhood of x and since $T^{-n}U$ is the open arc with end points $T^{-n}y$, $T^{-n+1}y$ we see that the sets $\{T^{-n}U\}_{n=0}^{\infty}$ are pairwise disjoint. Therefore x is wandering and $\Omega(T) = \{N, S\}$.

As mentioned before we show in Theorem 6.15 that if μ is a probability on the Borel subsets of X and invariant for T then $\mu(\Omega(T)) = 1$. For the north-south map we can use this to find all the invariant probabilities (see §6.4).

One can readily compute $\Omega(T)$ when $T: X_A \to X_A$ is a topological Markov chain. The calculation is like the division of a Markov chain into equivalent sets of states.

§5.4 Topological Transitivity

Topological transitivity is a weakening of minimality. Again X always denotes a compact metric space.

Definition 5.6

(i) A continuous transformation $T:X \to X$ is called *one-sided topologically transitive* if there exists some $x \in X$ with $\{T^n(x) \mid n \geq 0\}$ dense in X.

(ii) A homeomorphism $T:X \to X$ is called *topologically transitive* if there is some $x \in X$ with $0_T(x) = \{T^n(x) \mid n \in Z\}$ dense in X.

Both of these concepts make sense for a homeomorphism and we shall see how they are related after giving some equivalent forms of the definitions. Recall that a set which is the intersection of a countable collection of open sets is called a G_δ.

Theorem 5.8. *The following are equivalent for a homeomorphism $T:X \to X$ of a compact metric space.*

(i) *T is topologically transitive.*

(ii) *Whenever E is a closed subset of X and $TE = E$ then either $E = X$ or E is nowhere dense (or, equivalently, whenever U is an open subset of X with $TU = U$ then $U = \varnothing$ or U is dense).*

(iii) *Whenever U, V are non-empty open sets then there exists $n \in Z$ with*

$$T^n(U) \cap V \neq \varnothing.$$

(iv) *$\{x \in X : \overline{0_T(x)} = X\}$ is a dense G_δ.*

PROOF

(i) $\Rightarrow$ (ii). Suppose $\overline{0_T(x_0)} = X$ and let $E \neq \varnothing$, E closed and $TE = E$. Suppose U is open and $U \subset E$, $U \neq \varnothing$. Then there exists p with $T^p(x_0) \in U \subset E$ so that $0_T(x_0) \subset E$ and $X = E$. Therefore either E has no interior or $E = X$.

(ii) $\Rightarrow$ (iii). Suppose $U, V \neq \varnothing$ are open sets. Then $\bigcup_{n=-\infty}^{\infty} T^n U$ is an open T-invariant set, so it is necessarily dense by condition (ii). Thus $\bigcup_{n=-\infty}^{\infty} T^n U \cap V \neq \varnothing$.

(iii) $\Rightarrow$ (iv). Let $U_1, U_2, \ldots, U_n, \ldots$ be a countable base for X. Then $\{x \in X \mid \overline{0_T(x)} = X\} = \bigcap_{n=1}^{\infty} \bigcup_{m=-\infty}^{\infty} T^m U_n$ and $\bigcup_{m=-\infty}^{\infty} T^m(U_n)$ is clearly dense by condition (iii). Hence the result follows.

(iv) $\Rightarrow$ (i). This is clear.   □

For the analogous theorem for one-sided transitivity we assume $TX = X$. Note that if $E \subset X$ the condition $E \subset T^{-1}E$ is equivalent to $TE \subset E$.

Theorem 5.9. *The following are equivalent for a continuous transformation* $T: X \to X$ *with* $TX = X$.

(i) *T is one-sided topologically transitive.*

(ii) *Whenever E is a closed subset of X and $E \subset T^{-1}E$ then either $E = X$ or E is nowhere dense (equivalently, whenever U is an open subset of X and $T^{-1}U \subset U$ then $U = \varnothing$ or U is dense).*

(iii) *Whenever U, V are non-empty open sets there exists $n \geq 1$ with $T^{-n}U \cap V \neq \varnothing$.*

(iv) *The set of points x with $\{T^n(x) \mid n \geq 0\}$ dense in X is a dense G_δ.*

PROOF

(i) $\Rightarrow$ (ii). Suppose $\{T^n(x_0) \mid n \geq 0\}$ is dense in X, and suppose E is closed and $TE \subset E$. Suppose U is a non-empty open set with $U \subset E$. Then $T^p(x_0) \in U$ for some $p \geq 0$, so that $\{T^n(x_0) \mid n \geq p\} \subset E$. Therefore

$$\{x_0, T(x_0), \ldots, T^{p-1}(x_0)\} \cup E = X.$$

By applying T to each side we get $\{T(x_0), \ldots, T^{p-1}(x_0)\} \cup E = X$ so by repeated application of T we have $E = X$. So if E has interior then $E = X$.

(ii) $\Rightarrow$ (iii). Suppose U, V are non-empty open sets. Then $\bigcup_{n=1}^{\infty} T^{-n}U$ is open and $T^{-1}(\bigcup_{n=1}^{\infty} T^{-n}U) \subset \bigcup_{n=1}^{\infty} T^{-n}U$ so $\bigcup_{n=1}^{\infty} T^{-n}U$ is dense by (ii). Therefore $T^{-n}U \cap V \neq \varnothing$ for some $n \geq 1$.

(iii) $\Rightarrow$ (iv). If $\{U_n\}_{n=1}^{\infty}$ is a base for the topology then $\{x \mid \{T^n(x)\}_{n=0}^{\infty}$ is dense$\} = \bigcap_{n=1}^{\infty} \bigcup_{m=0}^{\infty} T^{-m}U_n$. By (iii) we know $\bigcup_{m=0}^{\infty} T^{-m}U_n$ is dense so the result follows.

(iv) $\Rightarrow$ (i). This is clear. $\square$

The assumption $TX = X$ was made because of the type of behaviour occurring in the following example. Let $X = \{1/n \mid n \geq 1\} \cup \{0\}$ with the induced topology from the real line. Define $T: X \to X$ by $T(0) = 0$ and $T(1/n) = 1/(n+1)$. Hence T moves each point $1/n$ to the next point on the left. Only the point 1 has a dense forward orbit so statements (i) and (iv) are not equivalent for this example. Also if $E = X \setminus \{1\}$ then $E \subset T^{-1}E$ and E is closed so (ii) is violated.

We now consider the connection between the two types of transitivity when T is a homeomorphism.

Theorem 5.10. *Let $T: X \to X$ be a homeomorphism. Then T is one-sided topologically transitive iff T is topologically transitive and $\Omega(T) = X$.*

PROOF. Suppose $\{T^n(x_0) \mid n \geq 0\}$ is dense in X. Clearly T is topologically transitive. If $\Omega(T) \neq X$ there is a non-empty open set U such that $\{T^nU \mid n \in Z\}$ are pairwise disjoint sets. For some $n_0 \geq 0$, $T^{n_0}(x_0) \in U$. Therefore $T^{n+n_0}(x_0) \in T^nU, n \geq 0$, so that only $\{x_0, T(x_0), \ldots, T^{n_0-1}(x_0)\}$ can belong to $\bigcup_{i=1}^{\infty} T^{-i}U$

contradicting the fact that each set $T^{-i}U$ must contain an element of the dense set $\{T^n(x_0)\mid n \geq 0\}$. Therefore $\Omega(T) = X$.

Now suppose T is topologically transitive and $\Omega(T) = X$. We use (iii) of Theorem 5.9 to show T is one-sided topologically transitive. Let U, V be non-empty open sets. We want to find some $k \geq 1$ with $T^{-k}U \cap V \neq \varnothing$. By (iii) of Theorem 5.8 we know there is some $N \in Z$ with $W = T^N U \cap V \neq \varnothing$ so we may as well suppose $N \geq 0$. Since $\Omega(T) = X$ Theorem 5.7 gives the existence of $n \geq N + 1$ with $T^{-n}W \cap W \neq \varnothing$. Then $T^{-(n-N)}U \cap V \supset T^{-n}W \cap W \neq \varnothing$ so we can take $k = n - N$. $\qquad \square$

Remarks

(1) An example of a topologically transitive homeomorphism which is not one-sided topologically transitive is the following. Let $X = \{0\} \cup \{1\} \cup \{(1/n)\mid n \geq 2\} \cup \{1 - (1/n)\mid n \geq 2\}$ with the induced topology as a subset of R. Define $T: X \to X$ by $T(0) = 0$, $T(1) = 1$ and T maps any other point to the next point on the left. Then if $x \notin \{0, 1\}$ the set $\{T^n(x)\mid n \in Z\}$ is dense, so that T is topologically transitive. Clearly T is not one-sided topologically transitive. Notice in this case that $\Omega(T) = \{0\} \cup \{1\}$.

(2) In an analogous way we could define a homeomorphism $T: X \to X$ to be one-sided minimal if $\{T^n(x)\mid n \geq 0\}$ is dense in X for each $x \in X$. Then one can show that T is one-sided minimal iff T is minimal. The "only if" part of this is trivial and the "if" part follows from the following. One-sided minimality is equivalent to $X = \bigcup_{n=0}^{\infty} T^{-n}U$ for each non-empty open set U. If U is open and non-empty then the minimality implies $\bigcup_{-\infty}^{\infty} T^{-k}U = X$. Since X is compact we have $X = T^{k_1}U \cup T^{k_2}U \cup \cdots \cup T^{k_r}U$ for some integers k_j. Choose $N > 0$ so that $N > |k_j|$, $1 \leq j \leq r$ and then $X = T^{-N}X = T^{-N+k_1}U \cup \cdots \cup T^{-N+k_r}U$ so $X = \bigcup_{n=0}^{\infty} T^{-n}U$. Hence minimality implies one-sided minimality. Of course $\Omega(T) = X$ when T is minimal, because it contradicts compactness of X to have an open set U with $\{T^n U\}_{-\infty}^{\infty}$ pairwise disjoint and $\bigcup_{n=-\infty}^{\infty} T^n U = X$. $\qquad \square$

Topologically transitive homeomorphisms enjoy some of the properties of minimal homeomorphisms and also allow other interesting things to occur; e.g., a dense set of periodic points. We know from Theorem 1.11 that an ergodic affine transformation of a compact, connected, metric abelian group is topologically transitive (even one-sided topologically transitive). We now see that some of them have a dense set of periodic points.

Theorem 5.11. *Let $A: K^n \to K^n$ be an ergodic automorphism of the n-torus K^n. The periodic points of A are exactly those points $(w_1, \ldots, w_n) \in K^n$ where each w_i is a root of unity. (In additive notation these are the points of R^n/Z^n of the form $(x_1, \ldots, x_n) + Z^n$ where each x_i is rational.) Even if A is not ergodic these points are periodic for A so every automorphism A has a dense set of periodic points.*

PROOF. Let A be an automorphism. Let $w = (w_1, \ldots, w_n) \in K^n$ be so that each w_i is a root of unity. There is some $k \geq 1$ with $w^k = e$, the identity element. For each fixed k the set $Y_k = \{z \in K^n : z^k = e\}$ is a finite subgroup of K^n and $AY_k = Y_k$. Hence each member of Y_k is a periodic point and so our original w is.

Now suppose A is ergodic. We shall use additive notation for this part of the proof. If $x + Z^n \in R^n/Z^n$ is fixed by A^k then $A^k x = x + p$ for some $p \in Z^n$. In matrix notation this equation becomes

$$([A]^k - I) \begin{pmatrix} x_1 \\ \vdots \\ x_n \end{pmatrix} = \begin{pmatrix} p_1 \\ \vdots \\ p_n \end{pmatrix}.$$

Since A is ergodic the matrix $[A]^k - I$ is an invertible matrix of integers and so its inverse has rational entries. Therefore each x_i is rational. Therefore each periodic point is of the form $x + Z^n$ where the coordinates of x are rational.   □

Theorem 5.12. *The two-sided and one-sided shifts have a dense set of periodic points. For the two-sided shift $\{x_n\}_{-\infty}^{\infty}$ is fixed by T^p iff $x_n = x_{n+p}$ $\forall n \in Z$. For the one-sided shift $\{x_n\}_0^{\infty}$ is fixed by T^p iff $x_n = x_{n+p}$ $\forall n \geq 0$.*

PROOF. We shall consider only the two-sided case. If $x = \{x_n\}_{-\infty}^{\infty}$ and $T^p x = x$ then $x_{p+i} = x_i$ for each i. In other words, the points fixed by T^p have the form $(\ldots, x_{p-1} \overset{*}{x_0} x_1, \ldots, x_{p-1} \overset{*}{x_0} x_1, \ldots, x_{p-1} x_0 x_1, \ldots, x_{p-1} x_0, \ldots)$ where we have free choice of $x_0, x_1, \ldots, x_{p-1}$. Therefore the periodic points are dense.   □

Parts (ii) and (iii) of Theorem 5.8 show that topological transitivity is (in some sense) a topological analogue of ergodicity. Also, topologically transitive homeomorphisms are "indecomposable;" i.e., we cannot write

$$X = \bigcup_\alpha E_\alpha, \qquad TE_\alpha = E_\alpha \quad \text{and } E_\alpha \text{ closed}$$

when T is topologically transitive. So it seems that topologically transitive homeomorphisms are better building blocks than minimal homeomorphisms. If T has a decomposition into minimal pieces then each piece is also topologically transitive. Not all homeomorphisms can be decomposed into topologically transitive pieces; see the example in Remark (1) above. Two important cases where a decomposition is possible are the following. A distal homeomorphism $T : X \to X$ (i.e. for every pair $x \neq y$ there exists $\delta = \delta(x, y) > 0$ with $d(T^n(x), T^n(y)) > \delta$ $\forall n \in Z$) can be decomposed into minimal pieces i.e. $X = \bigcup_{i \in I} X_i$ where I is some index set, the sets X_i are pairwise disjoint, closed, $TX_i = X_i$ and $T|_{X_i}$ is minimal and distal (Ellis [1]). If $T : M \to M$ is an Axiom A diffeomorphism of a compact manifold M then $T|_{\Omega(T)}$ is very important. It turns out that $\Omega(T) = \bigcup_{i=1}^r \Omega_i$ where the Ω_i are pairwise disjoint closed sets with $T\Omega_i = \Omega_i$ and $T|_{\Omega_i}$ is topologically transitive (Smale [1]).

The following gives a sufficient, but not necessary, condition for a topologically transitive homeomorphism to be minimal.

Theorem 5.13. *If X is a compact metrisable space, $T:X \to X$ a topologically transitive homeomorphism, and if there exists a metric on X making T an isometry, then T is minimal.*

PROOF. Suppose d is such a metric, i.e., $d(Tx, Ty) = d(x, y)$. Let $\overline{0_T(x_0)} = X$ and consider $x \in X$. We want to show that $\overline{0_T(x)} = X$. Let $y \in X$ and let $\varepsilon > 0$. There exist $n, m \in Z$ such that

$$d(x, T^m(x_0)) < \varepsilon, \qquad d(y, T^n(x_0)) < \varepsilon,$$

so that

$$\begin{aligned} d(y, T^{n-m}(x)) &\le d(y, T^n(x_0)) + d(T^n(x_0), T^{n-m}(x)) \\ &= d(y, T^n(x_0)) + d(T^m(x_0), x) \\ &< 2\varepsilon. \end{aligned}$$

Therefore $\overline{0_T(x)} = X$. $\qquad\qquad\qquad\qquad\qquad\qquad\qquad\qquad\qquad\qquad\qquad\square$

The following property of minimal homeomorphisms is enjoyed by topologically transitive ones.

Theorem 5.14. *If T is a topologically transitive homeomorphism or a one-sided topologically transitive continuous transformation then T has no non-constant invariant continuous function.*

PROOF. If $f \circ T = f$ then $f \circ T^n = f$ so f is constant on orbits of points. The result then follows. $\qquad\qquad\qquad\qquad\qquad\qquad\qquad\qquad\qquad\qquad\qquad\qquad\square$

Remark. If all the T-invariant continuous functions are constant then T need not be topologically transitive. The following is an example to illustrate this. Let

$$X = (K^2 \times \{0\}) \cup (K^2 \times \{1\})/(e, 0) \sim (e, 1)$$

i.e., two copies of the two-torus joined at the identity. Let $A:K^2 \to K^2$ be an ergodic automorphism and define $T:X \to X$ by

$$T(x, 0) = (Ax, 0), \qquad T(x, 1) = (Ax, 1).$$

Then T is not topologically transitive since T preserves $K^2 \times \{0\}$ and $K^2 \times \{1\}$. However, each continuous T-invariant function is constant since it must be constant on both $K^2 \times \{0\}$ and $K^2 \times \{1\}$, because A is ergodic, and these two constants must be the same because they must agree at the point $(e, 0) \equiv (e, 1)$.

The following theorems give many examples of topologically transitive homeomorphisms and one-sided topologically transitive maps.

Theorem 5.15. *Let* $T:X \to X$ *be a homeomorphism of the compact metric space* X *and let* m *be a probability measure on the Borel subsets of* X *giving non-zero measure to every non-empty open set. If* T *is an ergodic measure-preserving transformation with respect to* m, *then* $m(\{x \in X : \overline{0_T(x)} = X\}) = 1$. *In particular,* T *is topologically transitive.*

PROOF. Let $U_1, U_2, \ldots$ be a countable base for the topology. Then

$$\{x : \overline{0_T(x)} = X\} = \bigcap_{n=1}^{\infty} \bigcup_{k=-\infty}^{\infty} T^k U_n.$$

For each n the open set $\bigcup_{k=-\infty}^{\infty} T^k U_n$ is T-invariant so by ergodicity has measure 0 or 1. But U_n is contained in this set and $m(U_n) > 0$ since U_n is open. Therefore $m(\bigcup_{k=-\infty}^{\infty} T^k U_n) = 1$ and so $m\{x : \overline{0_T(x)} = X\} = 1$. ☐

Theorem 5.16. *Let* $T:X \to X$ *be continuous with* $TX = X$ *and let* m *be a probability on the Borel subsets of* X *giving non-zero measure to every non-empty open set. If* T *is an ergodic measure-preserving transformation with respect to* m *then* $m(\{x \in X : \{T^n(x)\}_0^{\infty} \text{ is dense}\}) = 1$. *In particular,* T *is one-sided topologically transitive.*

PROOF. If $\{U_n\}_1^{\infty}$ is a countable base for the topology then $\{x \mid \{T^n x\}_0^{\infty}$ is dense$\} = \bigcap_{n=1}^{\infty} \bigcup_{k=0}^{\infty} T^{-k} U_n$. For each n, $T^{-1}(\bigcup_{k=0}^{\infty} T^{-k} U_n) \subset \bigcup_{k=0}^{\infty} T^{-k} U_n$ and the ergodicity of T implies $m(\bigcup_{k=0}^{\infty} T^{-k} U_n) = 1$. The result then follows. ☐

We now see whether our examples are topologically transitive.

Clearly, the identity map on X is only topologically transitive when X has just one point. From Theorem 1.11 we know an affine transformation T is one-sided topologically transitive iff it is ergodic, and if T is invertible these conditions are equivalent to the topological transitivity of T. In particular a rotation $T(x) = ax$ of a compact group is topologically transitive iff $\{a^n \mid n \in Z\}$ is dense iff $\{a^n \mid n \geq 0\}$ is dense iff T is minimal.

Since the shifts are ergodic for Haar measure we know, by Theorem 5.16, they are one-sided topologically transitive. (Haar measure, in this case, is the product measure given by weights $(1/k, \ldots, 1/k)$, as can be seen by showing this measure is rotation invariant.) One can easily construct a point x with $\{T^n(x) \mid n \geq 0\}$ dense when T is the one-sided or two-sided shift.

The north–south map of K is not topologically transitive as the orbit of a point on the right half of the circle always lies in the right half.

For a two-sided topological Markov chain $T:X_A \to X_A$ one can readily show that T is one-sided topologically transitive iff the matrix A is irreducible (i.e. $\forall i, j \; \exists n > 0$ such that the (i, j)-th element of A^n is non-zero). A topologically transitive, but not one-sided topologically transitive, example is provided by the matrix $\left(\begin{smallmatrix} 1 & 0 \\ 1 & 1 \end{smallmatrix}\right)$.

§5.5 Topological Conjugacy and Discrete Spectrum

In this section we consider the topological analogue of the theory of transformations with discrete spectrum given in Chapter 3. To do this we need a notion of conjugacy for homeomorphisms. The following seems to be the most natural.

Definition 5.7. Let $T: X \to X$, $S: Y \to Y$ be homeomorphisms of compact spaces. We say T is *topologically conjugate* to S if there exists a homeomorphism $\phi: X \to Y$ such that $\phi T = S \phi$. The homeomorphism ϕ is called a *conjugacy*.

Remarks

(1) Topological conjugacy is an equivalence relation on the space of all homeomorphisms.

(2) If T and S are topologically conjugate then T is minimal iff S is minimal, and T is topologically transitive iff S is topologically transitive. If ϕ is a conjugacy as in Definition 5.7 then $\phi \Omega(T) = \Omega(S)$, and $T^n(x) = x$ iff $S^n \phi(x) = \phi(x)$.

Definition 5.8. Let X be a compact metric space, $T: X \to X$ a homeomorphism, and f a complex-valued continuous function on X which is not identically 0. We say that f is an *eigenfunction* for T if there exists $\lambda \in \mathbb{C}$ such that

$$f(Tx) = \lambda f(x) \quad \forall x \in X.$$

We then call λ the *eigenvalue* for T corresponding to the eigenfunction f.

Another way to write the above relationship is $U_T f = \lambda f$ or $f \circ T = \lambda f$. We have the following analogue of Theorem 3.1.

Theorem 5.17. *Let T be a homeomorphism of a compact metric space X and suppose T is topologically transitive. Then*

(i) *If $f \circ T = \lambda f$ where $0 \not\equiv f \in C(X)$, then $|\lambda| = 1$ and $|f|$ is constant.*

(ii) *If f, g are both eigenfunctions of T corresponding to the same eigenvalue then $f = cg$ where c is a constant.*

(iii) *A finite collection of eigenfunctions corresponding to distinct eigenvalues are linearly independent in $C(X)$.*

(iv) *The eigenvalues of T form a countable subgroup of K.*

PROOF

(i) Since $|f(Tx)| = |\lambda| |f(x)|$ we have

$$\sup_{x \in X} |f(Tx)| = |\lambda| \sup_{x \in X} |f(x)|.$$

Since $TX = X$ this gives

$$\sup_{x \in X} |f(x)| = |\lambda| \sup_{x \in X} |f(x)|$$

and hence $|\lambda| = 1$. Therefore $|f(Tx)| = |f(x)|$ and by Theorem 5.14 $|f(x)| =$ constant.

(ii) By (i) $|g(x)| > 0 \, \forall x \in X$. The function f/g is T-invariant and therefore constant by Theorem 5.14.

(iii) Let $f_n(Tx) = \lambda_n f_n(x)$ where $\{\lambda_n\}$ are all distinct for $n = 1, \ldots, k$. Suppose

$$a_1 f_1(x) + a_2 f_2(x) + \cdots + a_k f_k(x) = 0 \quad \forall x \in X,$$

where $a_i \in \mathbb{C}$ for $i = 1, \ldots, k$.

By applying the above equation to $T^i x$ instead of x, we get

$$a_1 \lambda_1^i f_1(x) + a_2 \lambda_2^i f_2(x) + \cdots + a_k \lambda_k^i f_k(x) = 0 \quad \forall x \in X.$$

Hence

$$
\begin{pmatrix}
1 & 1 & \cdots & 1 \\
\lambda_1 & \lambda_2 & \cdots & \lambda_k \\
\vdots & \vdots & & \vdots \\
\lambda_1^{k-1} & \lambda_2^{k-1} & & \lambda_k^{k-1}
\end{pmatrix}
\begin{pmatrix}
a_1 f_1(x) \\
a_2 f_2(x) \\
\vdots \\
a_k f_k(x)
\end{pmatrix}
=
\begin{pmatrix}
0 \\
0 \\
\vdots \\
0
\end{pmatrix}.
$$

All the λ_i's are distinct so the matrix is nonsingular. Therefore $a_i f_i(x) = 0$ $\forall x \in X$, $i = 1, \ldots, k$. Since f_i is not identically 0 we have $a_i = 0$ for each i. Hence $\{f_1, \ldots, f_k\}$ are linearly independent in $C(X)$.

(iv) The eigenvalues clearly form a subgroup of K. To check there are only countably many eigenvalues it suffices to show that if $h : X \to K$ is an eigenfunction corresponding to an eigenvalue $\tau \neq 1$ then $\|h - 1\| > \frac{1}{4}$. For then two eigenfunctions, with values in K, corresponding to different eigenvalues will be greater than distance $\frac{1}{4}$ apart in $C(X)$ and, since $C(X)$ has a countable dense set, there can only be countably many eigenvalues. So let $h(Tx) = \tau h(x)$, $\tau \neq 1$. Choose $x_0 \in X$ and $p \in Z$ so that $\tau^p h(x_0)$ is in the left-hand half of the unit circle. Then

$$\|h - 1\| = \sup_{x \in X} \|h(x) - 1\|$$

$$\geq \|h(T^p x_0) - 1\| = \|\tau^p h(x_0) - 1\| > \frac{1}{4}. \qquad \square$$

Remark. If T, S are topologically conjugate they have the same group of eigenvalues. This is because if ϕ is a homeomorphism with $\phi T = S\phi$ then $f \circ S = \lambda f$ iff $f \circ \phi \circ T = \lambda f \circ \phi$.

Definition 5.9. Let T be a homeomorphism of the compact metric space X. We say that T has *topological discrete spectrum* if the smallest closed linear subspace of $C(X)$ containing the eigenfunctions of T is $C(X)$, i.e., the eigenfunctions span $C(X)$.

From the above remarks we know that when T is topologically transitive and has topological discrete spectrum there is a countable collection $\{\lambda_n\}_{n=1}^\infty$ of eigenvalues and a linearly independent collection $\{f_n\}_{n=1}^\infty$ of functions $f_n : X \to K$ such that $\{f_n\}_1^\infty$ span $C(X)$ and $f_n \circ T = \lambda_n f_n$.

The following is a representation theorem for topologically transitive homeomorphisms with topological discrete spectrum. In the proof we shall use the Stone–Weierstrass Theorem: If A is a subalgebra of $C(X)$ which contains the constant functions, is closed under complex conjugation (i.e. $f \in A \Rightarrow \bar{f} \in A$), and separates the points of X (i.e. if $x \neq y$ then $f(x) \neq f(y)$ for some $f \in A$) then A is dense in $C(X)$ (Dunford and Schwartz [1], page 272).

Theorem 5.18 (Halmos and von Neumann). *The following are equivalent for a homeomorphism T of a compact metric space X.*

   (i) *T is topologically transitive and is an isometry for some metric on X.*

   (ii) *T is topologically conjugate to a minimal rotation on a compact abelian metric group.*

   (iii) *T is minimal and has topological discrete spectrum.*

   (iv) *T is topologically transitive and has topological discrete spectrum.*

PROOF

(i) $\Rightarrow$ (ii). Let ρ be a metric on X for which T is an isometry. Suppose $\overline{0_T(x_0)} = X$. Define a multiplication $*$ on $0_T(x_0)$ by $T^n x_0 * T^m x_0 = T^{n+m} x_0$. We have

$$\rho(T^n x_0 * T^m x_0, T^p x_0 * T^q x_0) = \rho(T^{n+m} x_0, T^{p+q} x_0)$$
$$\leq \rho(T^{n+m} x_0, T^{p+m} x_0) + \rho(T^{p+m} x_0, T^{p+q} x_0)$$
$$= \rho(T^n x_0, T^p x_0) + \rho(T^m x_0, T^q x_0).$$

Hence the map $* : 0_T(x_0) \times 0_T(x_0) \to 0_T(x_0)$ is uniformly continuous and therefore can be extended uniquely to a continuous map $* : X \times X \to X$.

Also, $\rho(T^{-n} x_0, T^{-m} x_0) = \rho(T^{m+n} T^{-n} x_0, T^{m+n} T^{-m} x_0) = \rho(T^m x_0, T^n x_0)$ and so the map

$$0_T(x_0) \xrightarrow{\text{inverse}} 0_T(x_0)$$

is uniformly continuous and can be uniquely extended to a continuous map of X. Thus we get that X is a topological group and is also abelian since it has a dense abelian subgroup $\{T^n x_0 : n \in Z\}$. Since $T(T^n x_0) = T^{n+1} x_0 = Tx_0 * T^n x_0$ we have $Tx = Tx_0 * x$ and so T is the rotation by Tx_0.

(ii) $\Rightarrow$ (iii). If T is a minimal rotation on a compact abelian group G then each character of G is an eigenfunction. Let A be the collection of all finite linear combinations of characters. Then A is a subalgebra of $C(X)$, contains the constants, is closed under complex conjugation, and separates points. Applying the Stone–Weierstrass Theorem we see that the topological closure of A is $C(X)$.

(iii) $\Rightarrow$ (iv) is trivial since minimality implies topological transitivity.

(iv) $\Rightarrow$ (i). We can choose eigenfunctions $f_n : X \to K$, $n \geq 1$, with $f_n(T) = \lambda_n f_n$ and where the f_n are linearly independent and span $C(X)$. Since the collection $\{f_n\}$ spans $C(X)$ it must separate the points of X, so

$$\rho(x, y) = \sum_{n=1}^{\infty} \frac{|f_n(x) - f_n(y)|}{2^n} \quad \text{defines a metric on } X.$$

Also

$$\rho(Tx, Ty) = \sum_{n=1}^{\infty} \frac{|\lambda_n f_n(x) - \lambda_n f_n(y)|}{2^n} = \rho(x, y), \quad \text{since } |\lambda_n| = 1.$$

It remains to check that ρ gives the topology on X. Let d be the original metric on X. It suffices to show the identity map from the compact metric space (X, d) to the metric space (X, ρ) is continuous, because a bijective continuous map from a compact space onto a Hausdorff space is a homeomorphism. Let $\varepsilon > 0$ and choose N so that $\sum_{n=N+1}^{\infty} (2/2^n) < \varepsilon/2$. By the continuity of the functions f_n $(n \leq N)$ there exists $\delta > 0$ such that $d(x, y) < \delta$ implies $|f_n(x) - f_n(y)| < \varepsilon/2$ for $1 \leq n \leq N$. Then $d(x, y) < \delta$ implies $\rho(x, y) < \sum_{n=1}^{N} (\varepsilon/2^{n+1}) + \varepsilon/2 < \varepsilon$. $\square$

Remark. If $Tx = ax$ is a minimal rotation of a compact metric abelian group G it is straightforward to show that the set of eigenvalues of T is $\{\gamma(a) : \gamma \in \hat{G}\}$ and every eigenfunction is a constant multiple of a character. This also follows from Theorem 3.5 since each continuous eigenfunction is an L^2-eigenfunction.

We have the following isomorphism theorem.

Theorem 5.19 (Topological Discrete Spectrum Theorem). *Two minimal homeomorphisms of compact metric spaces both having topological discrete spectrum are topologically conjugate iff they have the same eigenvalues.*

PROOF. If T, S are topologically conjugate they clearly have the same eigenvalues. We give the outline of two proofs of the converse.

(1) One proof is along the lines of the proof of Theorem 3.4, but instead of using Theorem 2.10 we use the Banach–Stone Theorem. This says that if X, Y are compact spaces, $\Phi : C(Y) \to C(X)$ is a bijective linear isometry,

and $\Phi(f \cdot g) = \Phi(f)\Phi(g)$, then there exists a homeomorphism $\phi: X \to Y$ such that $\Phi(f)(x) = f(\phi(x))$ (Dunford and Schwartz [1], p. 442).

(2) Another proof uses Theorem 5.18 and character theory. By Theorem 5.18 we can suppose T is a minimal rotation of a compact abelian group G, $Tx = ax$, and S is a minimal rotation of a compact abelian group H, $Sy = by$. We are assuming $\{\gamma(a): \gamma \in \hat{G}\} = \{\delta(b): \delta \in \hat{H}\}$. Define a map $\theta: \hat{H} \to \hat{G}$ by $(\theta(\delta))(a) = \delta(b)$. This is well-defined and a bijection. Moreover, θ is easily checked to be a group isomorphism and hence induces an isomorphism $C: G \to H$. It is easy to show that $CT = SC$.   □

Remarks

(1) Thus the theory of topologically transitive homeomorphisms with topological discrete spectrum is analogous to that of ergodic measure-preserving transformations with discrete spectrum.

(2) Just as in the case of measure-preserving transformations the theory of minimal homeomorphisms with discrete spectrum can be extended to an isomorphism theory of minimal homeomorphisms with quasi-discrete spectrum (Hahn and Parry [1], Hoare and Parry [1]).

(3) When studying some homeomorphisms it is desirable to consider notions of conjugacy weaker than topological conjugacy. Since the interesting "random" action of T takes place in $\Omega(T)$ one useful conjugacy notion is: T and S are Ω-conjugate if $T|_{\Omega(T)}$ and $S|_{\Omega(S)}$ are topologically conjugate. This is useful in studying the stability properties of Axiom A diffeomorphisms (Smale [1]). Since a minimal homeomorphism T satisfies $\Omega(T) = X$, Ω-conjugacy is the same as topological conjugacy for the class of homeomorphisms discussed in this section.

§5.6 Expansive Homeomorphisms

Expansive homeomorphisms are an important class of transformations. We shall see this in the study of topological entropy and the measure-theoretic entropies of homeomorphisms in Chapters 7 and 8. We begin by making a definition analogous to that of a generator (see §4.6).

Definition 5.10. Let X be a compact metrisable space and $T: X \to X$ a homeomorphism. A finite open cover α of X is a *generator* for T if for every bisequence $\{A_n\}_{-\infty}^{\infty}$ of members of α the set $\bigcap_{n=-\infty}^{\infty} T^{-n}\bar{A}_n$ contains at most one point of X. If this condition is replaced by "$\bigcap_{n=-\infty}^{\infty} T^{-n}A_n$ contains at most one point of X" then α is called a *weak generator*.

These concepts are due to Keynes and Robertson [1].

Theorem 5.20. *If $T: X \to X$ is a homeomorphism of a compact metrisable space then T has a generator iff T has a weak generator.*

PROOF. A generator is clearly a weak generator. Now suppose β is a weak generator for T, $\beta = \{B_1, \ldots, B_s\}$, and let δ be a Lebesgue number for β (see Theorem 0.20). Let α be a finite open cover by sets A_i having $\operatorname{diam}(\bar{A}_i) \leq \delta$. So if A_{i_n} is a bisequence in α then $\forall n \; \exists j_n$ with $\bar{A}_{i_n} \subseteq B_{j_n}$. Hence

$$\bigcap_{-\infty}^{\infty} T^{-n}\bar{A}_{i_n} \subseteq \bigcap_{-\infty}^{\infty} T^{-n}B_{j_n},$$

which is either empty or a single point. So α is a generator. $\qquad\square$

The following shows that a generator determines the topology on X. If α, β are open covers of X then $\alpha \vee \beta$ is the open cover of X by the sets $A \cap B$, $A \in \alpha$, $B \in B$. $T^{-1}\alpha$ is the open cover by the sets $T^{-1}A$, $A \in \alpha$.

Theorem 5.21. *Let $T: X \to X$ be a homeomorphism of a compact metric space (X, d). Let α be a generator for T. Then $\forall \varepsilon > 0 \; \exists N > 0$ such that each set in $\bigvee_{-N}^{N} T^{-n}\alpha$ has diameter less than ε. Conversely, $\forall N > 0 \; \exists \varepsilon > 0$ such that $d(x, y) < \varepsilon$ implies*

$$x, y \in \bigcap_{-N}^{N} T^{-n}A_n$$

for some $A_{-N}, \ldots, A_N \in \alpha$.

PROOF. Suppose the first part of the theorem does not hold. Then $\exists \varepsilon > 0$ such that $\forall j > 0 \; \exists x_j, y_j, d(x_j, y_j) > \varepsilon$ and $\exists A_{j,i} \in \alpha$, $-j \leq i \leq j$ with $x_j, y_j \in \bigcap_{i=-j}^{j} T^{-i}A_{j,i}$. There is a subsequence $\{j_k\}$ natural numbers such that $x_{j_k} \to x$ and $y_{j_k} \to y$ since X is compact. We have $x \neq y$. Consider the sets $A_{j_k,0}$. Infinitely many of them coincide since α is finite. Thus $x_{j_k}, y_{j_k} \in A_0$, say, for infinitely many k and hence $x, y \in \bar{A}_0$. Similarly, for each n, infinitely many $A_{j_k,n}$ coincide and we obtain $A_n \in \alpha$ with $x, y \in T^{-n}\bar{A}_n$. Thus

$$x, y \in \bigcap_{-\infty}^{\infty} T^{-n}\bar{A}_n,$$

contradicting the fact that α is a generator.

To prove the converse let $N > 0$ be given. Let $\delta > 0$ be a Lebesgue number for α. Choose $\varepsilon > 0$ such that $d(x, y) < \varepsilon$ implies $d(T^ix, T^iy) < \delta$ for $-N \leq i \leq N$. Hence if $d(x, y) < \varepsilon$ and $|i| \leq N$ then $T^ix, T^iy \in A_i$ for some $A_i \in \alpha$. Hence

$$x, y \in \bigcap_{-N}^{N} T^{-i}A_i. \qquad\square$$

Generators are connected with expansive homeomorphisms which were studied for several years before generators were introduced.

Definition 5.11. A homeomorphism T of a compact metric space (X, d) is said to be *expansive* if $\exists \delta > 0$ with the property that if $x \neq y$ then $\exists n \in Z$ with $d(T^nx, T^ny) > \delta$. We call δ an *expansive constant* for T.

Remark. Another way to give this definition is as follows. Consider $X \times X$ with $T \times T$ acting on it. Define a metric D on $X \times X$ by $D((u,v),(x,y)) = \max\{d(u,x), d(v,y)\}$. Then T is expansive iff $\exists \delta > 0$ such that if (x, y) is not an element of the diagonal, then some power of $T \times T$ takes (x, y) out of the δ-neighbourhood of the diagonal.

The following theorem is due to Reddy [1] and Keynes and Robertson [1].

Theorem 5.22. *Let T be a homeomorphism of a compact metric space (X, d). Then T is expansive iff T has a generator iff T has a weak generator.*

PROOF. By Theorem 5.20 it suffices to show T is expansive iff T has a generator.

Let δ be an expansive constant for T and α a finite cover by open balls of radius $\delta/2$. Suppose $x, y \in \bigcap_{-\infty}^{\infty} T^{-i}\bar{A}_n$ where $A_n \in \alpha$. Then $d(T^n x, T^n y) \leq \delta$ $\forall n \in Z$ so, by assumption $x = y$. Therefore α is a generator.

Conversely, suppose α is a generator. Let δ be a Lebesgue number for α. If $d(T^n x, T^n y) \leq \delta$ $\forall n$ then $\forall n \, \exists A_n \in \alpha$ with $T^n x, T^n y \in A_n$ and so,

$$x, y \in \bigcap_{-\infty}^{\infty} T^{-n}A_n.$$

Since this intersection contains at most one point we have $x = y$. Hence T is expansive. $\qquad\square$

Corollary 5.22.1

(i) *Expansiveness is independent of the metric as long as the metric gives the topology of X. (However the expansive constant does change.)*

(ii) *If $k \neq 0$ then T is expansive iff T^k is expansive.*

(iii) *Expansiveness is a topological conjugacy invariant i.e. if, for $i = 1, 2$, $T_i : X_i \to X_i$ is a homeomorphism of a compact metrisable space and if $\phi : X_1 \to X_2$ is a homeomorphism with $\phi T_1 = T_2 \phi$ then T_1 is expansive iff T_2 is expansive.*

PROOF

(i) This is because the concept of generator does not depend on the metric.

(ii) If α is a generator for T then

$$\alpha \vee T^{-1}\alpha \vee \cdots \vee T^{-(k-1)}\alpha$$

is a generator for T^k. Also any generator for T^k is a generator for T.

(iii) A cover α is a generator for T_2 iff $\phi^{-1}\alpha$ is a generator for T_1. $\qquad\square$

Remarks. We make some more comments on how expansiveness behaves relative to natural ways of getting a new homeomorphism from an old one.

(1) If $T : X \to X$ is expansive and Y is a closed subset of X with $TY = Y$ then $T|_Y$ is expansive (i.e. a subsystem of an expansive system is expansive).

(2) If $T_i: X_i \to X_i$, $i = 1, 2$, are expansive then so is $T_1 \times T_2: X_1 \times X_2 \to X_1 \times X_2$. This extends to finite products but not to infinite products.

(3) If $T_i: X_i \to X_i$, $i = 1, 2$ are homeomorphisms and if $\phi: X_1 \to X_2$ is a continuous map of X_1 onto X_2 with $\phi T_1 = T_2 \phi$ then T_2 is said to be a factor of T_1. It is clear that if T_1 is minimal or topologically transitive or has a dense set of periodic points then any factor T_2 also has the corresponding property. However expansiveness is not preserved under the operation of taking factors as the following examples show.

EXAMPLE 1. Consider the 2-torus K^2 and identity $(z, w) \in K^2$ with $(\bar{z}, \bar{w})$. i.e. an element of K^2 is identified with its group inverse. This identification is two-to-one except at the four points $(1, 1)$, $(1, -1)$, $(-1, 1)$ and $(-1, -1)$ which are their own group inverses. The identification space with the quotient topology is homeomorphic to the 2-sphere S^2. Let $\phi: K^2 \to S^2$ be the projection. Let $A: K^2 \to K^2$ be a continuous automorphism. Since A maps equivalence classes to equivalence classes it induces a homeomorphism $T: S^2 \to S^2$. Clearly T is a factor of A. We shall see later that A is expansive if the matrix $[A]$ has no eigenvalues of unit modulus. However the homeomorphism $T: S^2 \to S^2$, induced by such an expansive $A: K^2 \to K^2$, is not expansive. To see this let us use additive notation on K^2, so the identification means $(x, y) + Z^2$ is identified with $(-x, -y) + Z^2$. Let V_s, V_u be the eigenspaces in R^2 corresponding to the eigenvalues λ_s, λ_u of the linear transformation $\tilde{A}$ where $|\lambda_s| < 1$ and $|\lambda_u| > 1$. (s denotes stable and u denotes unstable.) Let $\varepsilon > 0$ be given and choose any point (x, y) in the Euclidean ball in R^2 with centre $(0, 0)$ and radius ε. Consider the parallelogram deter-

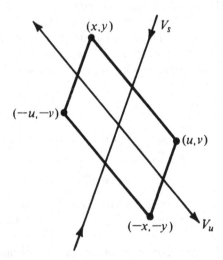

mined by the translates of V_s, V_u that go through (x, y) and those that pass through $(-x, -y)$. The other vertices have the form (u, v) and $(-u, -v)$ and all the vertices are contained in the Euclidean ball of radius $c\varepsilon$ for some constant c depending only on the slopes of V_s and V_u. Notice that $\|\tilde{A}^n(x, y) - \tilde{A}^n(u, v)\| = \lambda_u^n\|(x, y) - (u, v)\|$ if $n \le 0$ since $(u, v) - (x, y) \in V_u$, and $\|\tilde{A}^n(x, y) - \tilde{A}^n(-u, -v)\| = \lambda_s^n\|(x, y) - (-u, -v)\|$ if $n \ge 0$ since $(-u, -v) - (x, y) \in V_s$. So if d is the metric on the torus induced from $\|\cdot\|$, and if we write (x, y) rather than $(x, y) + Z^2$ as a point of K^2, we have $d(A^n(x, y), A^n(u, v)) < 2c\varepsilon$ $\forall n \le 0$ and $d(A^n(x, y), A^n(-u, -v)) < 2c\varepsilon$ $\forall n \ge 0$. If $T : S^2 \to S^2$ had a generator $\gamma = \{C_1, \ldots, C_k\}$ then $\phi^{-1}\gamma$ would have the property that each set $\bigcap_{n=-\infty}^{\infty} A^{-n}\phi^{-1}\bar{C}_{i_n}$ contains at most one equivalence class. Choose ε so that $2c\varepsilon$ is a Lebesgue number for the open cover $\phi^{-1}\gamma$ of K^2. The above shows that some set $\bigcap_{n=-\infty}^{\infty} A^{-n}\phi^{-1}\bar{C}_{i_n}$ contains the equivalence class of (x, y) and the equivalence class of (u, v), contradicting the fact that it contains at most one equivalence class.

EXAMPLE 2. Let $Tz = az$ be a minimal rotation of K. We shall represent T as a factor of a subset of the two-sided shift on two symbols. Consider the cover of K by the closed intervals (arcs) between -1 and 1 on K. Call one of them A_0 and the other A_1. If $z \in K\backslash\{a^n, -a^n : n \in Z\}$ we can uniquely asso-

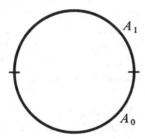

ciate a member of $\prod_{-\infty}^{\infty} \{0, 1\}$ to z by $z \to \{a_n\}_{-\infty}^{\infty}$ if $T^n z \in A_{a_n}$. Let Λ denote the subset of $\prod_{-\infty}^{\infty} \{0, 1\}$ arising in this way. Let $\psi : K\backslash\{a^n, -a^n : n \in Z\} \to \Lambda$ be the map defined above. We want to show ψ is injective and the inverse map can be extended to a continuous map $\phi : \bar{\Lambda} \to K$. We do this by proving for each $\varepsilon > 0$ there is an integer N such that if $x, y \in K\backslash\{a^n, -a^n : n \in Z\}$ and $(\psi(x))_n = (\psi(y))_n$ for $|n| \le N$ then $d(x, y) < \varepsilon$. Suppose $\varepsilon > 0$ is given. Choose $N > 0$ so that $\{1, a^{\pm 1}, a^{\pm 2}, \ldots, a^{\pm N}\}$ is $\varepsilon/2$-dense in K. Suppose $x, y \in K\backslash\{a^n, -a^n : n \in Z\}$ and $(\psi(x))_n = (\psi(y))_n$ for $|n| \le N$. We shall show $d(x, y) < \varepsilon$. The assumption $(\psi(x))_n = (\psi(y))_n$, $|n| \le N$ means that $a^n x$ and $a^n y$ belong to the same element of the cover for $|n| \le N$. If $y = -x$ then this clearly cannot happen. So suppose the counter-clockwise distance from y to x is smaller than the clockwise distance. For some n with $|n| \le N$ $a^n x$ is in the open interval of length ε starting at 1 and going counter-clockwise. Hence $a^n y$ must also be in the upper half of the circle and by the assumption about the relative positions of x and y, $a^n y$ must be between 1 and $a^n x$. Hence $d(a^n x, a^n y) < \varepsilon$ and so $d(x, y) < \varepsilon$.

Since $\psi T = S\psi$, where S denotes the shift, we have $\phi S(x) = T\phi(x)$ $\forall x \in \bar{\Lambda}$ where $\phi : \bar{\Lambda} \to K$ denotes the extension of ψ^{-1}. The continuous map ϕ is surjective because $\phi(\bar{A})$ is a closed subset of K containing the dense set $K \backslash \{a^n\}_{-\infty}^{\infty}$. We shall see later that S is an expansive homeonormphism whereas T, being an isometry, is not expansive.

We shall now show that every expansive homeomorphism is a factor of a subset of a two-sided shift.

We shall need to use the following result which is similar to half of Theorem 5.21. Recall that the diameter of a cover is the supremum of the diameters of its members.

Theorem 5.23. *Let T be an expansive homeomorphism of a compact metric space (X, d) and let δ be an expansive constant. Let γ be a finite cover of X (not necessarily an open cover) by sets $\{C_1, \ldots, C_r\}$ with $\mathrm{diam}(C_j) \leq \delta$, $1 \leq j \leq r$. Then $\mathrm{diam}(\bigvee_{j=-n}^{n} T^{-j}\gamma) \to 0$ as $n \to \infty$.*

PROOF. Suppose the conclusion is false. There exists $\varepsilon_0 > 0$ a subsequence $\{n_i\}$ of natural numbers, points x_i, y_i with $d(x_i, y_i) \geq \varepsilon_0$ and $x_i, y_i \in \bigcap_{j=-n_i}^{n_i} T^{-j}C_{i,j}$ where $C_{i,j} \in \gamma$. We can choose a subsequence $\{i_k\}$ of natural members such that $x_{i_k} \to x$ and $y_{i_k} \to y$. Hence $d(x, y) \geq \varepsilon_0$. Consider the sets $C_{i_k,0}$. Infinitely many of them coincide so for some $C_{l_0} \in \gamma$, $x_{i_k}, y_{i_k} \in C_{l_0}$ for infinitely many k. Therefore $x, y \in \bar{C}_{l_0}$. Similarly, for each j infinitely many of the sets $C_{i_k,j}$ coincide so there is some $C_{l_j} \in \gamma$ with $x, y \in T^{-j}\bar{C}_{l_j}$. Therefore $d(T^jx, T^jy) \leq \delta$ $\forall j \in Z$ and so $x = y$, contradicting $d(x, y) \geq \varepsilon_0$. $\qquad \square$

Theorem 5.24. *Let $T: X \to X$ be an expansive homeomorphism of a compact metric space. Then there is an integer $k > 0$, a closed subset Ω of*

$$X_k = \prod_{-\infty}^{\infty} \{0, 1, \ldots, k-1\}$$

such that $S\Omega = \Omega$, where S is the shift on X_k, and a continuous surjection $\phi: \Omega \to X$ such that

$$\phi S(y) = T\phi(y) \qquad y \in \Omega.$$

PROOF. The proof will resemble that of Example 2 above. Let δ be an expansive constant for T. We shall construct a cover $\gamma = \{C_0, \ldots, C_{k-1}\}$ of X by closed sets with $\mathrm{diam}(C_i) < \delta$ for each i, $C_i \cap C_j = \partial C_i \cap \partial C_j$ if $i \neq j$ and $\bigcup_{i=0}^{k-1} \partial C_i$ having no interior.

We can do this as follows. Take an open cover $\{B_0, \ldots, B_{k-1}\}$ by open balls of radius $\delta/3$. Let $C_0 = \bar{B}_0$, and for $n > 0$ let $C_n = \bar{B}_n \backslash (B_0 \cup \cdots \cup B_{n-1})$. Then if $i < j$ we have

$$C_i \cap C_j = \partial C_j \cap C_i \quad \text{(since } \mathrm{int}(C_j) \text{ is } B_j \backslash \overline{(B_0 \cup \cdots \cup B_{j-1})})$$
$$= \partial C_j \cap \partial C_i \quad \text{(since } \partial C_j \cap \mathrm{int}(C_i) \subset B_i \backslash (B_0 \cup \cdots \cup B_{j-1}) = \varnothing).$$

Also $\bigcup_{i=0}^{k-1} \partial C_i \subset \bigcup_{i=0}^{k-1} \partial B_i$ which has no interior.

Let $D = \bigcup_{i=0}^{k-1} \partial C_i$ and $D_\infty = \bigcup_{-\infty}^{\infty} T^n D$. Then D_∞ is a first category set so $X \backslash D_\infty$ is dense in X. For each $x \in X \backslash D_\infty$ we can assign, uniquely, a member of X_k by $x \to (a_n)_{-\infty}^{\infty}$ if $T^n x \in C_{a_n}$. Let Λ denote the collection of points of X_k arising in this way and let $\psi : X \backslash D_\infty \to \Lambda$ denote the map just defined. We want to show that ψ is injective and that the inverse of ψ can be extended to a continuous map $\phi : \bar{\Lambda} \to X$. This will follow if we show for each $\varepsilon > 0$ there is an integer N such that whenever $x, y \in X \backslash D_\infty$ and $(\psi(x))_n = (\psi(y))_n$ for all $|n| \leq N$ then $d(x, y) < \varepsilon$.

Let $\varepsilon > 0$ be given. Choose N so that diam $(\bigvee_{n=-N}^{N} T^n \gamma) < \varepsilon$, by Theorem 5.23. If $(\psi(x))_n = (\psi(y))_n$ for $|n| \leq N$ then x, y are in the same element of $\bigvee_{-N}^{N} T^n \gamma$ and so $d(x, y) < \varepsilon$.

Since $\psi T = S \psi$ we have $\phi S(y) = T \phi(y) \; \forall y \in \bar{\Lambda}$. The map ϕ is surjective since the dense set $X \backslash D_\infty$ is in its image. $\square$

The following gives many measure-theoretic generators for an expansive homeomorphism.

Theorem 5.25. *Let T be an expansive homeomorphism of a compact metric space (X, d) and let δ be an expansive constant for T. If $\xi = \{A_1, \ldots, A_k\}$ is a partition of X into Borel sets with diam$(A_j) \leq \delta$, $1 \leq j \leq k$, then $\bigvee_{n=-\infty}^{\infty} T^{-n} \mathscr{A}(\xi) = \mathscr{B}(X)$. Therefore, if μ is a probability measure on $(X, \mathscr{B}(X))$ for which T is measure-preserving then $h(T) = h(T, \mathscr{A}(\xi))$ (by Theorem 4.17).*

PROOF. Consider any open ball $B(x; \varepsilon)$. By Theorem 5.23 for each $n \geq 1$ choose N_n such that diam$(\bigvee_{i=-N_n}^{N_n} T^{-i} \xi) < 1/n$. Let E_n denote the union of all the members of $\bigvee_{i=-N_n}^{N_n} T^{-i} \xi$ that intersect $B(x; \varepsilon - 1/n)$. Also

$$B(x; \varepsilon - 1/n) \subset E_n \subset B(x; \varepsilon) \quad \text{so} \quad \bigcup_{n=1}^{\infty} E_n = B(x; \varepsilon).$$

Therefore $B(x; \varepsilon) \in \bigvee_{n=-\infty}^{\infty} T^{-n} \mathscr{A}(\xi)$, and since every open set is a countable union of open balls we see each open set belongs to $\bigvee_{-\infty}^{\infty} T^{-n} \mathscr{A}(\xi)$. Hence $\mathscr{B}(X) = \bigvee_{n=-\infty}^{\infty} T^{-n} \mathscr{A}(\xi)$. $\square$

The above result will be important in Chapter 8.

Let us examine some examples.

(1) Isometries are never expansive except on finite spaces. Therefore rotations on compact metrisable groups are not expansive if the group is infinite.

(2) Let A be an automorphism of the n-torus, and $[A]$ the corresponding matrix. Then A is expansive iff $[A]$ has no eigenvalues of modulus 1.

SKETCH OF PROOF. One first shows that A is expansive iff the linear map $\tilde{A}$ of R^n (that covers A) is expansive. (The definition of expansiveness does not need a compact space.) Then show that $\tilde{A}$ is expansive iff the complexification of $\tilde{A}$ is expansive. Then one shows that the complexification of $\tilde{A}$ is expansive iff the transformation given by the Jordan normal form is expansive. Lastly, one shows that the normal form is expansive iff there are no eigenvalues of modulus 1.

(*Note*: By Theorem 5.25 any partition of K^n into sufficiently small n-rectangles is a measure theoretic generator for an expansive automorphism of K^n.)

(3) The two-sided shift on k symbols is expansive. (and by Remark 1 so are all two-sided topological Markov chains).

PROOF (1). Let the state space be $\{0, 1, \ldots, k-1\}$. Let $A_i = \{\{x_n\}:x_0 = i\}$, $i = 0, 1, \ldots, k-1$. Then $A_0 \cup A_1 \cup \cdots \cup A_{k-1} = X$ and each A_i is open. The cover $\alpha = \{A_0, \ldots, A_{k-1}\}$ is a generator for the shift since if $x \in \bigcap_{-\infty}^{\infty} T^{-n} A_{i_n}$ where the $A_{i_n} \in \alpha$ then

$$x = (\ldots, i_{-2}, i_{-1}, \overset{*}{i_0}, i_1, i_2, \ldots).$$

We then use Theorem 5.22.   □

PROOF (2). Let d be the metric given by:

$$d(\{x_n\}, \{y_n\}) = \sum_{n=-\infty}^{\infty} \frac{|x_n - y_n|}{2^{|n|}}.$$

Suppose $\{x_n\} \neq \{y_n\}$. Then for some n_0, $x_{n_0} \neq y_{n_0}$ and

$$d(T^{n_0}\{x_n\}, T^{n_0}\{y_n\}) = \sum_{n=-\infty}^{\infty} \frac{1}{2^{|n|}} |x_{n+n_0} - y_{n+n_0}|$$

$$\geq |x_{n_0} - y_{n_0}| \geq 1.$$

Thus 1 is an expansive constant.   □

The last two examples show an expansive homeomorphism can have a dense set of periodic points. There are expansive homeomorphisms with no periodic points: in fact, there are expansive minimal homeomorphisms which can be chosen of the form $T|_E$ where E is a minimal set for an expansive homeomorphism T.

Expansiveness is not related to topological transitivity or the size of the non-wandering set. There is some restriction though on periodic points as the next result shows.

Theorem 5.26. *Let $T: X \to X$ be an expansive homeomorphism of a compact metric space. For each integer $p > 0$ the homeomorphism T^p has only a finite number of fixed points.*

PROOF. Let δ be an expansive constant for T^p. Suppose $T^p(x) = x$ and $T^p(y) = y$. Then either $x = y$ or $d(x, y) > \delta$.   □

The following shows there is some restriction on which spaces admit expansive homeomorphisms.

Theorem 5.27. *There are no expansive homeomorphisms of the unit circle K.*

PROOF. Suppose $T: K \to K$ is a homeomorphism. By replacing T by T^2, if necessary, we can assume T preserves orientation.

Case 1. Suppose T has a periodic point so that T^p has a fixed point for some $p > 0$. We know T is expansive iff T^p is. Suppose $T^p(w_1) = w_1$. If T^p has infinitely many fixed points then it is not expansive (Theorem 5.26). So assume T^p has only a finite number of fixed points and let w_2 be the first fixed point of T^p one reaches by going anticlockwise around the circle from w_1. (We could have $w_2 = w_1$.) Since T^p preserves orientation the anticlockwise interval from w_1 to w_2 is mapped to itself by T^p. Let z be a point in the interior of this interval and suppose $T^p(z)$ is anticlockwise from z (we use a similar proof if the opposite is true). Then $T^{np}(z) \to w_2$ as $n \to \infty$ and $T^{-np}z \to w_1$ as $n \to \infty$. Hence pairs of points in the interior of this interval which are close stay close under iteration by T^p. Therefore T^p is not expansive.

Case 2. Suppose T has no periodic points. We shall show in Theorem 6.18 that there is a continuous surjection $\phi: K \to K$ and a rotation $S: K \to K$ such that $\phi T = S \phi$ and for each $w \in K$ the set $\phi^{-1}(w)$ is either a point or closed interval. If each set $\phi^{-1}(w)$ is a point then ϕ is a homeomorphism and T is not expansive because the rotation S is not expansive. (See Example 1 above). Suppose for some w_0 the set $\phi^{-1}(w_0)$ is a closed interval of positive length. Since $\phi T = S \phi$ we know the sets $\{T^{-n}\phi^{-1}(w_0): n \in Z\}$ are mutually disjoint closed intervals. If $\delta > 0$ is given we can choose N so that if $|n| \geq N$ the length of $T^{-n}\phi(w_0)$ is less than δ. Then by continuity of T we can find two distinct points z_1, z_2 in $\phi^{-1}(w_0)$ such that $d(T^n z_1, T^n z_2) < \delta$ for $|n| \leq N$. Then $d(T^n z_1, T^n z_2) < \delta \ \forall n \in Z$ so that δ is not an expansive constant for T. Therefore T is not expansive. □

R. Mane [1] has proved that if $T: X \to X$ is an expansive homeomorphism of a compact metric space then X has finite covering dimension and any minimal set for T has zero covering dimension.

Remark. If $T: X \to X$ is a continuous map of a compact metric space (X, d) then we can define T to be positively expansive if $\exists \delta > 0$ so that $d(T^n(x), T^n(y)) \leq \delta \ \forall n \geq 0$ implies $x = y$. Similar theorems to ones above can be proved for positively expansive maps. One-sided shifts provided examples of positively expansive maps.

CHAPTER 6
Invariant Measures for Continuous Transformations

In this chapter X will denote a compact metrisable space and d will denote a metric on X. The σ-algebra of Borel subsets of X will be denoted by $\mathscr{B}(X)$. So $\mathscr{B}(X)$ is the smallest σ-algebra containing all open subsets of X and the smallest σ-algebra containing all closed subsets of X. We shall denote by $M(X)$ the collection of all probability measures defined on the measurable space $(X, \mathscr{B}(X))$. We call the members of $M(X)$ Borel probability measures on X. Each $x \in X$ determines a member δ_x of $M(X)$ defined by

$$\delta_x(A) = \begin{cases} 1 & \text{if } x \in A \\ 0 & \text{if } x \notin A. \end{cases}$$

So the map $x \to \delta_x$ imbeds X inside $M(X)$. Notice that $M(X)$ is a convex set where $pm + (1 - p)\mu$ is defined by $(pm + (1 - p)\mu)(B) = pm(B) + (1 - p)\mu(B)$ if $p \in [0, 1]$.

Our aim in this chapter is to study the invariant measures for a continuous transformation $T: X \to X$. In the first section we collect some standard facts about the set $M(X)$.

§6.1 Measures on Metric Spaces

Our first aim is to show that a member of $M(X)$ is determined by how it integrates continuous functions. This will follow from the following theorem which doesn't need the assumption of compactness of X.

146

Theorem 6.1. *A Borel probability measure m on a metric space X is regular (i.e., $\forall B \in \mathscr{B}(X)$ and $\forall \varepsilon > 0 \,\exists$ an open set U_ε and a closed set C_ε with $C_\varepsilon \subseteq B \subseteq U_\varepsilon$ and $m(U_\varepsilon \backslash C_\varepsilon) < \varepsilon$).*

PROOF. (The proof does not require X to be metric but only that each closed set be a G_δ.) Let $\mathscr{R}$ be the collection of all sets such that the regularity condition holds, i.e., $\mathscr{R} = \{A \in \mathscr{B} : \forall \varepsilon > 0 \,\exists$ open U_ε, closed C_ε with $C_\varepsilon \subseteq A \subseteq U_\varepsilon$ and $m(U_\varepsilon \backslash C_\varepsilon) < \varepsilon\}$. We show that $\mathscr{R}$ is a σ-algebra. Clearly $X \in \mathscr{R}$. Let $A \in \mathscr{R}$; we show that $X \backslash A \in \mathscr{R}$. If $\varepsilon > 0$ there are open U_ε, closed C_ε with $C_\varepsilon \subseteq A \subseteq U_\varepsilon$ such that $m(U_\varepsilon \backslash C_\varepsilon) < \varepsilon$. Thus $X \backslash U_\varepsilon \subseteq X \backslash A \subseteq X \backslash C_\varepsilon$ and $(X \backslash C_\varepsilon) \backslash (X \backslash U_\varepsilon) = U_\varepsilon \backslash C_\varepsilon$, so

$$m((X \backslash C_\varepsilon) \backslash (X \backslash U_\varepsilon)) = m(U_\varepsilon \backslash C_\varepsilon) < \varepsilon.$$

Therefore $A \backslash X \in \mathscr{R}$.

We now show $\mathscr{R}$ is closed under countable unions. Let $A_1, A_2, \ldots \in \mathscr{R}$ and let $A = \bigcup_{i=1}^\infty A_i$. Let $\varepsilon > 0$ be given. There exist open $U_{\varepsilon,n}$, closed $C_{\varepsilon,n}$ such that $C_{\varepsilon,n} \subseteq A_n \subseteq U_{\varepsilon,n}$ and $m(U_{\varepsilon,n} \backslash C_{\varepsilon,n}) < \varepsilon/3^n$. Let $U_\varepsilon = \bigcup_{n=1}^\infty U_{\varepsilon,n}$ (which is open), $\tilde{C}_\varepsilon = \bigcup_{n=1}^\infty C_{\varepsilon,n}$, and choose k such that $m(\tilde{C}_\varepsilon \backslash \bigcup_{n=1}^k C_{\varepsilon,n}) < \varepsilon/2$. Let $C_\varepsilon = \bigcup_{n=1}^k C_{\varepsilon,n}$ (which is closed). We have $C_\varepsilon \subseteq A \subseteq U_\varepsilon$. Also,

$$m(U_\varepsilon \backslash C_\varepsilon) \leq m(U_\varepsilon \backslash \tilde{C}_\varepsilon) + m(\tilde{C}_\varepsilon \backslash C_\varepsilon)$$

$$\leq \sum_{n=1}^\infty m(U_{\varepsilon,n} \backslash C_{\varepsilon,n}) + m(\tilde{C}_\varepsilon \backslash C_\varepsilon)$$

$$\leq \sum_{n=1}^\infty \frac{\varepsilon}{3^n} + \frac{\varepsilon}{2} = \varepsilon.$$

Therefore $\mathscr{R}$ is a σ-algebra.

To complete the proof we show that $\mathscr{R}$ contains all the closed subsets of X. Let C be a closed set and $\varepsilon > 0$. Define $U_n = \{x \in X : d(C, x) < 1/n\}$. This is an open set, $U_1 \supseteq U_2 \supseteq \cdots \supseteq U_n \supseteq \cdots$ and $\bigcap_{i=1}^\infty U_i = C$. Choose k such that $m(U_k \backslash C) < \varepsilon$ and let $C_\varepsilon = C$ and $U_\varepsilon = U_k$. This shows $C \in \mathscr{R}$. $\quad\square$

Corollary 6.1.1. *For a Borel probability measure m on a metric space X we have that for $B \in \mathscr{B}(X)$*

$$m(B) = \sup_{\substack{C \text{ closed} \\ C \subseteq B}} m(C), \quad \text{and} \quad m(B) = \inf_{\substack{U \text{ open} \\ U \supseteq B}} m(U).$$

The next result says that each $m \in M(X)$ is determined by how it integrates continuous functions.

Theorem 6.2. *Let m, μ be two Borel probability measures on the metric space X. If $\int_X f \, dm = \int_X f \, d\mu \; \forall f \in C(X)$ then $m = \mu$.*

PROOF. By the above corollary it suffices to show that $m(C) = \mu(C)$ for all closed sets $C \subseteq X$. Suppose C is closed and let $\varepsilon > 0$. By the regularity of m there exists an open set U with $C \subset U$ and $m(U\backslash C) < \varepsilon$.

Define $f : X \to R$ by

$$f(x) = \begin{cases} 0 & \text{if } x \notin U \\ \dfrac{d(x, X\backslash U)}{d(x, X\backslash U) + d(x, C)} & \text{if } x \in U. \end{cases}$$

This is well-defined since the denominator is not zero. Also f is continuous, $f = 0$ on $X\backslash U$, $f = 1$ on C, and $0 \le f(x) \le 1 \; \forall x \in X$. Hence

$$\mu(C) \le \int_X f \, d\mu = \int_X f \, dm \le m(U) < m(C) + \varepsilon.$$

Therefore $\mu(C) < m(C) + \varepsilon \quad \forall \varepsilon > 0$, so $\mu(C) \le m(C)$. By symmetry we get that $m(C) \le \mu(C)$. $\square$

The next theorem relates elements of $M(X)$ to linear functionals on $C(X)$.

Theorem 6.3 (Riesz Representation Theorem). *Let X be a compact metric space and $J : C(X) \to C$ a continuous linear map such that J is a positive operator (i.e., if $f \ge 0$ then $J(f) \ge 0$) and $J(1) = 1$. Then there exists $\mu \in M(X)$ such that $J(f) = \int_X f \, d\mu \; \forall f \in C(X)$.*

For the proof see Parthasarathy [2] p. 145.

Therefore the map $\mu \to J$ is a bijection between $M(X)$ and the collection of all normalised positive linear functionals on $C(X)$. (Injectivity follows from Theorem 6.2 and surjectivity by Theorem 6.3.) We shall denote the image of μ under this map by J_μ. Clearly this bijection is an affine map (i.e. $J_{p\mu + (1-p)m} = pJ_\mu + (1-p)J_m$, $p \in [0, 1]$, $m, \mu \in M(X)$) so $M(X)$ is identified with a convex subset of the unit ball in $C(X)^*$. This allows us to get a topology on $M(X)$ from the weak* topology on $C(X)^*$.

Definition 6.1. The *weak* topology* on $M(X)$ is the smallest topology making each of the maps $\mu \to \int_X f \, d\mu$ ($f \in C(X)$) continuous. A basis is given by the collection of all sets of the form $V_\mu(f_1, \ldots, f_k; \varepsilon) = \{m \in M(X) \, | \, |\int f_i \, dm - \int f_i \, d\mu| < \varepsilon, 1 \le i \le k\}$ where $\mu \in M(X)$, $k \ge 1$, $f_i \in C(X)$ and $\varepsilon > 0$.

Clearly this topology on $M(X)$ is independent of any metric chosen on X.

Theorem 6.4. *If X is a compact metrisable space then the space $M(X)$ is metrisable in the weak* topology. If $\{f_n\}_{n=1}^\infty$ is a dense subset of $C(X)$ then*

$$D(m, \mu) = \sum_{n=1}^\infty \frac{|\int f_n \, dm - \int f_n \, d\mu|}{2^n \|f_n\|}$$

is a metric on $M(X)$ giving the weak* topology.

PROOF. The function $D: M(X) \times M(X) \to R$ is clearly a metric. Consider the metric space $(M(X), D)$. For each fixed i the map $\mu \to \int f_i \, d\mu$ is clearly continuous on (X, D) because $|\int f_i \, dm - \int f_i \, d\mu| \leq 2^i \|f_i\| D(m, \mu)$. Since $\{f_i\}_1^\infty$ is dense in $C(X)$ it follows that for each $f \in C(X)$ the map $\mu \to \int f \, d\mu$ is continuous on $(M(X), D)$. Therefore every open set in the weak* topology is open in the metric space $(M(X), D)$. To show the converse it will suffice to prove each ball $\{m \in M(X) \,|\, D(m, \mu) < \varepsilon\}$ in $(M(X), D)$ contains a set $V_\mu(g_1, \ldots, g_k; \delta)$ where $k \geq 1, g_i \in C(X)\, 1 \leq i \leq k$, and $\delta > 0$. If $\mu \in M(X)$ and $\varepsilon > 0$ are given, choose N so that

$$\sum_{n=N+1}^{\infty} \frac{2}{2^n} < \frac{\varepsilon}{2}.$$

Let

$$\delta = \frac{\varepsilon}{2}\left(\sum_{n=1}^{N} \frac{1}{2^n \|f_n\|}\right)^{-1}.$$

Then $V_\mu(f_1, \ldots, f_N; \delta) \subset \{m \in M(X) \,|\, D(m, \mu) < \varepsilon\}$. $\qquad\square$

Remarks

(1) In the weak* topology $\mu_n \to \mu$ in $M(X)$ iff $\forall f \in C(X)\, \int f \, d\mu_n \to \int f \, d\mu$.
(2) The imbedding $X \to M(X)$ given by $x \to \delta_x$ is continuous.
(3) If $\mu_n, \mu \in M(X), n \geq 1$, one can prove the following are equivalent.

(i) $\mu_n \to \mu$ in the weak*-topology

(ii) For each closed subset F of X, $\limsup\limits_{n \to \infty} \mu_n(F) \leq \mu(F)$.

(iii) For each open subset U of X, $\liminf\limits_{n \to \infty} \mu_n(U) \geq \mu(U)$.

(iv) For every $A \in \mathcal{B}$ with $\mu(\partial A) = 0$, $\mu_n(A) \to \mu(A)$.

We shall want to use (i) $\Rightarrow$ (iv) so we give a proof. We shall in fact show that (i) $\Rightarrow$ (ii) and then (ii) $\Rightarrow$ (iii) and (iv). Let F be a closed subset of X and for $k \geq 1$ let $U_k = \{x \in X \,|\, d(x, F) < 1/k\}$. The sets U_k are open and decrease to F. Therefore $\mu(U_k) \to \mu(F)$. By Urysohn's lemma choose $f_k \in C(X)$ with $0 \leq f_k \leq 1, f_k = 1$ on F and $f_k = 0$ on $X \backslash U_k$. Then

$$\limsup_{n \to \infty} \mu_n(F) \leq \limsup_{n \to \infty} \int f_k \, d\mu_n = \int f_k \, d\mu \leq \mu(U_k)$$

so $\limsup_{n \to \infty} \mu_n(F) \leq \mu(F)$. Therefore (i) $\Rightarrow$ (ii). Suppose (ii) is true and let U be an open subset of X. Then

$$\limsup_{n \to \infty} \mu_n(X \backslash U) \leq \mu(X \backslash U) \quad \text{so } \liminf_{n \to \infty} \mu_n(U) \geq \mu(U).$$

Therefore (ii) $\Rightarrow$ (iii). Suppose (ii) is true and $\mu(\partial A) = 0$. Then $\mu(\text{int}(A)) = \mu(A) = \mu(\bar{A})$ and $\limsup_{n \to \infty} \mu_n(\bar{A}) \leq \mu(\bar{A}) = \mu(A)$ and $\liminf_{n \to \infty} \mu_n(\text{int}(A)) \geq \mu(\text{int}(A)) = \mu(A)$. Therefore $\mu_n(A) \to \mu(A)$. We have shown (i) $\Rightarrow$ (iv). $\qquad\square$

(See Parthasarathy [1] for the proofs of the other implications.)

The following important result is an easy consequence of the compactness of the unit ball of $C(X)^*$ in the weak*-topology, but we give a direct proof.

Theorem 6.5. *If X is a compact metrisable space then $M(X)$ is compact in the weak*-topology.*

PROOF. We shall write $\mu(f)$ instead of $\int f\,d\mu$. Let $\{\mu_n\}_1^\infty$ be a sequence in $M(X)$ and we shall show it has a convergent subsequence.

Choose $f_1, f_2, \ldots$ dense in $C(X)$. Consider the sequence of complex numbers $\{\mu_n(f_1)\}$. This is bounded by $\|f_1\|$, and so has a convergent subsequence, say $\{\mu_n^{(1)}(f_1)\}$. Consider the sequence of numbers $\{\mu_n^{(2)}(f_2)\}$; this is bounded and so has a convergent subsequence $\{\mu_n^{(2)}(f_2)\}$. Notice that $\{\mu_n^{(2)}(f_1)\}$ also converges. We proceed in this manner, and for each $i \geq 1$, construct a subsequence $\{\mu_n^{(i)}\}$ of $\{\mu_n\}$ such that $\{\mu_n^{(i)}\} \subseteq \{\mu_n^{(i-1)}\} \subseteq \cdots \subseteq \{\mu_n^{(1)}\} \subseteq \{\mu_n\}$, and so that $\{\mu_n^{(i)}(f)\}$ converges for $f = f_1, f_2, \ldots, f_i$. Consider the diagonal $\{\mu_n^{(n)}\}$. The sequence $\{\mu_n^{(n)}(f_i)\}$ converges for all i; thus $\{\mu_n^{(n)}(f)\}$ converges for all $f \in C(X)$ (by an easy approximation argument). Let $J(f) = \lim_{n\to\infty}\mu_n^{(n)}(f)$. Clearly $J:C(X) \to \mathbb{C}$ is linear and bounded, as $|J(f)| \leq \|f\|$. Also $J(1) = 1$, and if $f \geq 0$ then $J(f) \geq 0$. By Theorem 6.3, there exists a Borel probability measure μ on X such that $J(f) = \int_X f\,d\mu$ for all $f \in C(X)$, i.e.,

$$\int_X f\,d\mu_n^{(n)} \to \int_X f\,d\mu. \qquad \square$$

Hence $M(X)$ is a compact convex metrisable space and this will allow us to use the fixed point theorems valid for maps of such spaces.

§6.2. Invariant Measures for Continuous Transformations

Let $T:X \to X$ be a continuous transformation of the compact metrisable space X. We shall show in this section that there is always some $\mu \in M(X)$ for which T is a measure-preserving transformation of $(X, \mathscr{B}(X), \mu)$.

We first notice that $T^{-1}\mathscr{B}(X) \subset \mathscr{B}(X)$ (i.e. T is measurable) because $\{E \in \mathscr{B}(X) | T^{-1}E \in \mathscr{B}(X)\}$ is a σ-algebra and contains the open sets. Therefore we have a map $\tilde{T}:M(X) \to M(X)$ given by $(\tilde{T}\mu)(B) = \mu(T^{-1}B)$. We sometimes write $\mu \circ T^{-1}$ instead of $\tilde{T}\mu$. We shall need the following.

Lemma 6.6

$$\int f\,d(\tilde{T}\mu) = \int f \circ T\,d\mu \quad \forall f \in C(X).$$

PROOF. It suffices to deal with real-valued $f \in C(X)$. By definition of $\tilde{T}$ we have $\int \chi_B\,d(\tilde{T}\mu) = \int \chi_B \circ T\,d\mu \; \forall B \in \mathscr{B}(X)$. Therefore $\int h\,d(\tilde{T}\mu) = \int h \circ T\,d\mu$ if h is a simple function. The same formula holds when h is a non-negative measurable function, by choosing an increasing sequence of simple functions converging pointwise to h. Therefore the formula holds for any continuous $f: X \to R$ by considering the positive and negative part of f. □

Theorem 6.7. *The map* $\tilde{T}: M(X) \to M(X)$ *is continuous and affine.*

PROOF. If $f \in C(X)$ then $\int f \, d\tilde{T}\mu = \int f \circ T \, d\mu$. Therefore if $\mu_n \to \mu$ in $M(X)$ then $\int f \, d\tilde{T}\mu_n = \int f \circ T \, d\mu_n \to \int f \circ T \, d\mu = \int f \, d\tilde{T}\mu$ and so $\tilde{T}\mu_n \to \tilde{T}\mu$. This proves $\tilde{T}$ is continuous.

If $m, \mu \in M(X)$ and $p \in [0,1]$ then $\tilde{T}(pm + (1-p)\mu)(B) = pm(T^{-1}B) + (1-p)\mu(T^{-1}B) = (p\tilde{T}m + (1-p)\tilde{T}\mu)(B) \; \forall B \in \mathscr{B}(X)$. This shows $\tilde{T}$ is affine. □

We are interested in those members of $M(X)$ that are invariant measures for T.

Let $M(X, T) = \{\mu \in M(X) \,|\, \tilde{T}\mu = \mu\}$. This set consists of all $\mu \in M(X)$ making T a measure-preserving transformation of $(X, \mathscr{B}(X), \mu)$.

Theorem 6.8. *If* $T: X \to X$ *is continuous and* $\mu \in M(X)$ *then* $\mu \in M(X, T)$ *iff* $\int f \circ T \, d\mu = \int f \, d\mu \quad \forall f \in C(X)$.

PROOF. This is immediate from Lemma 6.6 and Theorem 6.2. □

Since $\tilde{T}: M(X) \to M(X)$ is a continuous affine map of a convex compact subset of $C(X)^*$ we could use the Markov-Kakutani theorem (see Dunford and Schwartz [1], p. 456) to show $\tilde{T}$ has a fixed point. However we will show directly that $M(X, T)$ is non-empty. The following gives us a method of constructing members of $M(X, T)$.

Theorem 6.9. *Let* $T: X \to X$ *be continuous. If* $\{\sigma_n\}_{n=1}^\infty$ *is a sequence in* $M(X)$ *and we form the new sequence* $\{\mu_n\}_{n=1}^\infty$ *by* $\mu_n = (1/n) \sum_{i=0}^{n-1} \tilde{T}^i \sigma_n$ *then any limit point* μ *of* $\{\mu_n\}$ *is a member of* $M(X, T)$. *(Such limit points exist by the compactness of* $M(X)$.)

PROOF. Let $\mu_{n_j} \to \mu$ in $M(X)$. Let $f \in C(X)$. Then

$$\left| \int f \circ T \, d\mu - \int f \, d\mu \right| = \lim_{j \to \infty} \left| \int f \circ T \, d\mu_{n_j} - \int f \, d\mu_{n_j} \right|$$

$$= \lim_{j \to \infty} \left| \frac{1}{n_j} \int \sum_{i=0}^{n_j - 1} (f \circ T^{i+1} - f \circ T^i) \, d\sigma_{n_j} \right|$$

$$= \lim_{j \to \infty} \left| \frac{1}{n_j} \int (f \circ T^{n_j} - f) \, d\sigma_{n_j} \right|$$

$$\leq \lim_{j \to \infty} \frac{2\|f\|}{n_j} = 0.$$

Therefore $\mu \in M(X, T)$. □

Corollary 6.9.1 (Krylov and Bogolioubov). *If* $T: X \to X$ *is a continuous transformation of a compact metric space* X *then* $M(X, T)$ *is non-empty.*

PROOF. We can make any choice for σ_n in Theorem 6.9; in particular choose $y \in X$ and put $\sigma_n = \delta_y$ for each n.   ☐

We have the following properties of $M(X, T)$.

Theorem 6.10. *If* T *is a continuous transformation of the compact metric space* X *then*

(i) $M(X, T)$ *is a compact subset of* $M(X)$.

(ii) $M(X, T)$ *is convex.*

(iii) μ *is an extreme point of* $M(X, T)$ *iff* T *is an ergodic measure-preserving transformation of* $(X, \mathscr{B}(X), \mu)$.

(iv) *If* $\mu, m \in M(X, T)$ *are both ergodic and* $m \neq \mu$ *then they are mutually singular.*

PROOF

(i) Suppose $\{\mu_n\}_1^\infty$ is a sequence of members of $M(X, T)$ and $\mu_n \to \mu$ in $M(X)$. Then $\int f d T \mu = \int f \circ T \, d\mu = \lim_{n \to \infty} \int f \circ T \, d\mu_n = \lim_{n \to \infty} \int f \, d\mu_n = \int f \, d\mu$, so $\mu \in M(X, T)$.

(ii) This is clear.

(iii) Suppose $\mu \in M(X, T)$ and μ is not ergodic. There exists a Borel set E such that $T^{-1}E = E$ and $0 < \mu(E) < 1$. Define measures μ_1 and μ_2 by

$$\mu_1(B) = \frac{\mu(B \cap E)}{\mu(E)} \quad \text{and} \quad \mu_2(B) = \frac{\mu(B \cap (X \backslash E))}{\mu(X \backslash E)}, \qquad B \in \mathscr{B}(X).$$

Note that μ_1 and μ_2 are in $M(X, T)$, $\mu_1 \neq \mu_2$, and

$$\mu(B) = \mu(E)\mu_1(B) + (1 - \mu(E))\mu_2(B).$$

Therefore μ is not an extreme point of $M(X, T)$.

Conversely, suppose $\mu \in M(X, T)$ is ergodic, and

$$\mu = p\mu_1 + (1 - p)\mu_2$$

where $\mu_1, \mu_2 \in M(X, T)$ and $p \in [0, 1]$. We must show $\mu_1 = \mu_2$. Clearly $\mu_1 \ll \mu$ (μ_1 is absolutely continuous with respect to μ) so that the Radon–Nikodym derivative $d\mu_1/d\mu$ exists, (i.e.,

$$\mu_1(E) = \int_E \frac{d\mu_1(x)}{d\mu} \, d\mu(x), \qquad \forall E \in \mathscr{B}(X).$$

See Theorem 0.10). We have $d\mu_1/d\mu \geq 0$. Let

$$E = \left\{ x \left| \frac{d\mu_1}{d\mu}(x) < 1 \right. \right\}.$$

We have

$$\int_{E\cap T^{-1}E} \frac{d\mu_1}{d\mu}\, d\mu + \int_{E\backslash T^{-1}E} \frac{d\mu_1}{d\mu}\, d\mu = \mu_1(E) = \mu_1(T^{-1}E)$$

$$= \int_{E\cap T^{-1}E} \frac{d\mu_1}{d\mu}\, d\mu + \int_{T^{-1}E\backslash E} \frac{d\mu_1}{d\mu}\, d\mu$$

so that

$$\int_{E\backslash T^{-1}E} \frac{d\mu_1}{d\mu}\, d\mu = \int_{T^{-1}E\backslash E} \frac{d\mu_1}{d\mu}\, d\mu.$$

Since $d\mu_1/d\mu < 1$ on $E\backslash T^{-1}E$ and $d\mu_1/d\mu \geq 1$ on $T^{-1}E\backslash E$ and since $\mu(T^{-1}E\backslash E) = \mu(T^{-1}E) - \mu(T^{-1}E \cap E) = \mu(E) - \mu(T^{-1}E \cap E) = \mu(E\backslash T^{-1}E)$ we have $\mu(E\backslash T^{-1}E) = 0 = \mu(T^{-1}E\backslash E)$. Therefore $\mu(T^{-1}E \triangle E) = 0$ so $\mu(E) = 0$ or 1. If $\mu(E) = 1$ then $\mu_1(X) = \int_E (d\mu_1/d\mu)\, d\mu < \mu(E) = 1$ contradicting $\mu_1(X) = 1$. Hence we must have $\mu(E) = 0$.

Similarly if $F = \{x | d\mu_1/d\mu > 1\}$ we have $\mu(F) = 0$ so that $d\mu_1/d\mu = 1$ a.e. (μ). Hence $\mu_1 = \mu$ and so μ is an extreme point of $M(X, T)$.

(iv) By the Lebesgue decomposition (Theorem 0.11) there are unique probability measures μ_1, μ_2 and a unique $p \in [0, 1]$ such that $\mu = p\mu_1 + (1 - p)\mu_2$ where $\mu_1 \ll m$ and μ_2 is singular with respect to m. But $\mu = \tilde{T}\mu = p\tilde{T}\mu_1 + (1 - p)\tilde{T}\mu_2$ and since $\tilde{T}\mu_1 \ll \tilde{T}m = m$ and $\tilde{T}\mu_2$ is singular with respect to $\tilde{T}m = m$ the uniqueness of the decomposition imply $\mu_1, \mu_2 \in M(X, T)$. Since μ is an extreme point we must have either $p = 0$ or $p = 1$.

If $p = 0$ then $\mu = \mu_2$ and so μ is singular relative to m. If $p = 1$ then $\mu \ll m$ and we can argue with $d\mu/dm$ as in (iii) to get $\mu = m$, a contradiction. □

Remarks

(1) The second part of the proof of (iii) shows that if $\mu_1, \mu \in M(X, T)$, $\mu_1 \ll \mu$ and μ is ergodic then $\mu_1 = \mu$.

(2) Since $M(X, T)$ is a compact convex set we can use the Choquet representation theorem to express each member of $M(X, T)$ in terms of the ergodic members of $M(X, T)$. If $E(X, T)$ denotes the set of extreme points of $M(X, T)$ then for each $\mu \in M(X, T)$ there is a unique measure τ on the Borel subsets of the compact metrisable space $M(X, T)$ such that $\tau(E(X, T)) = 1$ and $\forall f \in C(X)$

$$\int_X f(x)\, d\mu(x) = \int_{E(X,T)} \left(\int_X f(x)\, dm(x) \right) d\tau(m).$$

We write $\mu = \int_{E(X,T)} m\, d\tau(m)$ and call this the ergodic decomposition of μ. See Phelps [1]. Hence every $\mu \in M(X, T)$ is a generalised convex combination of ergodic measures. This is related to the decomposition of a measure-preserving transformation into ergodic transformations (see §1.5).

We shall now interpret ergodicity and the mixing conditions in terms of the weak*-topology on $M(X, T)$.

§6.3 Interpretation of Ergodicity and Mixing

Let $T: X \to X$ be a continuous transformation of a compact metric space. We say $\mu \in M(X, T)$ is ergodic or weak-mixing or strong mixing if the measure-preserving transformation T of the measure space $(X, \mathscr{B}(X), \mu)$ has the corresponding property.

Recall that μ is ergodic iff $\forall f, g \in L^2(\mu)$

$$\frac{1}{n} \sum_{i=0}^{n-1} \int f(T^i x) g(x) \, d\mu(x) \to \int f \, d\mu \int g \, d\mu.$$

We change this slightly for our needs.

Lemma 6.11. *Let $\mu \in M(X, T)$. Then*

(i) *μ is ergodic iff $\forall f \in C(X) \ \forall g \in L^1(\mu)$*

$$\frac{1}{n} \sum_{i=0}^{n-1} \int f(T^i x) g(x) \, d\mu(x) \to \int f \, d\mu \int g \, d\mu.$$

(ii) *μ is strong mixing iff $\forall f \in C(X) \ \forall g \in L^1(\mu)$*

$$\int f(T^i x) g(x) \, d\mu(x) \to \int f \, d\mu \int g \, d\mu.$$

(iii) *μ is weak mixing iff there is a set J of natural numbers of density zero such that $\forall f \in C(X) \ \forall g \in L^1(\mu)$*

$$\lim_{J \not\ni n \to \infty} \int f(T^i x) g(x) \, d\mu(x) \to \int f \, d\mu \int g \, d\mu.$$

PROOF

(i) Suppose the convergence condition holds and let $F, G \in L^2(\mu)$. Then

$$G \in L^1(\mu) \quad \text{so } \frac{1}{n} \sum_{i=0}^{n-1} \int f(T^i x) G(x) \, d\mu(x) \to \int f \, d\mu \int G \, d\mu \quad \forall f \in C(X).$$

Now approximate F in $L^2(\mu)$ by continuous functions to get

$$\frac{1}{n} \sum_{i=0}^{n-1} \int F(T^i x) G(x) \, d\mu(x) \to \int F \, d\mu \int G \, d\mu.$$

Now suppose μ is ergodic. Let $f \in C(X)$. Then $f \in L^2(\mu)$ so if $h \in L^2(\mu)$ we have

$$\frac{1}{n} \sum_{i=0}^{n-1} \int f(T^i x) h(x) \, d\mu(x) \to \int f \, d\mu \int h \, d\mu.$$

If $g \in L^1(\mu)$ then by approximating g in $L^1(\mu)$ by $h \in L^2(\mu)$ we obtain

$$\frac{1}{n} \sum_{i=0}^{n-1} \int f(T^i x) g(x) \, d\mu(x) \to \int f \, d\mu \int g \, d\mu.$$

The proofs of (ii) and (iii) are similar and use Theorem 1.23.   □

Theorem 6.12. *Let T be a continuous transformation of a compact metric space. Let $\mu \in M(X, T)$.*

(i) *μ is ergodic iff whenever $m \in M(X)$ and $m \ll \mu$ then*

$$\frac{1}{n} \sum_{i=0}^{n-1} \tilde{T}^i m \to \mu.$$

(ii) *μ is strong mixing iff whenever $m \in M(X)$ and $m \ll \mu$ then $\tilde{T}^n m \to \mu$.*

(iii) *u is weak mixing iff there exists a set J of natural numbers of density zero such that whenever $m \in M(X)$ and $m \ll \mu$ then $\lim_{J \not\ni n \to \infty} \tilde{T}^n m \to \mu$.*

PROOF

(i) We use the ergodicity condition of Lemma 6.11.

Let μ be ergodic and suppose $m \ll \mu$, $m \in M(X)$. Let $g = dm/d\mu \in L^1(\mu)$. If $f \in C(X)$ then

$$\int f d \left(\frac{1}{n} \sum_{i=0}^{n-1} \tilde{T}^i m \right) = \frac{1}{n} \sum_{i=0}^{n-1} \int f \circ T^i \, dm = \frac{1}{n} \sum_{i=0}^{n-1} \int f(T^i x) g(x) \, d\mu(x)$$

$$\to \int f \, d\mu \int g \, d\mu = \int f \, d\mu \int 1 \, dm = \int f \, d\mu.$$

Therefore $(1/n) \sum_{i=0}^{n-1} \tilde{T}^i m \to \mu$.

We now show the converse.

Suppose the convergence condition holds. Let $g \in L^1(\mu)$ and $g \geq 0$. Define $m \in M(X)$ by $m(B) = c \int_B g \, d\mu$ where $c = 1/(\int_X g \, d\mu)$. Then if $f \in C(X)$ we have, by reversing the above reasoning,

$$\frac{1}{n} \sum_{i=0}^{n-1} \int f(T^i x) g(x) \, d\mu(x) \to \int f \, d\mu \int g \, d\mu.$$

If $g \in L^1(\mu)$ is real-valued then $g = g_+ - g_-$ where $g_+, g_- \geq 0$ and we apply the above to g_+ and g_- to get the desired condition of Lemma 6.11 for g. The case of complex-valued g follows.

The proofs of (ii) and (iii) are similar to the above and use the corresponding parts of Lemma 6.11.   □

We now give an interpretation of the ergodic theorem for $\mu \in M(X, T)$.

Lemma 6.13. *If $T: X \to X$ is continuous and $\mu \in M(X, T)$ is ergodic then there exists $Y \in \mathscr{B}(X)$ with $\mu(Y) = 1$ such that*

$$\lim_{n \to \infty} \frac{1}{n} \sum_{i=0}^{n-1} f(T^i x) = \int f \, d\mu \quad \forall x \in Y, \quad \forall f \in C(X).$$ □

PROOF. Choose a countable dense subset $\{f_k\}_1^\infty$ of $C(X)$. By the ergodic theorem there is $X_k \in \mathscr{B}(X)$ such that $\mu(X_k) = 1$ and

$$\lim_{n \to \infty} \frac{1}{n} \sum_{i=0}^{n-1} f_k(T^i x) = \int f_k \, d\mu \quad \forall x \in X_k.$$

Put $Y = \bigcap_{k=1}^\infty X_k$. We have $\mu(Y) = 1$ and

$$\lim_{n \to \infty} \frac{1}{n} \sum_{i=0}^{n-1} f_k(T^i x) = \int f_k \, d\mu \quad \forall x \in Y$$

and each $k \geq 1$. The result now follows by approximating a given $f \in C(X)$ by members of $\{f_k\}_1^\infty$. $\qquad \square$

Theorem 6.14. *Let $T : X \to X$ be continuous and $\mu \in M(X, T)$. Then μ is ergodic iff*

$$\frac{1}{n} \sum_{i=0}^{n-1} \delta_{T^i x} \to \mu \text{ a.e.}$$

PROOF. If μ is ergodic then Lemma 6.13 says $(1/n) \sum_{i=0}^{n-1} \delta_{T^i x} \to \mu \ \forall x \in Y$, where $\mu(Y) = 1$.

Conversely suppose

$$\frac{1}{n} \sum_{i=0}^{n-1} \delta_{T^i x} \to \mu \quad \text{for } x \in Y \text{ and } \mu(Y) = 1.$$

Then

$$\frac{1}{n} \sum_{i=0}^{n-1} f(T^i x) \to \int f \, d\mu \quad \forall x \in Y \quad \forall f \in C(X).$$

If $x \in Y$, $f \in C(X)$ and $g \in L^1(\mu)$ then

$$\frac{1}{n} \sum_{i=0}^{n-1} f(T^i x) g(x) \to g(x) \int f \, d\mu$$

so applying the dominated convergence theorem yields the ergodicity condition of Lemma 6.11. $\qquad \square$

§6.4 Relation of Invariant Measures to Non-wandering Sets, Periodic Points and Topological Transitivity

We have mentioned earlier the following result about non-wandering sets.

Theorem 6.15. *Let $T : X \to X$ be a continuous transformation of a compact metrisable space X.*

(i) *For every $\mu \in M(X, T)$ we have $\mu(\Omega(T)) = 1$.*

(ii) *If there is some $\mu \in M(X, T)$ giving non-zero measure to every non-empty open set then $\Omega(T) = X$.*

PROOF

(i) Let $\{U_n\}_1^\infty$ be a base for the topology. Then $X\backslash\Omega(T)$ is the union of those U_n such that the sets $U_n, T^{-1}U_n, T^{-2}U_n, \ldots$ are pairwise disjoint. Such a set U_n must have measure zero for any invariant measure so $\mu(X\backslash\Omega(T)) = 0\ \forall\mu \in M(X, T)$.

(ii) If $U, T^{-1}U, T^{-2}U, \ldots$ are pairwise disjoint then $\mu(U) = 0\ \forall\mu \in M(X, T)$. Therefore if there is some $\mu \in M(X, T)$ giving non-zero measure to each non-empty open set then $\Omega(T) = X$. $\square$

Corollary 6.15.1. *If $\Omega_\infty(T)$ denotes the centre of T then $\mu(\Omega_\infty(T)) = 1$ for all $\mu \in M(X, T)$.*

PROOF. By Theorem 6.15(i) we can naturally identify $M(X, T)$ with $M(\Omega(T), T|_{\Omega(T)})$. Applying Theorem 6.15(i) to $T|_{\Omega(T)}$ gives $\mu(\Omega_2(T)) = 1\ \forall\mu \in M(X, T)$. By induction we have $\mu(\Omega_n(T)) = 1\ \forall\mu \in M(X, T)\ \forall n \geq 1$, so the result follows. $\square$

Theorem 6.15(i) and Corollary 6.15.1 help us to find $M(X, T)$ for some examples. When $T:K \to K$ is the north-south map we know $\Omega(T) = \{N, S\}$ so that we conclude $M(K, T) = \{p\delta_N + (1 - p)\delta_S | p \in [0, 1]\}$. The ergodic invariant measures are δ_N, δ_S. Theorem 6.15(ii) can sometimes be used to calculate $\Omega(T)$. We used it to show $\Omega(T) = G$ when $T:G \to G$ is an affine transformation of a compact metrisable group G. This is because T preserves Haar measure which is non-zero on non-empty open sets.

Remark. We can strengthen Theorem 6.15(i) as follows. A point $x \in X$ is said to be recurrent for T if there is a sequence of natural numbers $\{n_i\}$ with $n_i \nearrow \infty$ and $T^{n_i}(x) \to x$. If $R(T)$ denotes the collection of all recurrent points for T then $\mu(R(T)) = 1$ for all $\mu \in M(X, T)$. (It is clear that $R(T) \subset \Omega(T)$.) We can deduce this from the Poincaré recurrence theorem (Theorem 1.4). If $\{U_n\}_1^\infty$ is a base for the topology then $X\backslash R(T) = \bigcup_{n=1}^\infty (U_n \cap \bigcap_{k=1}^\infty T^{-k}(X\backslash U_n))$ and for each n and each $\mu \in M(X, T)$ we have $\mu(U_n \cap \bigcap_{k=1}^\infty T^{-k}(X\backslash U_n)) = 0$ by the Poincaré recurrence theorem.

We now consider the connection between periodic points and invariant measures. The following result implies that we can consider periodic points as contained in $M(X, T)$.

Theorem 6.16. *Let $T:X \to X$ be a continuous transformation of a compact metrisable space X. Let $N \geq 1$ and $x \in X$. Then $T^N(x) = x$ iff*

$$\frac{1}{N} \sum_{i=0}^{N-1} \delta_{T^i(x)} \in M(X, T).$$

If $\mu \in M(X, T)$ is purely atomic then μ is a (possibly countably infinite) convex combination of these "periodic orbit" atomic measures.

PROOF. If $\mu \in M(X)$ then $\mu \in M(X, T)$ iff $\int f \circ T \, d\mu = \int f \, d\mu \; \forall f \in C(X)$. Therefore

$$\frac{1}{N} \sum_{i=0}^{N-1} \delta_{T^i(x)} \in M(X, T) \quad \text{iff} \quad \frac{1}{N} \sum_{i=0}^{N-1} f(T^{i+1}x) = \frac{1}{N} \sum_{i=0}^{N-1} f(T^i x) \quad \forall f \in C(X).$$

This is equivalent to $f(T^N x) = f(x) \; \forall f \in C(X)$, which is equivalent to $T^N x = x$. The first part of the theorem follows from this. Now suppose $\mu \in M(X, T)$ is purely atomic. Then $\mu = \sum_{i=1}^{\infty} p_i \delta_{x_i}$ where $x_i \in X$ and $p_i > 0$, $\sum_1^{\infty} p_i = 1$. We can assume that $p_1 \geq p_2 \geq p_3 \geq \cdots$. Since $\int f \circ T \, d\mu = \int f \, d\mu \; \forall f \in C(X)$ we have $\sum_{i=1}^{\infty} p_i f(T(x_i)) = \sum_{i=1}^{\infty} p_i f(x_i)$ so that $T(x_1)$ is an atom with measure p_1. From this we get that $T^{N_1}(x_1) = x_1$ for some N_1 and each atom $T^i(x_1)$ has the same measure. The desired result follows from a simple induction. $\square$

Hence $M(X, T)$ can be considered to contain the periodic orbits. Some people like to think of an invariant probability for a continuous transformation as a generalisation of a periodic orbit.

Theorems 5.11 and 6.16 tell us that for an automorphism A of an n-torus K^n the set $M(K^n, A)$ contains many atomic measures. This is also true for the shift transformations (by Theorem 5.12).

We have already proved the following results (Theorems 5.15, 5.16) relating invariant measures to topological transitivity. If $T: X \to X$ is a homeomorphism of a compact metric space X and if there exists $\mu \in M(X, T)$ which is ergodic and gives non-zero measure to each non-empty open set then $\mu(\{x \in X \mid \overline{O_T(x)} = X\}) = 1$. Similarly, if such a $\mu \in M(X, T)$ exists for a continuous $T: X \to X$ with $TX = X$ then $\mu(\{x \in X \mid \{T^n(x)\}_0^{\infty}$ is dense in $X\}) = 1$. From these results we concluded that ergodic affine transformations are one-sided topologically transitive.

§6.5 Unique Ergodicity

In this section we study those transformations for which $M(X, T)$ is as small as possible i.e., contains only one member. It turns out that this is equivalent to strong behaviour in the ergodic theorem.

Definition 6.2. A continuous transformation $T: X \to X$ is a compact metrisable space X is called *uniquely ergodic* if there is only one T invariant Borel probability measure on X, i.e., $M(X, T)$ consists of one point.

If T is uniquely ergodic and $M(X, T) = \{\mu\}$ then μ is ergodic because it is an extreme point of $M(X, T)$ (Theorem 6.10(iii)).

Unique ergocity is connected to minimality by:

Theorem 6.17. *Suppose* $T: X \to X$ *is a homeomorphism of the compact metrisable space* X. *Suppose* T *is uniquely ergodic and* $M(X, T) = \{\mu\}$. *Then* T *is minimal iff* $\mu(U) > 0$ *for all non-empty open sets* U.

PROOF. Suppose T is minimal. If U is open, $U \neq \phi$, then $X = \bigcup_{n=-\infty}^{\infty} T^n(U)$, so if $\mu(U) = 0$ then $m(X) = 0$, a contradiction.

Conversely, suppose $\mu(U) > 0$ for all open non-empty U. Suppose also that T is not minimal. There exists a closed set K such that $TK = K$, $\emptyset \neq K \neq X$. The homeomorphism $T|_K$ has an invariant Borel probability measure μ_K on K by Corollary 6.9.1. Define $\tilde{\mu}$ on X by $\tilde{\mu}(B) = \mu_K(K \cap B)$ for all Borel sets B. Then $\tilde{\mu} \in M(X, T)$ and $\tilde{\mu} \neq \mu$ because $\mu(X \backslash K) > 0$, as $X \backslash K$ is non-empty and open, while $\tilde{\mu}(X \backslash K) = 0$. This contradicts the unique ergodicity of T. □

The map $T: K \to K$ given by $T(e^{2\pi i\theta}) = e^{2\pi i\theta^2}$, $\theta \in [0, 1]$, is an example of a uniquely ergodic homeomorphism which is not minimal. The point $1 \in K$ is a fixed point for T and $\Omega(T) = \{1\}$ so that $M(K, T) = \{\delta_1\}$.

We have the following result (due to H. Furstenberg [1]) about homeomorphisms of the unit circle K. We always write intervals on K anticlockwise so $[z, w]$ denotes the anticlockwise closed interval beginning at z and ending at w. We shall use the fact, proved in §6.6, that a minimal rotation of K is uniquely ergodic.

Theorem 6.18. *Let* $T: K \to K$ *be a homeomorphism with no periodic points. Then* T *is uniquely ergodic. Moreover*

(a) *there is a continuous surjection* $\phi: K \to K$ *and a minimal rotation* $S: K \to K$ *with* $\phi T = S\phi$. *The map* ϕ *has the property that for each* $z \in K$, $\phi^{-1}(z)$ *is either a point or a closed sub-interval of* K.

(b) *if* T *is minimal the map* ϕ *is a homeomorphism so that every minimal homeomorphism of* K *is topologically conjugate to a rotation.*

PROOF. Since T has no periodic points no member of $M(K, T)$ can give positive measure to a point of K. Let μ_1, $\mu_2 \in M(K, T)$ and put $v = \frac{1}{2}(\mu_1 + \mu_2) \in M(K, T)$. Define $\phi: K \to K$ by $\phi(z) = \exp(2\pi i v([1, z]))$. Since v has no points of positive measure we know that ϕ is continuous and is surjective.

For any three points z_1, z_2, z_3 of K we have $v([z_1, z_2]) + v([z_2, z_3]) = v([z_1, z_3])$ mod 1, so

$$\phi(T(z)) = \exp 2\pi i v([1, T(z)])$$
$$= \exp 2\pi i(v([1, T(1)]) + v([T(1), T(z)]))$$
$$= e^{2\pi i\alpha}\phi((z)) \quad \text{where } \alpha = \exp(2\pi i v([1, T(1)])).$$

In other words, if $S(z) = e^{2\pi i\alpha}z$ then $\phi T = S\phi$.

We now show α is irrational. If $\alpha^p = 1$ then $\phi(T^p z) = \phi(z)\ \forall z \in K$ so that $v([1, T^p(z)]) = v([1, z])\,\mathrm{mod}\,1\ \forall z \in K$. Hence $v([z, T^p z]) = 0\,\mathrm{mod}\,1\ \forall z \in K$. Since T^p has no fixed points there exists $\delta > 0$ such that $d(z, T^p z) > \delta\ \forall z \in K$. Therefore each point of K is the end point of an interval of length δ of zero v-measure. This is not possible. Therefore α is irrational.

We now know S is minimal and so Haar measure m is the only element of $M(K, S)$. Therefore $v \circ \phi^{-1}$, $\mu_1 \circ \phi^{-1}$, $\mu_2 \circ \phi^{-1}$ all equal m. Let $[a, b]$ be an interval in K. Then $\phi([a, b]) = [e^{2\pi i v([1, a])}, e^{2\pi i v([1, b])}]$ and $\phi^{-1}\phi([a, b]) = [c, d]$ where $c = \inf\{z : v([z, a]) = 0\}$ and $d = \sup\{w : v([b, w]) = 0\}$. Since $v(\phi^{-1}\phi([a, b]) \triangle [a, b]) = 0$ we have $\mu_i(\phi^{-1}\phi([a, b]) \triangle [a, b]) = 0$, $i = 1, 2$, so $\mu_i([a, b]) = \mu_i(\phi^{-1}\phi([a, b])) = m(\phi([a, b]))$. Hence $\mu_1([a, b]) = \mu_2([a, b])$ and so $\mu_1 = \mu_2$. Therefore T is uniquely ergodic.

Let $w \in K$ and consider the set $\phi^{-1}(w)$. Let $z_1 \in \phi^{-1}(w)$. Then $z_2 \in \phi^{-1}(w)$ iff $v([z_1, z_2]) = 0$ or 1 so that $\phi^{-1}(w)$ is the largest closed interval with zero v-measure which contains z_1.

It remains to prove (b). Suppose T is minimal. We need to show ϕ is a homeomorphism. From Theorem 6.17 we know that if $M(K, T) = \{v\}$ then $v(U) > 0$ for all non-empty open sets U. If $\phi(z) = \phi(w)$ then $v([z, w]) = 0$ or 1 so that either $v([z, w]) = 0$ or $v([w, z]) = 0$. This can only happen when $z = w$. ☐

H. Furstenberg [1] constructed an example of a minimal homeomorphism $T : K^2 \to K^2$ of the two dimensional torus which is not uniquely ergodic. The example preserves Haar measure and has the form $T(z, w) = (az, \phi(z)w)$ where $\{a^n\}_{-\infty}^{\infty}$ is dense in K and $\phi : K \to K$ is a well chosen continuous map.

Recall that if $\mu \in M(X, T)$ is ergodic then there is a $Y \in \mathscr{B}(X)$ such that $\mu(Y) = 1$ and

$$\frac{1}{n}\sum_{i=0}^{n-1} f(T^i x) \to \int f\,d\mu \quad \forall x \in Y,\quad \forall f \in C(X)$$

(Lemma 6.13). When T is uniquely ergodic we get much stronger behaviour of these ergodic averages.

Theorem 6.19. *Let* $T : X \to X$ *be a continuous transformation of a compact metrisable space* X. *The following are equivalent:*

(i) *For every* $f \in C(X)$ *$(1/n)\sum_{i=0}^{n-1} f(T^i x)$ converges uniformly to a constant.*
(ii) *For every* $f \in C(X)$ *$(1/n)\sum_{i=0}^{n-1} f(T^i x)$ converges pointwise to a constant.*
(iii) *There exists* $\mu \in M(X, T)$ *such that for all* $f \in C(X)$ *and all* $x \in X$,

$$\frac{1}{n}\sum_{i=0}^{n-1} f(T^i x) \to \int f\,d\mu.$$

(iv) *T is uniquely ergodic.*

PROOF

(i) $\Rightarrow$ (ii) holds trivially.

(ii) $\Rightarrow$ (iii). Define $k: C(X) \to \mathbb{C}$ by

$$k(f) = \lim_{n \to \infty} \frac{1}{n} \sum_{i=0}^{n-1} fT^i(x).$$

Observe that k is a linear operator and is continuous since

$$\left| \frac{1}{n} \sum_{i=0}^{n-1} fT^i(x) \right| \leq \|f\|.$$

Also $k(1) = 1$, and $f \geq 0$ implies $k(f) \geq 0$. Thus by the Riesz Representation Theorem there exists a Borel probability measure μ such that $k(f) = \int f \, d\mu$. But $k(fT) = k(f)$ and so $\int fT \, d\mu = \int f \, d\mu$. Hence $\mu \in M(X, T)$ by Theorem 6.8.

(iii) $\Rightarrow$ (iv). Suppose that $v \in M(X, T)$. We have

$$\frac{1}{n} \sum_{i=0}^{n-1} fT^i(x) \to f^* \quad \forall x \in X$$

where $f^* = \int f \, d\mu$. Integrating with respect to v, and using the bounded convergence theorem we get that

$$\int f \, dv = \int f^* \, dv = f^* = \int f \, d\mu \quad \forall f \in C(X).$$

Hence $v = \mu$ by Theorem 6.2. Therefore T is uniquely ergodic.

(iv) $\Rightarrow$ (i). If $(1/n) \sum_{i=0}^{n-1} fT^i(x)$ converges uniformly to a constant then this constant must be $\int f \, d\mu$, where $\{\mu\} = M(X, T)$. Suppose (i) is false. Then $\exists g \in C(X), \exists \varepsilon > 0$ such that $\forall N \, \exists n > N$ and $\exists x_n \in X$ with

$$\left| \frac{1}{n} \sum_{i=0}^{n-1} gT^i(x_n) - \int g \, d\mu \right| \geq \varepsilon.$$

If

$$\mu_n = \frac{1}{n} \sum_{i=0}^{n-1} \delta_{T^i x_n} = \frac{1}{n} \sum_{i=0}^{n-1} \tilde{T}^i \delta_{x_n}$$

then $|\int g \, d\mu_n - \int g \, d\mu| \geq \varepsilon$. Choose a convergence subsequence $\{\mu_{n_i}\}$ of $\{\mu_n\}$. If $\mu_{n_i} \to \mu_\infty$ then $\mu_\infty \in M(X, T)$ by Theorem 6.9. Also $|\int g \, d\mu_\infty - \int g \, d\mu| \geq \varepsilon$ so $\mu_\infty \neq \mu$. This contradicts the unique ergodicity of T. $\qquad \square$

Results about unique ergodicity known before 1952 are given in Oxtoby [1]. More recent results of Jewett [1] and Krieger [2] imply that any ergodic invertible measure-preserving transformation of a Lebesgue space is isomorphic in the sense of Chapter 2 to a minimal uniquely ergodic homeomorphism of a zero dimensional compact metrisable space. In particular there are minimal uniquely ergodic homeomorphisms with any prescribed non-negative real number for their entropy. Hahn and Katznelson [1] had previously found minimal uniquely ergodic transformations with arbitrarily large measure-theoretic entropy.

§6.6 Examples

We now investigate $M(X, T)$ for the examples listed in §5.1.

(1) The space of invariant measures for the identity map of X is the space $M(X)$ of all probabilities on $(X, \mathscr{B}(X))$.

(2) **Theorem 6.20.** *If $T(x) = ax$ is a rotation on the compact metrisable group G then T is uniquely ergodic iff T is minimal. The Haar measure is the only invariant measure.*

PROOF. If T is uniquely ergodic then T is minimal by Theorem 6.17, since Haar measure is non-zero on non-empty open sets. If T is minimal then $\{a^n\}_{-\infty}^{\infty}$ is dense in G. Suppose $\mu \in M(G, T)$. Then

$$\int f(a^n x)\, d\mu(x) = \int f(x)\, d\mu(x) \quad \forall f \in C(X) \quad \forall n \in Z.$$

If $b \in G$ there is a sequence a^{n_j} converging to b and by the dominated convergence theorem

$$\int f(bx)\, d\mu(x) = \lim_{j \to \infty} \int f(a^{n_j} x)\, d\mu(x) = \int f(x)\, d\mu(x) \quad \forall f \in C(X).$$

This shows μ is invariant for every rotation of G and is therefore Haar measure. □

(3) If $A:G \to G$ is a surjective endomorphism of a compact metrisable group G then $M(G, A)$ contains many measures. Two of the members of $M(G, A)$ are always Haar measure and δ_e where e is the identity element of G. When $A:K^n \to K^n$ is an automorphism of a torus Theorems 5.11 and 6.16 give us many atomic measures contained in $M(G, A)$. Clearly $A:G \to G$ can only be uniquely ergodic when $G = \{e\}$.

(4) When $T = a \cdot A:G \to G$ is an affine transformation of a compact metrisable group the set $M(G, T)$ is sometimes small (as in (2)) and sometimes large (as in (3)). When T is abelian we have that T is uniquely ergodic iff T is minimal. The 'only if' part follows by Theorem 6.17 and the fact that T preserves Haar measure. The "if" part can be proved by checking statement (i) of Theorem 6.19 holds. This was done by Hahn and Parry [1].

(5) The one-sided and two-sided shift maps have many invariant measures. For every probability vector $(p_0, \ldots, p_{k-1})$ on the state space Y the corresponding product measure belongs to $M(X, T)$. Other members of $M(X, T)$ are provided by Markov measures: if $P = (p_{ij})$ is a $k \times k$ stochastic matrix

and $\mathbf{p} = (p_0, \dots, p_{k-1})$ is a probability vector with $\sum_{i=0}^{k-1} p_i p_{ij} = p_j$ the corresponding Markov measure belongs to $M(X, T)$. The product measures are special examples of Markov measures $(p_{ij} = p_j)$. Atomic measures in $M(X, T)$ are provided by Theorems 5.12 and 6.16.

(6) When $T: K \to K$ is the north–south map we know $\Omega(T) = \{N, S\}$ so that, by Theorem 6.15, $M(K, T) = \{p\delta_N + (1 - p)\delta_S | p \in [0, 1]\}$.

(7) Suppose $T: X_A \to X_A$ is a two-sided topological Markov chain, where $A = (a_{ij})_{i,j=0}^{k-1}$ is a $k \times k$ matrix with $a_{ij} \in \{0, 1\}$. The set $M(X_A, T)$ depends very much on A; when $a_{ij} = 1 \; \forall i, j$ then we have Example 5, and when $A = I$ then T is the identity map on a space with k points. However if $P = (p_{ij})$ is a $k \times k$ stochastic matrix with $0 \le p_{ij} \le a_{ij}$ all i, j and $\mathbf{p} = (p_0, \dots, p_{k-1})$ is a probability vector with $\sum_{i=0}^{k-1} p_i p_{ij} = p_j$ the Markov measure determined by $\mathbf{p}$ and P is a member of $M(X_A, T)$ because it gives zero measure to $X \backslash X_A$. When A is irreducible (i.e. $\forall i, j$ there is some $n = n(i, j)$ such that A^n has (i, j)-th element non-zero) we can obtain such a member as follows. By the Perron-Frobenius theory of non-negative matrices (see §0.9) there is $\lambda > 0$ which is an eigenvalue of A and no other eigenvalue of A has larger absolute value. If A is irreducible then λ is a simple eigenvalue and the corresponding right and left eigenvectors have strictly positive entries. Suppose $\sum_{i=0}^{k-1} u_i a_{ij} = \lambda u_j$ and $\sum_{j=0}^{k-1} a_{ij} v_j = \lambda v_i$ where $u_i > 0$, $v_i > 0$ all i. Normalise $(u_1, \dots, u_n)$ and $(v_1, \dots, v_n)$ so that $\sum_{i=0}^{k-1} u_i v_i = 1$. Put $p_i = u_i v_i$ and $p_{ij} = a_{ij} v_j / \lambda v_i$. Then $P = (p_{ij})$ is a stochastic matrix, $0 \le p_{ij} \le a_{ij}$, and $\sum_{i=0}^{k-1} p_i p_{ij} = p_j$. Therefore the Markov measure determined by $(p_0, \dots, p_{k-1})$ and P is a member of $M(X_A, T)$. We shall see in Chapter 8 that this is a very important member of $M(X_A, T)$.

We now know that every continuous transformation $T: X \to X$ of a compact metric space has an invariant probability defined on the Borel subsets of X. For some transformations, such as toral automorphisms and shift homeomorphisms, the space $M(X, T)$ contains many elements. The question arises as to which are the important elements of $M(X, T)$ to study. It would be good if we could characterise certain members of $M(X, T)$ by "physical principles" (such as variational principles) and that these measures had strong ergodic properties (such as making T a Bernoulli automorphism). We shall see in Chapters 8 and 9 that such a variational principle exists (and is analogous to a well known variational principle in statistical mechanics). It turns out that for some transformations the measures picked out by the variational principle do have strong ergodic properties. The variational principle uses the idea of topological entropy which we discuss in the next chapter.

CHAPTER 7

Topological Entropy

Adler, Konheim, and McAndrew [1] introduced topological entropy as an invariant of topological conjugacy and also as an analogue of measure theoretic entropy. To each continuous transformation $T: X \to X$ of a compact topological space a non-negative real number or ∞, denoted by $h(T)$, is assigned. Later Dinaburg and Bowen gave a new, but equivalent, definition and this definition led to proofs of the results connecting topological and measure-theoretic entropies. Bowen defined the entropy of a uniformly continuous map of a (not necessarily compact) metric space and this leads to a geometric proof of the formula for the topological entropy of an auto-morphism of an n-torus. In the first section of this chapter we give the original definition of topological entropy and in §7.2 we give the other definition. In the last section we calculate the topological entropy of our examples.

We shall use natural logarithms because this will be more appropriate when we discuss topological pressure in Chapter 9.

§7.1 Definition Using Open Covers

Let X be a compact topological space. We shall be interested in open covers of X which we denote by $\alpha, \beta, \ldots$.

Definition 7.1. If α, β are open covers of X their *join* $\alpha \vee \beta$ is the open cover by all sets of the form $A \cap B$ where $A \in \alpha$, $B \in \mathscr{B}$. Similarly we can define the join $\bigvee_{i=1}^{n} \alpha_i$ of any finite collection of open covers of X.

Definition 7.2. An open cover β is a *refinement* of an open cover α, written $\alpha < \beta$, if every member of β is a subset of a member of α.

Hence $\alpha < \alpha \vee \beta$ for any open covers α, β. Also if β is a subcover of α then $\alpha < \beta$.

Definition 7.3. If α is an open cover of X and $T : X \to X$ is continuous then $T^{-1}\alpha$ is the open cover consisting of all sets $T^{-1} A$ where $A \in \alpha$.
We have

$$T^{-1}(\alpha \vee \beta) = T^{-1}(\alpha) \vee T^{-1}(\beta), \quad \text{and } \alpha < \beta \text{ implies } T^{-1}\alpha < T^{-1}\beta.$$

We shall denote $\alpha \vee T^{-1}\alpha \vee \cdots \vee T^{-(n-1)}\alpha$ by $\bigvee_{i=0}^{n-1} T^{-i}\alpha$.

Definition 7.4. If α is an open cover of X let $N(\alpha)$ denote the number of sets in a finite subcover of α with smallest cardinality. We define the entropy of α by $H(\alpha) = \log N(\alpha)$.

Remarks

(1) $H(\alpha) \geq 0$.
(2) $H(\alpha) = 0$ iff $N(\alpha) = 1$ iff $X \in \alpha$.
(3) If $\alpha < \beta$ then $H(\alpha) \leq H(\beta)$.

PROOF. Let $\{B_1, \ldots, B_{N(\beta)}\}$ be a subcover of β with minimal cardinality. For each $i \, \exists A_i \in \alpha$ with $A_i \supseteq B_i$. Therefore $\{A_1, \ldots, A_{N(\beta)}\}$ covers X and is a subcover of α. Thus $N(\alpha) \leq N(\beta)$. $\qquad\square$

(4) $H(\alpha \vee \beta) \leq H(\alpha) + H(\beta)$.

PROOF. Let $\{A_1, \ldots, A_{N(\alpha)}\}$ be a subcover of α of minimal cardinality, and $\{B_1, \ldots, B_{N(\beta)}\}$ be a subcover of β of minimal cardinality. Then

$$\{A_i \cap B_j : 1 \leq i \leq N(\alpha), 1 \leq j \leq N(\beta)\}$$

is a subcover of $\alpha \vee \beta$, so $N(\alpha \vee \beta) \leq N(\alpha)N(\beta)$. $\qquad\square$

(5) If $T : X \to X$ is a continuous map then $H(T^{-1}\alpha) \leq H(\alpha)$. If T is also surjective then $H(T^{-1}\alpha) = H(\alpha)$.

PROOF. If $\{A_1, \ldots, A_{N(\alpha)}\}$ is a subcover of α of minimal cardinality then $\{T^{-1}A_1, \ldots, T^{-1}A_{N(\alpha)}\}$ is a subcover of $T^{-1}\alpha$, so $N(T^{-1}\alpha) \leq N(\alpha)$. If T is surjective and $\{T^{-1}A_1, \ldots, T^{-1}A_{N(T^{-1}\alpha)}\}$ is a subcover of $T^{-1}\alpha$ of minimal cardinality then $\{A_1, \ldots, A_{N(T^{-1}\alpha)}\}$ also covers X so $N(\alpha) \leq N(T^{-1}\alpha)$. $\qquad\square$

Theorem 7.1. *If α is an open cover of X and $T : X \to X$ is continuous then* $\lim_{n \to \infty} (1/n)H(\bigvee_{i=0}^{n-1} T^{-i}\alpha)$ *exists.*

PROOF. Recall that if we set

$$a_n = H\left(\bigvee_{i=0}^{n-1} T^{-i}\alpha\right)$$

then by Theorem 4.9 it suffices to show that

$$a_{n+k} \leq a_n + a_k \quad \text{for } k, n \geq 1.$$

We have

$$a_{n+k} = H\left(\bigvee_{i=0}^{n+k-1} T^{-i}\alpha\right)$$

$$\leq H\left(\bigvee_{i=0}^{n-1} T^{-i}\alpha\right) + H\left(T^{-n} \bigvee_{j=0}^{k-1} T^{-j}\alpha\right) \text{ by Remark (4)}$$

$$\leq a_n + a_k \text{ by Remark (5).} \qquad \square$$

Definition 7.5. If α is an open cover of X and $T: X \to X$ is a continuous map then the *entropy of T relative to α* is given by:

$$h(T, \alpha) = \lim_{n \to \infty} \frac{1}{n} H\left(\bigvee_{i=0}^{n-1} T^{-i}\alpha\right).$$

Remarks

(6) $h(T, \alpha) \geq 0$ by (1).

(7) $\alpha < \beta$ then $h(T, \alpha) \leq h(T, \beta)$.

PROOF. If $\alpha < \beta$ then $\bigvee_{i=0}^{n-1} T^{-i}\alpha < \bigvee_{i=0}^{n-1} T^{-i}\beta$, so by (3) we have that $H(\bigvee_{i=0}^{n-1} T^{-i}\alpha) \leq H(\bigvee_{i=0}^{n-1} T^{-i}\beta)$. Hence $h(T, \alpha) \leq h(T, \beta)$. $\qquad \square$

Note that if β is a finite subcover of α then $\alpha < \beta$ so then $h(T, \alpha) \leq h(T, \beta)$.

(8) $h(T, \alpha) \leq H(\alpha)$.

PROOF. By (4) we have

$$H\left(\bigvee_{i=0}^{n-1} T^{-i}\alpha\right) \leq \sum_{i=0}^{n-1} H(T^{-i}\alpha)$$

$$\leq n \cdot H(\alpha) \quad \text{by (5).} \qquad \square$$

Definition 7.6. If $T: X \to X$ is continuous, the *topological entropy* of T is given by:

$$h(T) = \sup_{\alpha} h(T, \alpha)$$

where α ranges over all open covers of X.

Remarks

(9) $h(T) \geq 0$.

(10) In the definition of $h(T)$ one can take the supremum over finite open covers of X. This follows from (7).

(11) $h(I) = 0$ where I is the identity map of X.

(12) If Y is a closed subset of X and $TY = Y$ then $h(T|Y) \le h(T)$.

The next result shows that topological entropy is an invariant of topological conjugacy.

Theorem 7.2. *If X_1, X_2 are compact spaces and $T_i:X_i \to X_i$ are continuous for $i = 1$, 2, and if $\phi:X_1 \to X_2$ is a continuous map with $\phi X_1 = X_2$ and $\phi T_1 = T_2 \phi$ then $h(T_1) \ge h(T_2)$. If ϕ is a homeomorphism then $h(T_1) = h(T_2)$.*

PROOF. Let α be an open cover of X_2. Then

$$h(T_2, \alpha) = \lim \frac{1}{n} H\left(\bigvee_{i=0}^{n-1} T_2^{-i}\alpha\right)$$

$$= \lim_n \frac{1}{n} H\left(\phi^{-1} \bigvee_{i=0}^{n-1} T_2^{-i}\alpha\right) \quad \text{by (5)}$$

$$= \lim_n \frac{1}{n} H\left(\bigvee_{i=0}^{n-1} \phi^{-1} T_2^{-i}\alpha\right)$$

$$= \lim_n \frac{1}{n} H\left(\bigvee_{i=0}^{n-1} T_1^{-i}\phi^{-1}\alpha\right)$$

$$= h(T_1, \phi^{-1}\alpha).$$

Hence $h(T_2) \le h(T_1)$. If ϕ is a homeomorphism then $\phi^{-1}T_2 = T_1\phi^{-1}$ so, by the above, $h(T_1) \le h(T_2)$. □

In the next section we shall give a definition of $h(T)$ that does not require X to be compact and we shall prove properties of $h(T)$ in this more general setting. However, one result that is false when X is not compact is the following.

Theorem 7.3. *If $T:X \to X$ is a homeomorphism of a compact space X then $h(T) = h(T^{-1})$.*

PROOF

$$h(T, \alpha) = \lim \frac{1}{n} H\left(\bigvee_{i=0}^{n-1} T^{-i}\alpha\right)$$

$$= \lim \frac{1}{n} H\left(T^{n-1}\left(\bigvee_{i=0}^{n-1} T^{-i}\alpha\right)\right) \quad \text{by Remark (5)}$$

$$= \lim \frac{1}{n} H\left(\bigvee_{i=0}^{n-1} T^{i}\alpha\right)$$

$$= h(T^{-1}, \alpha). \qquad \qquad □$$

§7.2 Bowen's Definition

In this section we give the definition of topological entropy using separating and spanning sets. This was done by Dinaburg and by Bowen, but Bowen also gave the definition when the space X is not compact and this will prove useful later. We shall give the definition when X is a metric space but the definition can easily be formulated when X is a uniform space.

In this section (X, d) is a metric space, not necessarily compact. The open ball centre x radius r will be denoted by $B(x; r)$, and the closed ball by $\bar{B}(x; r)$. We shall define topological entropy for uniformly continuous maps $T: X \to X$. The space of all uniformly continuous maps of the metric space (X, d) will be denoted by $UC(X, d)$. Our definitions will depend on the metric d on X; we shall see later what the dependence on d is.

Throughout this section T will denote a fixed member of $UC(X, d)$. If n is a natural number we can define a new metric d_n on X by $d_n(x, y) = \max_{0 \leq i \leq n-1} d(T^i(x), T^i(y))$. (The notation does not show the dependence on T.) The open ball centre x and radius r in the metric d_n is $\bigcap_{i=0}^{n-1} T^{-i} B(T^i x; r)$.

Definition 7.7. Let n be a natural number, $\varepsilon > 0$ and let K be a compact subset of X. A subset F of X is said to (n, ε) *span* K with respect to T if $\forall x \in K$ $\exists y \in F$ with $d_n(x, y) \leq \varepsilon$. (i.e.

$$K \subset \bigcup_{y \in F} \bigcap_{i=0}^{n-1} T^{-i} \bar{B}(T^i y; \varepsilon)).$$

Definition 7.8. If n is a natural number, $\varepsilon > 0$ and K is a compact subset of X let $r_n(\varepsilon, K)$ denote the smallest cardinality of any (n, ε)-spanning set for K with respect to T. (If we need to emphasise T we shall write $r_n(\varepsilon, K, T)$.)

Remarks

(1) Clearly $r_n(\varepsilon, K) < \infty$ because the compactness of K implies the covering of K by the open sets $\bigcap_{i=0}^{n-1} T^{-i} B(T^i x; \varepsilon)$, $x \in X$, has a finite subcover.

(2) If $\varepsilon_1 < \varepsilon_2$ then $r_n(\varepsilon_1, K) \geq r_n(\varepsilon_2, K)$.

Definition 7.9. If $\varepsilon > 0$ and K is a compact subset of X let $r(\varepsilon, K, T) = \limsup_{n \to \infty} (1/n) \log r_n(\varepsilon, K)$. We write $r(\varepsilon, K, T, d)$ if we wish to emphasise the metric d.

Remarks

(3) If $\varepsilon_1 < \varepsilon_2$ then $r(\varepsilon_1, K, T) \geq r(\varepsilon_2, K, T)$ (by Remark 2).

(4) The value of $r(\varepsilon, K, T)$ could be ∞. (An example is given in Remark 14.)

Definition 7.10. If K is a compact subset of X let $h(T; K) = \lim_{\varepsilon \to 0} r(\varepsilon, K, T)$. The *topological entropy* of T is $h(T) = \sup_K h(T; K)$, where the supremum is taken over the collection of all compact subsets of X. We sometimes write $h_d(T)$ to emphasise the dependence on d.

Before giving any interpretations or explanations of this definition we shall give an equivalent but "dual" definition. This definition will use the idea of separated sets which is dual to the notion of spanning sets.

Definition 7.11. Let n be a natural number, $\varepsilon > 0$ and K be a compact subset of X. A subset E of K is said to be (n, ε) *separated* with respect to T if $x, y \in E, x \neq y$, implies $d_n(x, y) > \varepsilon$. (i.e., for $x \in E$ the set $\bigcap_{i=0}^{n-1} T^{-i}\bar{B}(T^i x; \varepsilon)$ contains no other point of E).

Definition 7.12. If n is a natural number, $\varepsilon > 0$ and K is a compact subset of X let $s_n(\varepsilon, K)$ denote the largest cardinality of any (n, ε) separated subset of K with respect to T. (We write $s_n(\varepsilon, K, T)$ to emphasise T if we need to.)

Remarks

(5) We have $r_n(\varepsilon, K) \leq s_n(\varepsilon, K) \leq r_n(\varepsilon/2, K)$ and hence $s_n(\varepsilon, K) < \infty$.

PROOF. If E is an (n, ε) separated subset of K of maximum cardinality then E is an (n, ε) spanning set for K. Therefore $r_n(\varepsilon, K) \leq s_n(\varepsilon, K)$. To show the other inequality suppose E is an (n, ε) separated subset of K and F is an $(n, \varepsilon/2)$ spanning set for K. Define $\phi: E \to F$ by choosing, for each $x \in E$, some point $\phi(x) \in F$ with $d_n(x, \phi(x)) \leq \varepsilon/2$. Then ϕ is injective and therefore the cardinality of E is not greater than that of F. Hence $s_n(\varepsilon, K) \leq r_n(\varepsilon/2, K)$. □

(6) If $\varepsilon_1 < \varepsilon_2$ then $s_n(\varepsilon_1 K) \geq s_n(\varepsilon_2, K)$.

Definition 7.13. If $\varepsilon > 0$ and K is a compact subset of X put $s(\varepsilon, K, T) = \limsup_{n \to \infty} (1/n) \log s_n(\varepsilon, K)$. We sometimes write $s(\varepsilon, K, T, d)$ when we need to emphasise the metric d.

Remarks

(7) We have $r(\varepsilon, K, T) \leq s(\varepsilon, K, T) \leq r(\varepsilon/2, K, T)$ by Remark (5).
(8) If $\varepsilon_1 < \varepsilon_2$ then $s(\varepsilon_1, K, T) \geq s(\varepsilon_2, K, T)$.
(9) We have $h(T; K) = \lim_{\varepsilon \to 0} s(\varepsilon, K, T)$, by Remark (7), so that $h(T) = \sup_K \lim_{\varepsilon \to 0} s(\varepsilon, K, T)$.

Hence $h(T)$ can be defined using either spanning or separating sets.

Remarks

(10) If $r'_n(\varepsilon, K)$ denotes the smallest cardinality of a subset of K that (n, ε) spans K then the proof of Remark (5) gives $r'_n(\varepsilon, K) \leq s_n(\varepsilon, K) \leq r'_n(\varepsilon/2, K)$

so we also have

$$h(T) = \sup_{K} \lim_{\varepsilon \to 0} \limsup_{n \to \infty} \frac{1}{n} \log r'_n(\varepsilon, K).$$

We now make some comments about the definition of $h(T)$.

Remarks

(11) If T is an isometry of (X, d) then clearly $d_n = d$ for all n so that $s_n(\varepsilon, K) = s_1(\varepsilon, K)$ and $h_d(T) = 0$.

(12) For $r_n(\varepsilon, K)$ to increase with n the mapping T needs to increase distances between some points. We can think of $h_d(T)$ as a measure of the expansion of T relative to the metric d.

(13) The ideas for the definition come from the work of Kolmogorov on the size of a metric space (see Kolmogorov and Tihomirov [1]). If (X, ρ) is a metric space then a subset F is said to ε-*span* X if $\forall x \in X \; \exists y \in F$ with $\rho(x, y) \le \varepsilon$, and a subset E is said to be ε-*separated* if whenever $y, z \in E$, $y \ne z$, then $\rho(y, z) > \varepsilon$. The ε-*entropy* of (X, ρ) is then the logarithm of the minimum number of elements of an ε-spanning set and the ε-*capacity* is the logarithm of the maximum number of elements in an ε-separated set. So in the above definitions we are considering the metric spaces (K, d_n) and $r'_n(\varepsilon, K)$ (see Remark 10) is the ε-entropy of (K, d_n) and $s_n(\varepsilon, K)$ is the ε-capacity of (K, d_n). Therefore

$$h(T; K) = \lim_{\varepsilon \to 0} \limsup_{n \to \infty} \frac{1}{n} (\varepsilon\text{-entropy of } (K, d_n))$$

$$= \lim_{\varepsilon \to 0} \limsup_{n \to \infty} \frac{1}{n} (\varepsilon\text{-capacity of } (K, d_n)).$$

(14) The following is an example when $r(\varepsilon, K, T)$ can be ∞ (see Remark (4)). Consider the real line R with the Euclidean metric and let $T(x) = x^2$. Let $K = [3, 4]$. If $x, y > 2$ we have $d_n(x, y) \le \varepsilon$ iff $|x^{2^{n-1}} - y^{2^{n-1}}| \le \varepsilon$. By the mean-value theorem $|x^{2^{n-1}} - y^{2^{n-1}}| = 2^{n-1} z^{2^{n-1}-1} |x - y|$ for some z between x and y, so that $d_n(x, y) \le \varepsilon$ implies

$$|x - y| \le \frac{\varepsilon}{2^{n-1} 2^{2^{n-1}-1}}.$$

Therefore a (n, ε)-spanning set for $[3, 4]$ contains at least

$$\frac{2^{n-2} 2^{2^{n-1}-1}}{\varepsilon}$$

points so that

$$r_n(\varepsilon, K) \ge \frac{2^{n-2} 2^{2^{n-1}-1}}{\varepsilon} \quad \text{and} \quad r(\varepsilon, K, T) = \infty.$$

We now investigate the dependence of $h_d(T)$ on the metric d and then we shall consider $h_d(T)$ when X is compact. In this case the definition has a geometric interpretation and we shall show it coincides with the definition given in §7.1.

Definition 7.14. Two metrics d and d' on X are *uniformly equivalent* if

$$\text{id.}:(X, d) \to (X, d') \quad \text{and id.}:(X, d') \to (X, d)$$

are both uniformly continuous.

In this case, $T \in UC(X, d)$ iff $T \in UC(X, d')$.

Theorem 7.4. *If d and d' are uniformly equivalent and $T \in UC(X, d)$ then $h_d(T) = h_{d'}(T)$.*

PROOF. Let $\varepsilon_1 > 0$. Choose $\varepsilon_2 > 0$ such that

$$d'(x, y) < \varepsilon_2 \Rightarrow d(x, y) < \varepsilon_1$$

and choose $\varepsilon_3 > 0$ such that

$$d(x, y) < \varepsilon_3 \Rightarrow d'(x, y) < \varepsilon_2.$$

Let K be compact. Then

$$r_n(\varepsilon_1, K, d) \le r_n(\varepsilon_2, K, d') \quad \text{and}$$
$$r_n(\varepsilon_2, K, d') \le r_n(\varepsilon_3, K, d).$$

Hence $r(\varepsilon_1, K, T, d) \le r(\varepsilon_2, K, T, d') \le r(\varepsilon_3, K, T, d)$. If $\varepsilon_1 \to 0$, then $\varepsilon_2 \to 0$, and $\varepsilon_3 \to 0$ so we have

$$h_d(T, K) = h_{d'}(T, K). \qquad \square$$

Remarks

(15) The following is an example of two equivalent, but not uniformly equivalent metrics which give different values of entropy for some transformation. Let $X = (0, \infty)$. Define $T:(0, \infty) \to (0, \infty)$ by $T(x) = 2x$. Let d be the Euclidean metric on $(0, \infty)$. Then $T \in UC(X, d)$ and one can easily show $h_d(T) \ge \log(2)$ by estimating the value of $r_n(\varepsilon, [1, 2])$. Let d' be the metric which coincides with d on $[1, 2]$ but is so that T is an isometry for d' i.e. use the fact that the intervals $(2^{n-1}, 2^n]$, $n \in Z$, partition X and $T((2^{n-1}, 2^n]) = (2^n, 2^{n+1}]$. Then $h_{d'}(T) = 0$ by Remark 11 since T is an isometry for d'. The metrics d, d' are equivalent but not uniformly equivalent.

If X is compact and if d and d' are equivalent metrics then they are uniformly equivalent. Also each continuous map $T:X \to X$ is uniformly continuous. Therefore if X is a compact metrisable space the entropy of T does not depend on the metric chosen on X (provided that metric induces the topology of X).

The following will be useful later and it will allow us to simplify the definition of $h(T)$ when X is compact.

Theorem 7.5. *Let* (X, d) *be a metric space and* $T \in UC(X, d)$. *If* $K \subset K_1 \cup \cdots \cup K_m$ *are all compact subsets of* X *then* $h(T; K) \leq \max_{1 \leq i \leq m} h(T; K_i)$.

PROOF. Certainly $s_n(\varepsilon, K) \leq s_n(\varepsilon, K_1) + \cdots + s_n(\varepsilon, K_m)$. Fix $\varepsilon > 0$. For each n choose $K_{i(n,\varepsilon)}$ such that $s_n(\varepsilon, K_{i(n,\varepsilon)}) = \max_j s_n(\varepsilon, K_j)$ Then $s_n(\varepsilon, K) \leq m \cdot s_n(\varepsilon, K_{i(n,\varepsilon)})$ and so,

$$\log s_n(\varepsilon, K) \leq \log m + \log s_n(\varepsilon, K_{i(n,\varepsilon)}).$$

Choose $n_j \to \infty$ such that

$$\frac{1}{n_j} \log s_{n_j}(\varepsilon, K) \to \limsup_{n \to \infty} \frac{1}{n} \log s_n(\varepsilon, K)$$

and so that $K_{i(n_j,\varepsilon)}$ does not depend on j (i.e., $K_{i(n_j,\varepsilon)} = K_{i(\varepsilon)}$ $\forall j$). Therefore $s(\varepsilon, K, T) \leq s(\varepsilon, K_{i(\varepsilon)}, T)$. Choose $\varepsilon_q \to 0$ so that $K_{i(\varepsilon_q)}$ is constant ($= K_{i_0}$, say). Then $h(T; K) \leq h(T; K_{i_0}) \leq \max_j h(T, K_j)$. $\square$

Corollary 7.5.1. *Let* (X, d) *be a metric space and* $T \in UC(X, d)$. *Let* $\delta > 0$. *In order to compute* $h_d(T)$ *is suffices to take the supremum of* $h(T; K)$ *over those compact sets of diameter less than* δ.

PROOF. If K is compact it can be covered by a finite number of balls $B_1, \ldots, B_m$ of diameter $\delta/2$ and hence $h(T; K) \leq \max_{1 \leq i \leq m} h(T; K \cap \bar{B}_i)$. $\square$

Corollary 7.5.2. *If* X *is a compact metrisable space and* d *is any metric on* X *then* $h(T) = h_d(T) = h(T; X)$.

PROOF. If K is a compact subset of X then $h(T; K) \leq h(T; X)$. It follows from Theorem 7.4 that $h_d(T)$ does not depend on d. $\square$

When X is compact we can use Corollary 7.5.2 to simplify the definition of $h(T)$. Take any metric d giving the topology of X. Then

$$h(T) = \lim_{\varepsilon \to 0} \limsup_{n \to \infty} \frac{1}{n} \log r_n(\varepsilon, X) = \lim_{\varepsilon \to 0} \limsup_{n \to \infty} \frac{1}{n} \log s_n(\varepsilon, X).$$

We can give the following interpretation of these expressions. Suppose we want to count the number of orbits of length n (an orbit of length n is a set $\{x, T(x), \ldots, T^{n-1}(x)\}$) but we can only measure to an error ε. Then $r_n(\varepsilon, X)$ and $s_n(\varepsilon, X)$ both can be interpreted as the number of orbits of length n up to error ε. So as $\varepsilon \to 0$ $h(T)$ is a measurement of the growth rate in n of the number of orbits of length n up to error ε.

We shall now prove that the definition of $h(T)$ in this section coincides with that given in §7.1, when T is a continuous map of a compact metrisable space. For the moment let us denote by $h^*(T)$ and $h^*(T, \alpha)$ the numbers occurring in the definition of topological entropy using open covers. In a metric space (X, d) we define the diameter of a cover to be $\text{diam}(\alpha) = \sup_{A \in \alpha} \text{diam}(A)$, where $\text{diam}(A)$ denotes the diameter of the set A. If α, γ are open covers of X and $\text{diam}(\alpha)$ is less than a Lebesgue number for γ then $\gamma < \alpha$. The following result is often useful for calculating $h^*(T)$.

Theorem 7.6. *Let (X, d) be a compact metric space. If $\{\alpha_n\}_1^\infty$ is a sequence of open covers of X with $\text{diam}(\alpha_n) \to 0$ then if $h^*(T) < \infty$ $\lim_{n \to \infty} h^*(T, \alpha_n)$ exists and equals $h^*(T)$, and if $h^*(T) = \infty$ then $\lim_{n \to \infty} h^*(T, \alpha_n) = \infty$.*

PROOF. Suppose $h^*(T) < \infty$. Let $\varepsilon > 0$ be given and choose an open cover γ with $h^*(T, \gamma) > h^*(T) - \varepsilon$. Let δ be a Lebesgue number for γ. Choose N so that $n \geq N$ implies $\text{diam}(\alpha_n) < \delta$. Then $\gamma < \alpha_n$ so $h^*(T, \gamma) \leq h^*(T, \alpha_n)$ when $n \geq N$. Hence $n \geq N$ implies $h^*(T) \geq h^*(T, \alpha_n) > h^*(T) - \varepsilon$ so $\lim_{n \to \infty} h^*(T, \alpha_n) = h^*(T)$. If $h^*(T) = \infty$ and $a > 0$ choose an open cover γ with $h^*(T, \gamma) > a$ and proceed as above to show $\lim h^*(T, \alpha_n) = \infty$. □

Corollary 7.6.1. *We have $h^*(T) = \lim_{\delta \to 0} \{\sup h^*(T, \alpha) | \text{diam}(\alpha) < \delta\}$.*

The next result gives the basic relationship between the two ways of defining topological entropy.

Theorem 7.7. *Let $T : X \to X$ be a continuous map of a compact metric space (X, d).*

(i) *If α is an open cover of X with Lebesgue number δ then*

$$N\left(\bigvee_{i=0}^{n-1} T^{-i}\alpha\right) \leq r_n(\delta/2, X) \leq s_n(\delta/2, X).$$

(ii) *If $\varepsilon > 0$ and γ is an open cover with $\text{diam}(\gamma) \leq \varepsilon$ then*

$$r_n(\varepsilon, X) \leq s_n(\varepsilon, X) \leq N\left(\bigvee_{i=0}^{n-1} T^{-i}\gamma\right).$$

PROOF. We know from Remark 5 that $r_n(\varepsilon, X) \leq s_n(\varepsilon, X) \; \forall \varepsilon > 0$.

(i) Let F be a $(n, \delta/2)$ spanning set for X of cardinality $r_n(\delta/2, T)$. Then

$$X = \bigcup_{x \in F} \bigcap_{i=0}^{n-1} T^{-i}\bar{B}(T^i x; \delta/2)$$

and since for each i $\bar{B}(T^i x; \delta/2)$ is a subset of a member of α we have $N(\bigvee_{i=0}^{n-1} T^{-i}\alpha) \leq r_n(\delta/2, X)$.

(ii) Let E be a (n, ε) separated set of cardinality $s_n(\varepsilon, X)$. No member of the cover $\bigvee_{i=0}^{n-1} T^{-i}\gamma$ can contain two elements of E so $s_n(\varepsilon, X) \leq N(\bigvee_{i=0}^{n-1} T^{-i}\gamma)$. $\qquad \square$

Corollary 7.7.1. *Let* $T\colon X \to X$ *be a continuous map of a compact metric space* (X, d). *Let* $\varepsilon > 0$. *Let* α_ε *be the cover of* X *by all open balls of radius* 2ε *and let* γ_ε *be any cover of* X *by open balls of radius* $\varepsilon/2$. *Then*

$$N\left(\bigvee_{i=0}^{n-1} T^{-i}\alpha_\varepsilon\right) \leq r_n(\varepsilon, X) \leq s_n(\varepsilon, X) \leq N\left(\bigvee_{i=0}^{n-1} T^{-i}\gamma_\varepsilon\right).$$

This leads directly to

Theorem 7.8. *If* $T\colon X \to X$ *is a continuous map of the compact metric space* (X, d) *then* $h(T) = h^*(T)$ *i.e. the two definitions of topological entropy coincide.*

PROOF. If $\varepsilon > 0$ and $\alpha_\varepsilon, \gamma_\varepsilon$ are as in Corollary 7.7.1 then $h^*(T, \alpha_\varepsilon) \leq r(\varepsilon, X, T) \leq s(\varepsilon, X, T) \leq h^*(T, \gamma_\varepsilon)$. If we put $\varepsilon = 1/n$ and let $n \to \infty$ the two end terms converge to $h^*(T)$ by Theorem 7.6 and the middle terms to $h(T)$. $\qquad \square$

Remark. If we had set up the definition of this section on a uniform space we could have proved Theorem 7.8, for a compact Haudsdorff space.

Corollary 7.7.1 also gives us

Theorem 7.9. *If* $T\colon X \to X$ *is a continuous map of a compact metric space* (X, d) *then*

$$h(T) = \lim_{\varepsilon \to 0} \liminf_{n \to \infty} \frac{1}{n} \log r_n(\varepsilon, X) = \lim_{\varepsilon \to 0} \liminf_{n \to \infty} \frac{1}{n} \log s_n(\varepsilon, X).$$

(We know by Corollary 7.5.2 that these formulae hold with "lim inf" replaced by "lim sup".)

PROOF. Corollary 7.7.1 gives

$$h^*(T, \alpha_\varepsilon) \leq \liminf_{n \to \infty} \frac{1}{n} \log r_n(\varepsilon, X) \leq \liminf_{n \to \infty} \frac{1}{n} \log s_n(\varepsilon, X) \leq h^*(T, \gamma_\varepsilon)$$

and then put $\varepsilon = 1/k$ and let $k \to \infty$ and use Theorem 7.6. $\qquad \square$

We now turn to some more properties of topological entropy.

Theorem 7.10

(i) *If* (X, d) *is a metric space,* $T \in UC(X, d)$ *and* $m > 0$ *then* $h_d(T^m) = m \cdot h_d(T)$.
(ii) *Let* (X_i, d_i), $i = 1, 2$, *be a metric space and* $T_i \in UC(X_i, d_i)$. *Define a metric* d *on* $X_1 \times X_2$ *by* $d((x_1, x_2), (y_1, y_2)) = \max\{d_1(x_1, y_1), d_2(x_2, y_2)\}$. *Then*

$T_1 \times T_2 \in UC(X_1 \times X_2, d)$ and $h_d(T_1 \times T_2) \le h_{d_1}(T_1) + h_{d_2}(T_2)$. If either X_1 or X_2 is compact then $h_d(T_1 \times T_2) = h_{d_1}(T_1) + h_{d_2}(T_2)$.

PROOF

(i) Since $r_n(\varepsilon, K, T^m) \le r_{mn}(\varepsilon, K, T)$ we have

$$\frac{1}{n} \log r_n(\varepsilon, K, T^m) \le \frac{m}{mn} \log r_{mn}(\varepsilon, K, T)$$

and therefore $h_d(T^m) \le m \cdot h_d(T)$.

Since T is uniformly continuous, $\forall \varepsilon > 0 \ \exists \delta > 0$ such that

$$d(x, y) < \delta \quad \text{implies} \quad \max_{0 \le j \le m-1} d(T^j x, T^j y) < \varepsilon.$$

So an (n, δ)-spanning set for K with respect to T^m is also an (nm, ε)-spanning set for K with respect to T. Hence $r_n(\delta, K, T^m) \ge r_{mn}(\varepsilon, K, T)$, so $mr(\varepsilon, K, T) \le r(\delta, K, T^m)$. Therefore

$$m \cdot h_d(T, K) \le h_d(T^m, K).$$

(ii) Let $K_i \subseteq X_i$ be compact, $i = 1, 2$. If F_i is an (n, ε)-spanning set for K_i with respect to T_i then $F_1 \times F_2$ is an (n, ε)-spanning set for $K_1 \times K_2$ with respect to $T_1 \times T_2$. Hence

$$r_n(\varepsilon, K_1 \times K_2, T_1 \times T_2) \le r_n(\varepsilon, K_1, T_1) \cdot r_n(\varepsilon, K_2, T_2)$$

which implies

$$r(\varepsilon, K_1 \times K_2, T_1 \times T_2) \le r(\varepsilon, K_1, T_1) + r(\varepsilon, K_2, T_2).$$

Therefore

$$h_d(T_1 \times T_2, K_1 \times K_2) \le h_{d_1}(T_1, K_1) + h_{d_2}(T_2, K_2).$$

Let $\pi_i : X_1 \times X_2 \to X_i$, $i = 1, 2$ be the projection map. If $K \subseteq X_1 \times X_2$ is compact then $K_1 = \pi_1(K)$ and $K_2 = \pi_2(K)$ are compact and $K \subseteq K_1 \times K_2$. Hence

$$h_d(T_1 \times T_2, K) \le h_d(T_1 \times T_2, K_1 \times K_2).$$

Therefore

$$\begin{aligned}
h_d(T_1 \times T_2) &= \sup_{\substack{K \subseteq X_1 \times X_2 \\ \text{compact}}} h_d(T_1 \times T_2, K) \\
&= \sup_{\substack{K_1 \subseteq X_1 \\ K_2 \subseteq X_2 \\ \text{cpt.}}} h_d(T_1 \times T_2, K_1 \times K_2) \\
&\le \sup_{\substack{K_1 \subseteq X_1 \\ \text{cpt.}}} h_{d_1}(T_1, K_1) + \sup_{\substack{K_2 \subseteq X_2 \\ \text{cpt.}}} h_{d_2}(T_2, K_2) \\
&= h_{d_1}(T_1) + h_{d_2}(T_2).
\end{aligned}$$

Now suppose X_1 is compact. (The proof is similar if X_1 is not compact but X_2 is compact.) Since any compact subset of $X_1 \times X_2$ is a subset of $X_1 \times K_2$ for some compact subset K_2 of X_2, we have $h_d(T_1 \times T_2) = \sup\{h_d(T_1 \times T_2, X_1 \times K_2) | K_2$ is a compact subset of $X_2\}$. Let K_2 be a compact subset of X_2. Let $\delta > 0$. If E_1 is a (n, δ) separated subset of X_1 and E_2 is a (n, δ) separated subset of K_2 then $E_1 \times E_2$ is a (n, δ) separated subset of $X_1 \times K_2$. Therefore $s_n(\delta, X_1 \times K_2, T_1 \times T_2) \geq s_n(\delta, X_1, T_1) \cdot s_n(\delta, K_2, T_2)$ so

$$s(\delta, X_1 \times K_2, T_1 \times T_2) \geq \limsup_{n \to \infty} \frac{1}{n}[\log s_n(\delta, X_1, T_1) + \log s_n(\delta, K_2, T_2)]$$

$$\geq \liminf_{n \to \infty} \frac{1}{n}\log s_n(\delta, X_1, T_1) + \limsup_{n \to \infty} \frac{1}{n}\log s_n(\delta, K_2, T_2).$$

Letting $\delta \to 0$ we get by Theorem 7.9

$$h_d(T_1 \times T_2, X_1 \times K_2) \geq h_{d_1}(T_1) + h_{d_2}(T_2). \qquad \square$$

There are examples of homeomorphisms $T_i : X_i \to X_i$ $(i = 1, 2)$ of non-compact metric spaces for which $h_d(T_1 \times T_2) < h_{d_1}(T_1) + h_{d_2}(T_2)$. From the end of the proof of Theorem 7.10 one can see that one needs

$$\limsup_{n \to \infty} \frac{1}{n}[\log s_n(\delta, K_1, T_1) + \log s_n(\delta, K_2, T_2)]$$

$$< \limsup_{n \to \infty} \frac{1}{n}\log s_n(\delta, K_1, T_1) + \limsup_{n \to \infty} \frac{1}{n}\log s_n(\delta, K_2, T_2).$$

P. Hulse has shown how to obtain such an example where each X_i is the real line equipped with a special metric d_i and each T_i is the map $x \to x + 1$. The idea is that d_1 differs from the Euclidean metric on some intervals $[n, n + 1]$ and d_2 differs from the Euclidean metric on $[n, n + 1]$ for different values of n.

Remark

(16) If T is a homeomorphism with $T, T^{-1} \in UC(X, d)$ then $h_d(T^{-1})$ can differ from $h_d(T)$. If $T : R \to R$ is given by $Tx = 2x$ and d is the usual Euclidean metric then we shall see later that $h_d(T) = \log 2$ (it is easy to show $h_d(T) \geq \log 2$ by estimating $s_n(\varepsilon, [0, 1])$). However T^{-1} decreases distances so $h_d(T^{-1}) = 0$.

§7.3 Calculation of Topological Entropy

Theorem 7.6 provided the only method we have given so far for calculating the topological entropy of examples. The following is an analogue of the Kolmogorov-Sinai theorem, and provides a method of calculating topological entropy for some examples.

Theorem 7.11. Let $T: X \to X$ be an expansive homeomorphism of the compact metric space (X, d).

(i) If α is a generator for T then $h(T) = h(T, \alpha)$.

(ii) If δ is an expansive constant for T then $h(T) = r(\delta_0, T) = s(\delta_0, T)$ for all $\delta_0 < \delta/4$.

PROOF.

(i) Let β be any open cover. Let δ be a Lebesgue number for β. By Theorem 5.21 choose $N > 0$ so that each member of $\bigvee_{-N}^{N} T^{-n}\alpha$ has diameter less than δ. Then $\beta < \bigvee_{-N}^{N} T^{-n}\alpha$, and so,

$$h(T, \beta) \leq h\left(T, \bigvee_{n=-N}^{N} T^{-n}\alpha\right)$$

$$= \lim_{k \to \infty} \frac{1}{k} H\left(\bigvee_{i=0}^{k-1} T^{-i}\left(\bigvee_{n=-N}^{N} T^{-n}\alpha\right)\right)$$

$$= \lim_{k \to \infty} \frac{1}{k} H\left(\bigvee_{n=-N}^{N+k-1} T^{-n}\alpha\right)$$

$$= \lim_{k \to \infty} \frac{1}{k} H\left(\bigvee_{n=0}^{2N+k-1} T^{-n}\alpha\right)$$

$$= \lim_{k \to \infty} \frac{2N+k-1}{k} \cdot \frac{1}{2N+k-1} H\left(\bigvee_{0}^{2N+k-1} T^{-n}\alpha\right)$$

$$= h(T, \alpha).$$

Therefore $h(T, \beta) \leq h(T, \alpha)$ for all open covers β and hence

$$h(T) = h(T, \alpha).$$

(ii) Let $\delta_0 < \delta/4$. Choose $x_1, \ldots, x_k$ such that $X = \bigcup_{i=1}^{k} B(x_i; (\delta/2) - 2\delta_0)$. The cover $\alpha = \{B(x_i; \delta/2) | 1 \leq i \leq k\}$ has $2\delta_0$ for a Lebesgue number so by Theorem 7.7 $h(T, \alpha) \leq r(\delta_0, X) \leq s(\delta_0, X) \leq h(T)$. The result follows by part (i) since α is a generator.   □

Corollary 7.11.1. An expansive homeomorphism has finite topological entropy.

We now apply Theorem 7.11 to some examples.

Theorem 7.12. The two-sided shift on $X = \prod_{-\infty}^{\infty} Y$, where $Y = \{0, 1, \ldots, k-1\}$, has topological entropy $\log(k)$.

PROOF. Let $\alpha = \{A_0, \ldots, A_{k-1}\}$ be the natural generator, i.e.

$$A_j = \{\{x_n\}_{-\infty}^{\infty} | x_0 = j\}.$$

Then by Theorem 7.11,

$$h(T) = h(T, \alpha) = \lim_{n \to \infty} \frac{1}{n} \log N \left(\bigvee_{i=0}^{n-1} T^{-i} \alpha \right)$$

$$= \lim_{n \to \infty} \frac{1}{n} \log(k^n) = \log(k). \qquad \square$$

In a similar way, using the idea of a one-sided generator for positively expansive maps, we get the entropy of the one-sided shift is $\log(k)$, when the the state space has k points.

We now generalise the previous theorem.

Theorem 7.13. *Let $T : X \to X$ be the two-sided shift on $X = \prod_{-\infty}^{\infty} Y$ where $Y = \{0, 1, \ldots, k-1\}$.*

(i) *If X_1 is a closed subset of X with $TX_1 = X_1$ then $h(T|_{X_1}) = \lim_{n \to \infty} (1/n) \log \theta_n(X_1)$, where $\theta_n(X_1)$ is the number of n-tuples $[i_0, i_1, \ldots, i_{n-1}]$ such that the set $\{\{x_n\}_{-\infty}^{\infty} \in X_1 | x_0 = i_0, \ldots, x_{n+1} = i_{n-1}\}$ is non-empty.*

(ii) *Let $T_A : X_A \to X_A$ denote the topological Markov chain given by an irreducible $k \times k$ matrix A whose entries belong to $\{0, 1\}$. Then $h(T_A) = \log \lambda$ where λ is the largest positive eigenvalue of A (see Theorem 0.16).*

PROOF

(i) Let α be the natural generator for $T : X \to X$, as in Theorem 7.12. Then α is a generator for $T|_{X_1}$ and $\theta_n(X_1) = N(\bigvee_{i=0}^{n-1} T_1^{-i} \alpha)$. The result follows from Theorem 7.11.

(ii) The set $\{\{x_n\}_{-\infty}^{\infty} \in X_1 | x_0 = i_0, \ldots, x_{n-1} = i_{n-1}\}$ is non-empty iff $a_{i_0 i_1} a_{i_1 i_2} \cdots a_{i_{n-2} i_{n-1}} = 1$. Therefore

$$\theta_n(X_A) = \sum_{i_0, \ldots, i_{n-1} = 0}^{k-1} a_{i_0 i_1} a_{i_1 i_2} \cdots a_{i_{n-2} i_{n-1}} = \sum_{i_0, i_{n-1} = 0}^{k-1} (A^{n-1})_{i_0 i_{n-1}}$$

where $(A^{n-1})_{ij}$ denotes the (i, j)th entry of A^{n-1}. If we define a norm on $k \times k$ matrices by $\|(b_{ij})\| = \sum_{i, j=0}^{k-1} |b_{ij}|$ then we have

$$\theta_n(X_A) = \|A^{n-1}\|, \quad \text{so } \frac{1}{n} \log \theta_n(X_A) = \log(\|A^{n-1}\|^{1/n}) \to \log \lambda$$

by the spectral radius formula. The result follows by (i). $\qquad \square$

The corresponding one-sided results are true. Part (ii) holds also when A is reducible by arranging the matrix A in lower diagonal block form as in the theory of Markov chains.

We now give a collection of transformations which will show there is a transformation with topological entropy equal to any given positive real number. Let $\beta > 1$ be given and we describe a transformation with entropy $\log \beta$. Suppose $\beta \notin Z$. (We know the k-shift has entropy $\log k$.)

Consider the expansion of 1 in powers of β^{-1}, i.e. $1 = \sum_{n=1}^{\infty} a_n \beta^{-n}$ where $a_1 = [\beta]$ and $a_n = [\beta^n - \sum_{i=1}^{n-1} a_i \beta^{n-i}]$. Here $[x]$ denotes the integral part of $x \in R$. Let $k = [\beta] + 1$. Then $0 \leq a_n \leq k - 1$ for all n so we can consider $a = \{a_n\}_1^{\infty}$ as a point in the space $X = \prod_{n=1}^{\infty} Y$ where $Y = \{0, 1, \ldots, k - 1\}$. Consider the lexicographical ordering on X, i.e. $x = \{x_n\}_1^{\infty} < y = \{y_n\}_1^{\infty}$ if $x_j < y_j$ for the smallest j with $x_j \neq y_j$. Let $T : X \to X$ denote the one-sided shift transformation. Note that $T^n a \leq a \, \forall n \geq 0$.

Let $X_\beta = \{x = \{x_n\}_1^{\infty} \mid x \in X \text{ and } T^n x \leq a \, \forall n \geq 0\}$. Then X_β is a closed subset of X and $TX_\beta = X_\beta$. We shall show $h(T|_{X_\beta}) = \log \beta$ by using the formula $h(T|_{X_\beta}) = \lim_{n \to \infty} (1/n) \log \theta_n(X_\beta)$ (Theorem 7.13).

We shall denote $\theta_n(X_\beta)$ by θ_n. Recall that θ_n is the number of n-blocks in X_β. A n-block $(b_1, \ldots, b_n)$ occurs in X_β iff for all $k \in \{1, \ldots, n\}$ we have $(b_k, \ldots, b_n) \leq (a_1, \ldots, a_{n-k+1})$. Put $\theta_0 = 1$ and $a_0 = 0$. We claim the formula $\theta_n = 1 + a_0 \theta_n + a_1 \theta_{n-1} + \cdots + a_n \theta_0$ holds for all $n \geq 0$. Indeed, if $(b_1, \ldots, b_n)$ occurs then either: (1) $b_1 < a_1$ and $(b_2, \ldots, b_n)$ occurs (there are $a_1 \theta_{n-1}$ such possibilities), or (2) $b_1 = a_1$ and $(b_2, \ldots, b_n) \leq (a_2, \ldots, b_n)$. This means either $b_2 < a_2$ and $(b_3, \ldots, b_n)$ occurs (there are $a_2 \theta_{n-2}$ such possibilities), or (3) $b_2 = a_3$ and $(b_3, \ldots, b_n) \leq (a_3, \ldots, a_n)$. Finally we get either $b_{n-1} < a_{n-1}$ and (b_n) occurs (there are $a_{n-1} \theta_1$ such possibilities), or $b_{n-1} = a_{n-1}$ and $b_n \leq a_n$ (there are $a_n + 1$ such possibilities). Therefore

$$\beta^{-n} \theta_n = \beta^{-n} + \beta^{-1} a_1 \beta^{-n-1} \theta_{n-1} + \cdots + \beta^{-n} a_n \theta_0$$

so by the renewal theorem (Theorem 0.18) we have $\lim_{n \to \infty} \beta^{-n} \theta_n = (\sum_0^{\infty} \beta^{-n})$ $(\sum_0^{\infty} n a_n \beta^{-n})^{-1} > 0$. ($\sum_0^{\infty} n a_n \beta^{-n}$ converges by the root test.) Therefore $h(T|_{X_\beta}) = \lim_{n \to \infty} (1/n) \log \theta_n = \log \beta$. The transformation $T|_{X_\beta}$ is called the *one-sided β-shift*. We can obtain the *two-sided β-shift* by letting $\tilde{X}_\beta = \{x = \{x_n\}_{-\infty}^{\infty} \mid x \in \prod_{-\infty}^{\infty} Y \text{ and } (x_i, x_{i+1}, \ldots) \in X_\beta \text{ for all } i \in Z\}$. Then $\tilde{X}_\beta$ is a closed subspace of $\prod_{-\infty}^{\infty} Y$ invariant under the two-sided shift

$$\tilde{T} : \prod_{-\infty}^{\infty} Y \to \prod_{-\infty}^{\infty} Y.$$

Therefore $\tilde{T}|_{\tilde{X}_\beta}$ is a homeomorphism and, since $\theta_n(\tilde{X}_\beta) = \theta_n(X_\beta)$, we have $h(\tilde{T}|_{\tilde{X}_\beta}) = \log \beta$.

We already know that a rotation T of a compact metric group G has zero topological entropy because there is a metric on G making T an isometry. In the next chapter we shall calculate the topological entropy of a torus automorphism. The north-south map of K has zero entropy. In fact we now show any homeomorphism of K has zero entropy.

Theorem 7.14. *If $T : K \to K$ is a homeomorphism of the unit circle then $h(T) = 0$.*

PROOF. We know T maps intervals to intervals because the intervals are the connected subsets of K. Suppose the circle has length 1. Choose $\varepsilon > 0$ such that

$$d(x, y) \leq \varepsilon \quad \text{implies} \quad d(T^{-1}x, T^{-1}y) \leq \tfrac{1}{4}.$$

Consider spanning sets for K with respect to T. Clearly $r_1(\varepsilon, K) \leq [1/\varepsilon] + 1$, where $[1/\varepsilon]$ denotes the integer part of $1/\varepsilon$. We shall show $r_n(\varepsilon, K) \leq n([1/\varepsilon] + 1)$.

Suppose we have a $(n-1, \varepsilon)$-spanning set F of minimal cardinality $r_{n-1}(\varepsilon, K)$. Consider the points of $T^{n-1}F$ and the intervals they determine. Add points to this set so that the new intervals have length less than ε. We have added at most $[1/\varepsilon] + 1$ points. If E denotes the collection of new points, put

$$F' = F \cup T^{-(n-1)}E.$$

We claim that F' is an (n, ε)-spanning set for K. Let $x \in K$. Then $\exists y \in F$ with

$$\max_{0 \leq i \leq n-2} d(T^i x, T^i y) \leq \varepsilon.$$

If $d(T^{n-1}x, T^{n-1}y) \leq \varepsilon$ then our claim is proved. If there is no $y \in F$ with both these properties, choose $y \in F$ with

$$\max_{0 \leq i \leq n-2} d(T^i x, T^i y) \leq \varepsilon.$$

There are two closed intervals with end points $T^{n-1}(x)$ and $T^{n-1}(y)$. Choose the one, and call it I, which is mapped by T^{-1} to the interval, I', with end points $T^{n-2}(x)$ and $T^{n-2}(y)$ and length less than or equal to ε. Choose a point $T^{n-1}(z) \in I$, $z \in F'$, with $d(T^{n-1}x, T^{n-1}z) \leq \varepsilon$. Then $T^{n-2}(z) \in I'$ so that $d(T^{n-2}(x), T^{n-2}(z)) \leq \varepsilon$. The interval I' is mapped by T^{-1} to an interval I'' with end points $T^{n-3}(x)$ and $T^{n-3}(y)$ and length less than $\frac{1}{4}$. Therefore the length of I'' is less than or equal to ε. Since $T^{n-3}(z) \in I''$ we have $d(T^{n-3}x, T^{n-3}z) \leq \varepsilon$. By induction

$$d(T^i x, T^i z) \leq \varepsilon \quad \text{for } 0 \leq i \leq n-1.$$

Thus F' is an (n, ε)-spanning set for K. So,

$$r_n(\varepsilon, K) \leq r_{n-1}(\varepsilon, K) + [1/\varepsilon] + 1$$

and hence

$$r_n(\varepsilon, K) \leq n([1/\varepsilon] + 1).$$

Therefore $r(\varepsilon, X) = 0$, and so

$$h(T) = 0. \qquad \qquad \square$$

Corollary 7.14.1. *Any homeomorphism of* $[0, 1]$ *has zero topological entropy.*

PROOF. $T : [0, 1] \to [0, 1]$ has either $T(0) = 0$ and $T(1) = 1$, or $T(0) = 1$ and $T(1) = 0$. In both cases T^2 fixes both 0 and 1. Let S be any homeomorphism of $[0, 1]$ which fixes both 0 and 1. Let $\phi : [0, 1] \to K$ be the continuous map $\phi(t) = e^{2\pi i t}$. The map ϕ is injective on $(0, 1)$ and $\phi S \phi^{-1}$ is a homeomorphism of K which fixes $1 \in K$. Let α_n be the open cover of $[0, 1]$ by the intervals $[0, 1/n), (1 - 1/n, 1]$ and $(k/2n, (k+2)/2n), 1 \leq k \leq 2n - 3$. Then the open arcs $\phi((k/2n), (k+2)/2n)), 1 \leq k \leq 2n - 3$, together with $\phi([0, 1/n) \cup (1 - 1/n, 1])$

from an open cover β_n of K. We have $N(\bigvee_{j=0}^{p-1} S^j \alpha_n) \leq 2^p N(\bigvee_{j=0}^{p-1} (\phi S \phi^{-1})^j \beta_n)$, and so $h(S, \alpha_n) \leq \log 2 + h(\phi S \phi^{-1}, \beta_n)$. When $n \to \infty$ Theorem 7.6 gives $h(S) \leq \log 2 + h(\phi S \phi^{-1}) = \log 2$. This formula holds for any homeomorphism S of $[0, 1]$ which fixes 0 and 1. We can put $S = T^{2q}$ and get $2qh(T) \leq \log 2$. Hence $h(T) = 0$.   □

We shall obtain an upper bound for the entropy of a differentiable map of a finite-dimensional Riemannian manifold. Suppose M is a p-dimensional Riemannian manifold, not necessarily compact. Let $\tau_x M$ denote the tangent space to M at x and let $\|\cdot\|$ denote the norm induced on $\tau_x M$ by the Riemannian metric. Let $T: M \to M$ be a differentiable map and let $\tau_x T: \tau_x M \to \tau_{Tx} M$ denote the linear transformation which is the tangent to T at x. The norm of this linear transformation, calculated using the norms on $\tau_x M$ and $\tau_{Tx} M$ induced by the Riemannian metric, will be denoted by $\|\tau_x T\|$. Let d denote the metric on M induced by the Riemannian metric.

Theorem 7.15. *For a differentiable transformation $T: M \to M$ of a p-dimensional Riemannian manifold M we have $h_d(T) \leq \max\{0, p \log(\sup_{x \in M} \|\tau_x T\|)\}$.*

PROOF. Let $a = \sup_{x \in X} \|\tau_x T\|$. If $a = \infty$ there is nothing to prove. If $a \leq 1$ the mean-value theorem implies T satisfies $d(Tx, Ty) \leq d(x, y) \ \forall x, y \in M$ so that $h_d(T) = 0$.

Suppose $1 < a < \infty$. By the mean value theorem $d(Tx, Ty) \leq ad(x, y)$. Suppose K is a compact subset of M and $\varepsilon > 0$. We shall show $r(\varepsilon, K, T) \leq p \log a$. We shall select convenient charts on M that cover K. Let $\|\|\cdot\|\|$ denote the norm on R^p given by $\|\|u\|\| = \max |u_i|$ if $u = (u_1, \dots, u_p) \in R^p$ and let $B(0; r)$ denote the open ball in R^p with centre 0 and radius r in this norm. Choose differentiable maps $f_j: B(0, 3) \to M$, $1 \leq j \leq q$, such that $K \subset \bigcup_{j=1}^q f_j(B(0; 1))$. Let $b > 0$ be so that $d(f_j(u), f_j(v)) \leq b \|\| u - v \|\| \ \forall u, v \in B(0; 2)$, $1 \leq j \leq q$. For any $\delta \in (0, 1)$ let $E(\delta) = \{(k_1\delta, \dots, k_p\delta) \in R^p \,|\, k_i \in Z\} \cap B(0; 2)$. The cardinality of $E(\delta)$ is at most $(4/\delta)^p$. Each point of $B(0; 2)$ is within distance δ of a point of $E(\delta)$. Consider $F(\delta) = \bigcup_{j=1}^q f_j E(\delta)$. This set is clearly a $(n, a^n b\delta)$ spanning set for K with respect to T. If we put $\delta = \varepsilon(a^n b)^{-1}$ then $r_n(\varepsilon, K) \leq q(4a^n b\varepsilon^{-1})^p = a^{np}(q4^p b^{-p}\varepsilon^{-p})$. Therefore $r(\varepsilon, K, T) \leq p \log a$.   □

This result was proved by Bowen [1]. A. G. Kushnirenko had been the first to show a C^1 transformation of a compact manifold has finite entropy. In Chapter 8 we shall calculate exactly the entropy of a linear transformation of R^p. Theorem 7.15 gives us an inequality for the entropy of a linear transformation $L: R^p \to R^p$. Suppose d is the Euclidean metric. By putting $T = L^n$ in Theorem 7.15 we get $n \, h_d(L) \leq \max(0, p \log \|L^n\|)$ so $h_d(L) \leq \max(0, \log(\text{spectral radius of } L))$.

Relationship Between Topological Entropy and Measure-Theoretic Entropy

In this chapter we study a fixed continuous transformation $T: X \to X$ and how the measure-theoretic entropy $h_\mu(T)$, where $\mu \in M(X, T)$, varies with μ. We shall prove that the supremum of $h_\mu(T)$, as μ varies over $M(X, T)$, is equal to the topological entropy of T. For some transformations T there is a unique member m of $M(X, T)$ with $h_m(T) = h(T)$ and this is an important natural way of choosing a member of $M(X, T)$.

We also calculate the topological entropy of affine transformations of a finite-dimensional torus.

§8.1 The Entropy Map

Let (X, d) be a compact metric space and let $T: X \to X$ be continuous. The σ-algebra of Borel subsets of X is denoted by $\mathscr{B}(X)$ and $M(X, T)$ denotes the space of all probability measures, on the measurable space $(X, \mathscr{B}(X))$, which are preserved by T. We know that $M(X, T)$ is a non-empty convex set which is compact in the weak*-topology. If $\mu \in M(X, T)$ then T is a measure-preserving transformation of the space $(X, \mathscr{B}(X), \mu)$ and hence has a measure-theoretic entropy that we shall denote by $h_\mu(T)$.

Definition 8.1. The *entropy map* of the continuous transformation $T: X \to X$ is the map $\mu \to h_\mu(T)$ which is defined on $M(X, T)$ and has values in $[0, \infty]$.

We now investigate how the entropy map ties in with the convexity structure and topological structure of $M(X, T)$. We first consider the connection with the convexity of $M(X, T)$.

Theorem 8.1. *Let $T: X \to X$ be a continuous map of a compact metric space. The entropy map of T is affine i.e., if μ, $m \in M(X, T)$ and $p \in [0, 1]$ then $h_{p\mu + (1-p)m}(T) = ph_\mu(T) + (1 - p)h_m(T)$.*

PROOF. We shall use $H_\mu(\xi)$ and $h_\mu(T, \xi)$ to denote the dependence on μ of the quantities that are used to define entropy.

Since $\phi(x) = x \log(x)$ is concave (Theorem 4.2) we know that if $B \in \mathscr{B}(X)$ then

$$
\begin{aligned}
0 &\geq \phi(p\mu(B) + (1 - p)m(B)) - p\phi(\mu(B)) - (1 - p)\phi(m(B)) \\
&= (p\mu(B) + (1 - p)m(B))\log(p\mu(B) + (1 - p)m(B)) - p\mu(B)\log\mu(B) \\
&\quad - (1 - p)m(B)\log m(B) \\
&= p\mu(B)[\log(p\mu(B) + (1 - p)m(B)) - \log(p\mu(B))] \\
&\quad + (1 - p)m(B)[\log(p\mu(B) + (1 - p)m(B)) - \log((1 - p)m(B))] \\
&\quad + p\mu(B)[\log(p\mu(B)) - \log(\mu(B))] \\
&\quad + (1 - p)m(B)[\log((1 - p)m(B)) - \log(m(B))] \\
&\geq 0 + 0 + p\mu(B)\log p + (1 - p)m(B)\log(1 - p) \quad \text{because log is increasing.}
\end{aligned}
$$

Therefore if ξ is any finite partition of $(X, \mathscr{B}(X))$

$$
\begin{aligned}
0 &\leq H_{p\mu + (1-p)m}(\xi) - pH_\mu(\xi) - (1 - p)H_m(\xi) \\
&\leq -(p \log p + (1 - p)\log(1 - p)) \\
&\leq \log 2.
\end{aligned}
$$

If η is any finite partition of $(X, \mathscr{B}(X))$ then by putting $\xi = \bigvee_{i=0}^{n-1} T^{-i}\eta$ in the above we have

$$
h_{p\mu + (1-p)m}(T, \eta) = ph_\mu(T, \eta) + (1 - p)h_m(T, \eta).
$$

Clearly

$$
h_{p\mu + (1-p)m}(T) \leq ph_\mu(T) + (1 - p)h_m(T).
$$

We now show the opposite inequality. Let $\varepsilon > 0$. Choose η_1 so that

$$
h_\mu(T, \eta_1) > \begin{cases} h_\mu(T) - \varepsilon & \text{if } h_\mu(T) < \infty \\ 1/\varepsilon & \text{if } h(T) = \infty \end{cases}
$$

and choose η_2 so that

$$
h_m(T, \eta_2) > \begin{cases} h_m(T) - \varepsilon & \text{if } h_m(T) < \infty \\ 1/\varepsilon & \text{if } h_m(T) = \infty. \end{cases}
$$

Putting $\eta = \eta_1 \vee \eta_2$ in the equality above gives

$$
h_{p\mu + (1-p)m}(T, \eta) > \begin{cases} ph_\mu(T) + (1 - p)h_m(T) - \varepsilon & \text{if } h_\mu(T), h_m(T) < \infty \\ 1/\varepsilon & \text{if either } h_\mu(T) = \infty \text{ or } h_m(T) = \infty \end{cases}
$$

so that $h_{p\mu + (1-p)m}(T) \geq ph_\mu(T) + (1 - p)h_m(T)$. $\qquad\square$

Remark. The first part of the proof shows that if μ, $m \in M(X)$, $p \in [0, 1]$, and ξ is a finite partition then $H_{p\mu + (1-p)m}(\xi) \geq pH_\mu(\xi) + (1-p)H_m(\xi)$. This implies the corresponding inequality for an arbitrary finite convex combination of members of $M(X)$. We shall use this in §8.2.

We now discuss the continuity properties of the entropy map of T.

The entropy map of T need not be continuous. It is easy to show this for many transformations including toral automorphisms and shifts. We shall do it when $T: X \to X$ is the two-sided shift on the space $X = \prod_{-\infty}^{\infty} \{0, 1\}$. The points of X which are fixed by T^p have the form $(\ldots, x_{p-1}\overset{*}{x_0}, x_1, \ldots, x_{p-1}, x_0, x_1, \ldots, x_{p-1}, x_0, \ldots)$ where we have free choice of $x_0, x_1, \ldots, x_{p-1}$ (Theorem 5.12). Let μ_p be the atomic measure that gives each of these points measure $1/2^p$. Note that $\mu_p \in M(X, T)$ and $h_{\mu_p}(T) = 0$ because μ_p is concentrated on a finite set of points. Let μ denote the $(\frac{1}{2}, \frac{1}{2})$-product measure. We know $h_\mu(T) = \log 2$. We now show $\mu_p \to \mu$ as $p \to \infty$. The collection of functions that depend on only a finite number of coordinates forms a dense subset $F(X)$ of $C(X)$ (by the Stone–Weierstrass theorem). It suffices to show $\int f \, d\mu_p \to \int f \, d\mu \, \forall f \in F(X)$. If $f \in F(X)$ there exists N such that $\int f \, d\mu_p = \int f \, d\mu$ if $p \geq N$. Therefore $\mu_p \to \mu$, and so the entropy map is not continuous at μ.

The next example shows that sometimes the entropy map of T is not even upper semi-continuous. Suppose $Y = \{0\} \cup \{(1/n) \, | \, n \geq 1\}$ with topology as a subset of R. Let $X = \prod_{-\infty}^{\infty} Y$ and let $T: X \to X$ be the shift homeomorphism. Let μ_j be the product measure obtained from the measure on Y that gives the points $1/(j-1)$ and $1/j$ each measure $\frac{1}{2}$. Then the measure-preserving transformation T on $(X, \mathscr{B}(X), \mu_j)$ is conjugate to the two-sided $(\frac{1}{2}, \frac{1}{2})$-shift and hence $h_{\mu_j}(T) = \log 2$. However $\mu_j \to \mu$ where μ is the atomic measure on X that gives measure 1 to the point $(\ldots, 0, 0, \overset{*}{0}, 0, 0, \ldots)$. Clearly $h_\mu(T) = 0$ so that the entropy map of T is not upper semi-continuous. One shows $\mu_j \to \mu$ by checking $\int f \, d\mu_j \to \int f \, d\mu$ for the dense subset of $C(X)$ consisting of these continuous functions that only depend on a finite number of coordinates in the product $\prod_{-\infty}^{\infty} Y$.

However, the following gives a family of homeomorphisms for which the entropy map is upper semi-continuous.

Theorem 8.2. *When $T: X \to X$ is an expansive homeomorphism of a compact metric space the entropy map of T is upper semi-continuous, i.e., if $\mu \in M(X, T)$ and $\varepsilon > 0$ there is a neighbourhood U of μ in $M(X, T)$ such that $m \in U$ implies $h_m(T) < h_\mu(T) + \varepsilon$.*

PROOF. In the proof we shall use the simple fact that if $\sum_{i=1}^m a_i = 1 = \sum_{i=1}^m b_i$ and if there exists $c > 0$ with $a_i - b_i < c \, \forall i$ then $|a_i - b_i| < cm, \forall i$ (because $b_i - a_i = \sum_{j \neq i}(a_j - b_j) < mc$). Let δ be an expansive constant for T. Let $\mu \in M(X, T)$ and $\varepsilon > 0$. Let $\gamma = \{C_1, \ldots, C_k\}$ be any partition of X into Borel sets with diam $(C_j) < \delta$. By Theorem 5.25 we know $h_\mu(T) = h_\mu(T, \gamma)$. Choose

N so that

$$\frac{1}{N} H_\mu \left(\bigvee_{j=0}^{N-1} T^{-j}\gamma \right) < h_\mu(T) + \frac{\varepsilon}{2}.$$

Fix $\varepsilon_1 > 0$ to be chosen later. Since μ is regular, choose compact sets $K(i_0, \ldots, i_{N-1}) \subset \bigcap_{j=0}^{N-1} T^{-j}C_{i_j}$ with $\mu(\bigcap_{j=0}^{N-1} T^{-j}C_{i_j}\backslash K(i_0, \ldots, i_{N-1})) < \varepsilon_1$. Then $C_i \supset L_i = \bigcup_{j=0}^{N-1} \{T^j K(i_0, \ldots, i_{N-1}) | i_j = i\}$. The sets $L_1, \ldots, L_k$ are compact and pairwise disjoint so there is a partition $\gamma' = \{C'_1, \ldots, C'_k\}$ with diam $(C'_j) < \delta$ and $L_j \subset \text{int}(C'_j)$ all j. We have

$$K(i_0, \ldots, i_{N-1}) \subset \text{int}\left(\bigcap_{j=0}^{N-1} T^{-j}C'_{i_j} \right).$$

By Urysohn's lemma choose $f_{i_0,\ldots,i_{N-1}} \in C(X)$ with $0 \le f_{i_0,\ldots,i_{N-1}} \le 1$ which vanishes on $X\backslash\text{int}(\bigcap_{j=0}^{N-1} T^{-j}C'_{i_j})$ and equals 1 on $K(i_0,\ldots, i_{N-1})$. Let

$$U(i_0, \ldots, i_{N-1}) = \left\{m \in M(X,T) \left| \left| \int f_{i_0,\ldots i_{N-1}} \, dm - \int f_{i_0,\ldots i_{N-1}} \, d\mu \right| < \varepsilon_1 \right. \right\}.$$

The set $U(i_0, \ldots, i_{N-1})$ is an open subset of $M(X,T)$ and if $m \in U(i_0, \ldots, i_{N-1})$ then $m(\bigcap_{j=0}^{N-1} T^{-j}C'_{i_j}) \ge \int f_{i_0,\ldots,i_{N-1}} \, dm > \int f_{i_0,\ldots,i_{N-1}} \, d\mu - \varepsilon_1 \ge \mu(K(i_0,\ldots, i_{N-1})) - \varepsilon_1$. Hence $m \in U(i_0, \ldots, i_{N-1})$ implies

$$\mu\left(\bigcap_{j=0}^{N-1} T^{-j}C_{i_j} \right) - m\left(\bigcap_{j=0}^{N-1} T^{-j}C'_{i_j} \right) < 2\varepsilon_1.$$

Let $U = \bigcap_{i_0,\ldots i_{N-1}=1}^k U(i_0, \ldots, i_{N-1})$. If $m \in U$ then, for any choice of $i_0, i_1, \ldots, i_{N-1}$

$$\left| m\left(\bigcap_{j=0}^{N-1} T^{-j}C'_{i_j} \right) - \mu\left(\bigcap_{j=0}^{N-1} T^{-j}C_{i_j} \right) \right| < 2\varepsilon_1 k^N$$

by the result about probability vectors mentioned at the start of the proof.

So if $m \in U$ and ε_1 is small enough the continuity of $x \log x$ gives

$$\frac{1}{N} H_m \left(\bigvee_{j=0}^{N-1} T^{-j}\gamma' \right) < \frac{1}{N} H_\mu \left(\bigvee_{j=0}^{N-1} T^{-j}\gamma \right) + \frac{\varepsilon}{2}.$$

Hence if $m \in U$ and ε_1 is small enough

$$h_m(T) = h_m(T, \gamma') \quad \text{by Theorem 5.25 since } \text{diam}(\gamma') < \delta$$

$$\le \frac{1}{N} H_m \left(\bigvee_{j=0}^{N-1} T^{-j}\gamma' \right) \quad \text{by Theorem 4.10}$$

$$< \frac{1}{N} H_\mu \left(\bigvee_{j=0}^{N-1} T^{-j}\gamma \right) + \frac{\varepsilon}{2} \quad \text{by the above}$$

$$< h_\mu(T) + \varepsilon.$$

Therefore the entropy map is upper semi-continuous. $\qquad\qquad\qquad\square$

Since an upper semi-continuous real-valued function of a compact space attains its supremum this result gives us that if T is an expansive homeomorphism there is some $m \in M(X, T)$ with $h_m(T) = \sup\{h_\mu(T) | \mu \in M(X, T)\}$. This will be useful in the next section when we show $\sup(h_\mu(T) | \mu \in M(X, T)\}$ is the topological entropy $h(T)$.

We shall use the following result in the proof of the next theorem. It gives a way of calculating measure-theoretic entropy for a continuous map. When $\xi = \{A_1, \ldots, A_k\}$ is partition recall that $\operatorname{diam}(\xi)$ denotes $\max_{1 \le i \le k} \operatorname{diam}(A_i)$.

Theorem 8.3. *Let $T: X \to X$ be a continuous map of a compact metric space. Let $(\xi_n)_{n=1}^\infty$ be a sequence of partitions X such that $\operatorname{diam}(\xi_n) \to 0$. For every $\mu \in M(X, T)$ $h_\mu(T) = \lim_{n \to \infty} h_\mu(T, \xi_n)$.*

PROOF. Let $\mu \in M(X, T)$. Let $\varepsilon > 0$. Choose a finite partition $\xi = \{A_1, \ldots, A_r\}$ such that $h_\mu(T, \xi) > h_\mu(T) - \varepsilon$ if $h_\mu(T) < \infty$, or $h_\mu(T, \xi) > 1/\varepsilon$ if $h_\mu(T) = \infty$. Choose $\delta > 0$ to correspond to ε and r as in Lemma 4.15. Choose compact sets $K_i \subset A_i$, $1 \le i \le r$, with $\mu(A_i \backslash K_i) < \delta/(r + 1)$. Let $\delta' = \inf_{i \ne j} d(K_i, K_j)$ and choose n with $\operatorname{diam}(\xi_n) < \delta'/2$.

For $1 \le i < r$ let $E_n^{(i)}$ be the union of all the elements of ξ_n that intersect K_i, and let $E_n^{(r)}$ be the union of the remaining elements of ξ_n. Since $\operatorname{diam}(\xi_n) < \delta'/2$ each $C \in \xi_n$ can intersect at most one K_i. Then $\xi_n' = \{E_n^{(1)}, \ldots, E_n^{(r)}\}$ is so that $\xi_n' \le \xi_n$ and $\mu(E_n^{(i)} \triangle A_i) = \mu(E_n^{(i)} \backslash A_i) + \mu(A_i \backslash E_n^{(i)}) \le \mu(X \backslash \bigcup_{j=1}^r K_j) + \mu(A_i \backslash K_i) < \delta$. By Lemma 4.15 we have $H_\mu(\xi/\xi_n') < \varepsilon$. Therefore if n is such that $\operatorname{diam}(\xi_n) < \delta'/2$ then

$$h_\mu(T, \xi) \le h_\mu(T, \xi_n') + \varepsilon \quad \text{by Theorem 4.12(iv)}$$
$$\le h_\mu(T, \xi_n) + \varepsilon.$$

Therefore $\operatorname{diam}(\xi_n) < \delta'/2$ implies $h_\mu(T, \xi_n) > h_\mu(T) - 2\varepsilon$ if $h_\mu(T) < \infty$ or $h_\mu(T, \xi_n) > (1/\varepsilon) - \varepsilon$ if $h_\mu(T) = \infty$. Therefore $\lim_{n \to \infty} h_\mu(T, \xi_n)$ exists and equals $h_\mu(T)$. $\square$

Recall from §6.2 that each $\mu \in M(X, T)$ has a unique ergodic decomposition $\mu = \int_{E(X,T)} m \, d\tau(m)$ where τ is a probability measure on the Borel subsets of $M(X, T)$ and $\tau(E(X, T)) = 1$ where $E(X, T)$ denotes the collection of ergodic members of $M(X, T)$. This implies that if $F: M(X, T) \to R$ is affine and upper semi-continuous then $F(\mu) = \int_{E(X,T)} F(m) \, d\tau(m)$, since such a function F is the limit of a decreasing sequence of continuous affine functions. We have the following relationship for entropy.

Theorem 8.4 (Jacobs). *Let $T: X \to X$ be a continuous map of a compact metrisable space. If $\mu \in M(X, T)$ and $\mu = \int_{E(X,T)} m \, d\tau(m)$ is the ergodic decomposition of μ then we have:*

  (i) *if ξ is a finite partition of $(X, \mathscr{B}(X))$ then $h_\mu(T, \xi) = \int_{E(X,T)} h_m(T, \xi) \, d\tau(m)$.*
  (ii) *$h_\mu(T) = \int_{E(X,T)} h_m(T) \, d\tau(m)$ (both sides could be ∞).*

PROOF

(i) Let $\xi = \{A_1, \ldots, A_k\}$. Let $\Sigma = \prod_{-\infty}^{\infty} Y$ where $Y = \{1, 2, \ldots, k\}$. Define $\phi : X \to \Sigma$ by $\phi(x) = \{i_n\}_{-\infty}^{\infty}$ if $T^n(x) \in A_{i_n}$. We have $\phi T = S\phi$ where $S : \Sigma \to \Sigma$ is the two-sided shift. The map ϕ is measurable and if $m \in M(X, T)$ $\tilde{\phi}m = m \circ \phi^{-1}$ is a measure on $(\Sigma, \mathscr{B}(\Sigma))$ and $\tilde{\phi}m \in M(\Sigma, S)$. If m is ergodic for T then clearly $\tilde{\phi}m$ is ergodic for S. Therefore if $\mu = \int_{E(X,T)} m \, d\tau(m)$ is the ergodic decomposition of $\mu \in M(X, T)$ then $\tilde{\phi}\mu = \int_{E(\Sigma,S)} p \, d\tau \circ \tilde{\phi}^{-1}(p)$ is the ergodic decomposition of $\tilde{\phi}\mu$. Since $S : \Sigma \to \Sigma$ is expansive the entropy map of S is an upper-semi continuous affine real-valued function of $M(\Sigma, S)$ and so

$$h_{\tilde{\phi}\mu}(S) = \int_{E(\Sigma,S)} h_p(S) \, d\tau \circ \tilde{\phi}^{-1}(p) = \int_{E(X,T)} h_{\tilde{\phi}(m)}(S) \, d\tau(m).$$

Since $\xi = \phi^{-1}\eta$, where η is the natural generator of S, we have $h_{\tilde{\phi}m}(S) = h_{\tilde{\phi}m}(S, \eta) = h_m(T, \xi) \, \forall m \in M(X, T)$. Therefore

$$h_\mu(T, \xi) = \int_{E(X,T)} h_m(T, \xi) \, d\tau(m).$$

(ii) Choose finite partitions ξ_q, $q \geq 1$, of $(X, \mathscr{B}(X))$ with $\xi_q \leq \xi_{q+1}$ for all q and $\mathrm{diam}(\xi_q) \to 0$. Then $\lim_{q \to \infty} h_m(T, \xi_q) = h_m(T) \, \forall m \in M(X, T)$, by Theorem 8.3, so by the monotone convergence theorem for the measure τ we have

$$h_\mu(T) = \lim_{q \to \infty} h_\mu(T, \xi_q) = \lim_{q \to \infty} \int_{E(X,T)} h_\mu(T, \xi_q) \, d\tau(m)$$

$$= \int_{E(X,T)} h_m(T) \, d\tau(m). \qquad \square$$

§8.2 The Variational Principle

In this section we prove the basic relationship between topological entropy and measure-theoretic entropy: if T is a continuous map of a compact metric space then $h(T) = \sup\{h_\mu(T) | \mu \in M(X, T)\}$. The inequality $\sup\{h_\mu(T) | \mu \in M(X, T)\} \leq h(T)$ was proved by L. W. Goodwyn in 1968. In 1970 E. I. Dinaburg proved equality when X has finite covering dimension and later in 1970 T. N. T. Goodman proved equality in the general case. The elegant proof we present is due to M. Misiurewicz.

We shall need the following simple lemma, where we use ∂B to denote the boundary of a set B ($\partial B = \bar{B} \backslash \mathrm{int}\,(B)$).

Lemma 8.5. *Let X be a compact metric space and $\mu \in M(X)$.*

(i) *If $x \in X$ and $\delta > 0$ there exists $\delta' < \delta$ such that $\mu(\partial B(x; \delta')) = 0$.*

(ii) *If $\delta > 0$ there is a finite partition $\xi = \{A_1, \ldots, A_k\}$ of $(X, \mathscr{B}(X))$ such that $\mathrm{diam}(A_j) < \delta$ and $\mu(\partial A_j) = 0$ for each j.*

PROOF

(i) This is clear since we cannot have an uncountable collection of disjoint sets of positive measure.

(ii) By (i) there is a finite open cover $\beta = \{B_1, \ldots, B_r\}$ of X by balls of radius less than $\delta/2$ with $\mu(\partial B_j) = 0$ for all j. Let $A_1 = \bar{B}_1$ and for $n > 1$ let $A_n = \bar{B}_n \backslash (\bar{B}_1 \cup \bar{B}_2 \cup \cdots \cup \bar{B}_{n-1})$. Then $\xi = \{A_1, \ldots, A_r\}$ is a partition of $(X, \mathscr{B}(X))$, $\operatorname{diam}(A_n) < \delta$, and since $\partial A_n \subset \bigcup_{i=1}^{n} \partial B_i$ we have $\mu(\partial A_n) = 0$ for all n. $\qquad\square$

We now collect together some results we will use in the proof of the variational principle. In this section X will always denote a compact metric space and $\mathscr{B}(X)$ the σ-algebra of Borel subsets.

Remarks

(1) If $\mu_i \in M(X)$, $1 \le i \le n$, and $p_i \ge 0$, $\sum_{i=1}^{n} p_i = 1$ then

$$H_{\sum_{i=1}^{n} p_i \mu_i}(\xi) \ge \sum_{i=1}^{n} p_i H_{\mu_i}(\xi)$$

for any finite partition ξ of $(X, \mathscr{B}(X))$. (For the proof see the remark following Theorem 8.1.)

(2) Suppose q, n are natural numbers and $1 < q < n$. For $0 \le j \le q - 1$ put $a(j) = [(n - j)/q]$. Here $[b]$ denotes the integer part of $b > 0$. We have the following facts.

(i) $a(0) \ge a(1) \ge \cdots \ge a(q - 1)$.
(ii) Fix $0 \le j \le q - 1$. Then

$$\{0, 1, \ldots, n - 1\} = \{j + rq + i | 0 \le r \le a(j) - 1, 0 \le i \le q - 1\} \cup S$$

where

$$S = \{0, 1, \ldots, j - 1, j + a(j)q, j + a(j)q + 1, \ldots, n - 1\}.$$

Since $j + a(j)q \ge j + [((n - j)/q) - 1]q = n - q$, we have the cardinality of S is at most $2q$.

(iii) For each $0 \le j \le q - 1$, $(a(j) - 1)q + j \le [((n - j)/q) - 1]q + j = n - q$. The numbers $\{j + rq | 0 \le j \le q - 1, 0 \le r \le a(j) - 1\}$ are all distinct and are all no greater than $n - q$.

(3) If $\mu \in M(X, T)$ and if $\mu(\partial A_i) = 0$, $0 \le i \le n - 1$, then

$$\mu\left(\partial\left(\bigcap_{i=0}^{n-1} T^{-i} A_i\right)\right) = 0 \quad \text{since} \quad \partial\left(\bigcap_{i=0}^{n-1} T^{-i} A_i\right) \subset \bigcup_{i=0}^{n-1} T^{-i} \partial A_i.$$

Theorem 8.6. *Let* $T: X \to X$ *be a continuous map of a compact metric space* X. *Then* $h(T) = \sup\{h_\mu(T) | \mu \in M(X, T)\}$.

PROOF

(1) Let $\mu \in M(X, T)$. We show in this part that $h_\mu(T) \leq h(T)$. Let $\xi = \{A_1, \ldots, A_k\}$ be a finite partition of $(X, \mathscr{B}(X))$. Choose $\varepsilon > 0$ so that $\varepsilon < 1/(k \log k)$. Since μ is regular there exist compact sets $B_j \subset A_j$, $1 \leq j \leq k$, with $\mu(A_j \backslash B_j) < \varepsilon$. Let η be the partition $\eta = \{B_0, B_1, \ldots, B_k\}$ where $B_0 = X \backslash \bigcup_{j=1}^k B_j$. We have $\mu(B_0) < k\varepsilon$, and

$$H_\mu(\xi/\eta) = -\sum_{i=0}^k \sum_{j=1}^k \mu(B_i)\phi\left(\frac{\mu(B_i \cap A_j)}{\mu(B_i)}\right)$$

$$= -\mu(B_0)\sum_{j=1}^k \phi\left(\frac{\mu(B_0 \cap A_j)}{\mu(B_0)}\right) \quad \text{since if } i \neq 0, \frac{\mu(B_i \cap A_j)}{\mu(B_i)} = 0 \text{ or } 1$$

$$\leq \mu(B_0)\log(k) \quad \text{by Corollary 4.2.}$$

$$< k\varepsilon \log(k) < 1.$$

The reason we can bring in topological entropy is that for each $i \neq 0$, $B_0 \cup B_i = X \backslash \bigcup_{j \neq i} B_j$ is an open set so $\beta = \{B_0 \cup B_1, \ldots, B_0 \cup B_k\}$ is an open cover of X. We have, if $n \geq 1$, $H_\mu(\bigvee_{i=0}^{n-1} T^{-i}\eta) \leq \log N(\bigvee_{i=0}^{n-1} T^{-i}\eta)$, by Corollary 4.2, where $N(\bigvee_{i=0}^{n-1} T^{-i}\eta)$ denotes the number of non-empty sets in the partition $\bigvee_{i=0}^{n-1} T^{-i}\eta$; so $H_\mu(\bigvee_{i=0}^{n-1} T^{-i}\eta) \leq \log(N(\bigvee_{i=0}^{n-1} T^{-i}\beta) \cdot 2^n)$. Therefore

$$h_\mu(T, \eta) \leq h(T, \beta) + \log 2 \leq h(T) + \log 2,$$

so

$$h_\mu(T, \xi) \leq h_\mu(T, \eta) + H_\mu(\xi/\eta) \quad \text{by Theorem 4.12(iv)}$$
$$\leq h(T) + \log 2 + 1.$$

This gives $h_\mu(T) \leq h(T) + \log 2 + 1$ for any continuous map T with $\mu \in M(X, T)$. It therefore holds for T^n so $nh_\mu(T) \leq nh(T) + \log 2 + 1$ by Theorems 4.13(i) and 7.10(i). Hence $h_\mu(T) \leq h(T)$.

(2) Let $\varepsilon > 0$ be given. We shall find some $\mu \in M(X, T)$ with $h_\mu(T) \geq s(\varepsilon, X, T)$, and this clearly implies $\sup \{h_\mu(T) | \mu \in M(X, T)\} \geq h(T)$.

Let E_n be a (n, ε) separated set for X of cardinality $s_n(\varepsilon, X)$. Let $\sigma_n \in M(X)$ be the atomic measure concentrated uniformly on the points of E_n i.e. $\sigma_n = (1/s_n(\varepsilon, X))\sum_{x \in E_n} \delta_x$. Let $\mu_n \in M(X)$ be defined by $\mu_n = (1/n)\sum_{i=0}^{n-1} \sigma_n \circ T^{-i}$. Since $M(X)$ is compact we can choose a subsequence $\{n_j\}$ of natural numbers such that $\lim_{j \to \infty} (1/n_j)\log s_{n_j}(\varepsilon, X) = s(\varepsilon, X, T)$ and $\{\mu_{n_j}\}$ converges in $M(X)$ to some $\mu \in M(X)$. By Theorem 6.9 we know $\mu \in M(X, T)$. We shall show $h_\mu(T) \geq s(\varepsilon, X, T)$.

By Lemma 8.5 choose a partition $\xi = \{A_1, \ldots, A_k\}$ of $(X, \mathscr{B}(X))$ so that $\text{diam}(A_i) < \varepsilon$ and $\mu(\partial A_i) = 0$ for $1 \leq i \leq k$. Then $H_{\sigma_n}(\bigvee_{i=0}^{n-1} T^{-i}\xi) = \log s_n(\varepsilon, X)$ since no member of $\bigvee_{i=0}^{n-1} T^{-i}\xi$ can contain more than one member of E_n and so $s_n(\varepsilon, X)$ members of $\bigvee_{i=0}^{n-1} T^{-i}\xi$ each have σ_n-measure $1/s_n(\varepsilon, X)$ and the others have σ_n-measure zero. Fix natural numbers q, n with $1 < q < n$ and, as in Remark (2), define $a(j)$, for $0 \leq j \leq q - 1$, by $a(j) = [(n-j)/q]$.

Fix $0 \leq j \leq q - 1$. From Remark 2(ii) we have

$$\bigvee_{i=0}^{n-1} T^{-i}\xi = \bigvee_{r=0}^{a(j)-1} T^{-(rq+j)} \bigvee_{i=0}^{q-1} T^{-i}\xi \vee \bigvee_{l \in S} T^{-l}\xi$$

and S has cardinality at most $2q$. Therefore

$$\log s_n(\varepsilon, X) = H_{\sigma_n}\left(\bigvee_{i=0}^{n-1} T^{-i}\xi\right)$$

$$\leq \sum_{r=0}^{a(j)-1} H_{\sigma_n}\left(T^{-(rq+j)} \bigvee_{i=0}^{q-1} T^{-i}\xi\right) + \sum_{k \in S} H_{\sigma_n}(T^{-k}\xi)$$

$$\text{by Theorem 4.3(viii)}$$

$$\leq \sum_{r=0}^{a(j)-1} H_{\sigma_n \circ T^{-(rq+j)}}\left(\bigvee_{i=0}^{q-1} T^{-i}\xi\right) + 2q\log(k) \quad \text{by Corollary 4.2.}$$

Sum this inequality over j from 0 to $q - 1$ and use Remark 2(iii) to get

$$q \log s_n(\varepsilon, X) \leq \sum_{p=0}^{n-1} H_{\sigma_n \circ T^{-p}}\left(\bigvee_{i=0}^{q-1} T^{-i}\xi\right) + 2q^2 \log(k).$$

If we divide by n and use Remark (1) we get

$$\frac{q}{n} \log s_n(\varepsilon, X) \leq H_{\mu_n}\left(\bigvee_{i=0}^{q-1} T^{-i}\xi\right) + \frac{2q^2}{n}\log(k). \qquad (*)$$

By Remark 3 we know the members of $\bigvee_{i=0}^{q-1} T^{-i}\xi$ have boundaries of μ-measure zero, so $\lim_{j \to \infty} \mu_{n_j}(B) = \mu(B)$ for each member B of $\bigvee_{i=0}^{q-1} T^{-i}\xi$ (Remark 3 §6.1) and therefore $\lim_{j \to \infty} H_{\mu_{n_j}}(\bigvee_{i=0}^{q-1} T^{-i}\xi) = H_\mu(\bigvee_{i=0}^{q-1} T^{-i}\xi)$. Therefore replacing n by n_j in $(*)$ and letting j go to ∞ we have $qs(\varepsilon, X, T) \leq H_\mu(\bigvee_{i=0}^{q-1} T^{-i}\xi)$. We can divide by q and let q go to ∞ to get $s(\varepsilon, X, T) \leq h_\mu(T, \xi) \leq h(T)_\mu$. $\qquad \square$

Corollary 8.6.1. *Let* $T: X \to X$ *be a continuous map of a compact metric space. Then*

(i) $h(T) = \sup\{h_\mu(T) | \mu \in E(X, T)\}$.

(ii) $h(T) = h(T|_{\Omega(T)})$.

(iii) $h(T) = h(T|_{\bigcap_{n=0}^{\infty} T^n X})$.

(iv) *If, for* $i = 1, 2$, $T_i: X_i \to X_i$ *is a continuous map of a compact metric space and if there is a bijection* $\phi: X_1 \to X_2$ *which is bimeasurable (i.e.* ϕ *and* ϕ^{-1} *are measurable) and* $\phi T_1 = T_2 \phi$ *then* $h(T_1) = h(T_2)$. *(This generalises the fact that topological entropy is an invariant of topological conjugacy.)*

PROOF

(i) Let $\varepsilon > 0$ be given. Choose $\mu \in M(X, T)$ such that

$$h_\mu(T) > \begin{cases} h(T) - \varepsilon & \text{if } h(T) < \infty \\ 1/\varepsilon & \text{if } h(T) = \infty. \end{cases}$$

If $\mu = \int_{E(X,T)} m \, d\tau(m)$ is the ergodic decomposition of μ then, by Theorem 8.4, $h_\mu(T) = \int_{E(X,T)} h_m(T) \, d\tau(m)$ and so $h_m(T) > h_\mu(T) - \varepsilon$ for some $m \in E(X,T)$. Hence

$$h_m(T) > \begin{cases} h(T) - 2\varepsilon & \text{if } h(T) < \infty \\ 1/\varepsilon - \varepsilon & \text{if } h(T) = \infty \end{cases}$$

so that $\sup\{h_\mu(T) \,|\, m \in E(X,T)\} = h(T)$.

(ii) By Theorem 6.15 we have $\mu(\Omega(T)) = 1 \ \forall \mu \in M(X,T)$ so that

$$\sup(h_\mu(T) \,|\, \mu \in M(X,T)\} = \sup\{h_\mu(T) \,|\, \mu \in M(\Omega(T), T|_{\Omega(T)})\}.$$

(iii) If $\mu \in M(X,T)$ then $\mu(T^n X) = \mu(T^{-n}T^n X) = \mu(X) = 1$. Therefore $\mu(\bigcap_{n=1}^\infty T^n X) = 1 \ \forall \mu \in M(X,T)$ so that we can identify $M(X,T)$ and $M(\bigcap_0^\infty T^n X, T)$. The result follows from the variational principle.

(iv) We have $\mu \in M(X_1, T_1)$ iff $\mu \circ \phi^{-1} \in M(X_2, T_2)$. Also $h_\mu(T_1) = h_{\mu \circ \phi^{-1}}(T_2)$ so the result follows from the variational principle. $\square$

§8.3 Measures with Maximal Entropy

The variational principle gives a natural way to pick out some members of $M(X,T)$.

Definition 8.2. Let $T: X \to X$ be a continuous transformation on a compact metric space X. A member μ of $M(X,T)$ is called a *measure of maximal entropy* for T if $h_\mu(T) = h(T)$.

Let $M_{\max}(X,T)$ denote the collection of all measures with maximal entropy for T. After the next theorem, which gives the properties of $M_{\max}(X,T)$, we shall give an example where $M_{\max}(X,T)$ is empty.

Theorem 8.7. *Let* $T: X \to X$ *be a continuous transformation of a compact metrisable space. Then*

(i) $M_{\max}(X,T)$ *is convex.*

(ii) *If* $h(T) < \infty$ *the extreme points of* $M_{\max}(X,T)$ *are precisely the ergodic members of* $M_{\max}(X,T)$.

(iii) *If* $h(T) < \infty$ *and* $M_{\max}(X,T) \neq \varnothing$ *then* $M_{\max}(X,T)$ *contains an ergodic measure.*

(iv) *If* $h(T) = \infty$ *then* $M_{\max}(X,T) \neq \varnothing$.

(v) *If the entropy map is upper semi continuous then* $M_{\max}(X,T)$ *is compact and non-empty.*

PROOF

(i) This follows since the entropy map is affine (Theorem 8.1).

(ii) If $\mu \in M_{\max}(X,T)$ is ergodic then it is an extreme point of $M(X,T)$ (Theorem 6.10(iii)) and hence of $M_{\max}(X,T)$. Now suppose $\mu \in M_{\max}(X,T)$

is an extreme point of $M_{max}(X, T)$ and $\mu = p\mu_1 + (1 - p)\mu_2$ for some $p \in [0, 1]$, $\mu_1, \mu_2 \in M(X, T)$. Then, since $h(T) = h_\mu(T) = ph_{\mu_1}(T) + (1 - p)h_{\mu_2}(T)$ (Theorem 8.1) and $h_{\mu_1}(T)$, $h_{\mu_2}(T) \leq h(T)$ (Theorem 8.6), we must have $\mu_1, \mu_2 \in M_{max}(X, T)$. Hence $\mu = \mu_1 = \mu_2$ and μ is an extreme point of $M(X, T)$ and hence ergodic (Theorem 6.10).

(iii) Let $\mu \in M_{max}(X, T)$ and let $\mu = \int_{E(X,T)} m\, d\tau(m)$ be the ergodic decomposition of μ. By Theorem 8.4 $h(T) = h_\mu(T) = \int_{E(X,T)} h_m(T)\, d\tau(m)$ and since $h_m(T) \leq h(T)$ (Theorem 8.6) we have $m \in M_{max}(X, T)$ for τ-almost all m.

(iv) By Theorem 8.6 choose $\mu_n \in M(X, T)$ with $h_{\mu_n}(T) > 2^n$. Let

$$\mu = \sum_{n=1}^{\infty} \frac{1}{2^n} \mu_n \in M(X, T).$$

Since

$$\mu = \sum_{n=1}^{N} \frac{1}{2^n} \mu_n + \frac{1}{2^N} v \quad \text{for some } v \in M(X, T)$$

we have

$$h_\mu(T) \geq \sum_{n=1}^{N} \frac{1}{2^n} h_{\mu_n}(T) > N \quad \text{for each } N \text{ (Theorem 8.1).}$$

Hence $\mu \in M_{max}(X, T)$.

(v) The set $M_{max}(X, T)$ is non-empty because an upper semi-continuous function on a compact space attains its supremum. The upper semi continuity also implies $M_{max}(X, T)$ is compact because if $\mu_n \in M_{max}(X, T)$ and $\mu_n \to \mu \in M(X, T)$ then $h_\mu(T) \geq \limsup_{n \to \infty} h_{\mu_n}(T) = h(T)$ so that $\mu \in M_{max}(X, T)$.   □

Remarks

(1) There is a minimal homeomorphism with $h(T) = \infty$ but $h_\mu(T) < \infty$ $\forall \mu \in E(X, T)$, showing that statement (iii) fails if $h(T) = \infty$ (Grillenberger [1]).

(2) Part (v) together with Theorem 8.2, shows that $M_{max}(X, T)$ is non-empty when T is expansive. This also follows from Theorem 7.11 and the proof of the second part of the variational principle which gives a description of a measure with maximal entropy as a limit of atomic measures on separated sets.

The first example of a homeomorphism with $M_{max}(X, T) = \emptyset$ was given by Gurevič. We now give an example.

Choose numbers β_n such that $1 < \beta_n < 2$ but $\beta_n \nearrow 2$. Let $T_n : X_n \to X_n$ denote the two-sided β_n-shift (see §7.3). We know $h(T_n) = \log \beta_n$. Suppose d_n is a metric on X_n and we can suppose $d_n(x, y) \leq 1$ $\forall x, y \in X_n$. We define a new space X which will be the disjoint union of the X_n together with a "compactification" point x_∞ and we shall put a metric on X so that the subsets X_n converge to x_∞. Define the metric ρ an X by $\rho(x, y) = (1/n^2)d_n(x, y)$

if $x, y \in X_n, \rho(y, x) = \rho(x, y) = \sum_{i=n}^{p} 1/i^2$ if $x \in X_n, y \in X_p$ and $n < p, \rho(x_\infty, x) = \sum_{i=n}^{\infty} 1/i^2$ if $x \in X_n$. Then (X, ρ) is a compact metric space. The transformation $T: X \to X$ with $T|_{X_n} = T_n$ and $T(x_\infty) = x_\infty$ is a homeomorphism. If $\mu \in M(X, T)$ then $\mu = \sum_{n=1}^{\infty} p_n \mu_n + (1 - \sum_{n=1}^{\infty} p_n) \delta_{x_\infty}$ where $\mu_n \in M(X_n, T_n)$ and $p_n \geq 0, \sum_{n=1}^{\infty} p_n \leq 1$. Hence if $\mu \in E(X, T)$ then either $\mu \in E(X_n, T_n)$ for some n or $\mu = \delta_{x_\infty}$. Therefore $h(T) = \sup\{h_\mu(T) \mid \mu \in E(X, T)\} = \sup_{n \geq 1} \sup\{h_{\mu_n}(T_n) \mid \mu_n \in E(X_n, T_n)\} = \sup_{n \geq 1} h(T_n) = \log 2$. If $M_{\max}(X, T) \neq \varnothing$ then by Theorem 8.7(iii) $M_{\max}(X, T)$ contains some ergodic measure μ. Then $\mu \in M(X_n, T_n)$ for some n so $h_\mu(T) = \log \beta_n < \log 2$. Therefore $M_{\max}(X, T) = \varnothing$.

There are minimal homeomorphisms with $h(T) < \infty$ and $M_{\max}(X, T) = \varnothing$ (Grillenberger [1]). There are also diffeomorphisms of compact manifolds with $M_{\max}(X, T) = \varnothing$ (Misiurewicz [1]). Note that if $h(T) = 0$ then $M_{\max}(X, T) = M(X, T)$.

There is a discussion in §20 of Denker, Grillenberger and Sigmund [1] of necessary and sufficient conditions for $M_{\max}(X, T) \neq \varnothing$. In particular the following result of Denker is proved. The conditions $h(T) < \infty$ and $M_{\max}(X, T) \neq \varnothing$ are equivalent to the existence of a sequence $\{\alpha_n\}_1^\infty$ of finite open covers of X with $\sum_{n=1}^{\infty} h(T, \alpha_n) < \infty$ and $\lim_{k \to \infty} h(T, \bigvee_{n=1}^{k} \alpha_n) = h(T)$.

The following is an entropy analogue of unique ergodicity.

Definition 8.3. A continuous transformation $T: X \to X$ of a compact metric space is said to have a *unique measure with maximal entropy* if $M_{\max}(X, T)$ consists of exactly one member. Such transformations are also called intrinsically ergodic (Weiss [1]).

Remarks

(1) If T is uniquely ergodic and $M(X, T) = \{\mu\}$ then T has a unique measure with maximal entropy, because the variational principle gives $h_\mu(T) = h(T)$ in this case.

(2) If $h(T) = \infty$ and T has a unique measure with maximal entropy then T is uniquely ergodic, because if $M_{\max}(X, T) = \{\mu\}$ and $m \in M(X, T)$ then $h_{\mu/2 + m/2}(T) = \infty$ so $m = \mu$.

(3) If $M_{\max}(X, T) = \{\mu\}$ then μ is ergodic. If $h(T) = \infty$ this follows from (2) and if $h(T) < \infty$ it follows from Theorem 8.7(iii).

(4) There are two ways that T can fail to have a unique measure with maximal entropy; either $M_{\max}(X, T) = \varnothing$ or $M_{\max}(X, T)$ has at least two members. One can easily obtain examples of the second type by taking a disjoint union of two compact spaces on which homeomorphisms act. There are however minimal homeomorphisms of the second type. Furstenberg's example of a minimal homeomorphism T of K^2 which is not uniquely ergodic (see §6.5) provides such an example because $h(T) = 0$ and therefore $M_{\max}(X, T) = M(X, T)$.

One way that unique measures with maximal entropy are useful is in constructing isomorphisms.

Theorem 8.8. *Let* $T_i : X_i \to X_i$ $(i = 1, 2)$ *be a continuous transformation of a compact metrisable space and suppose* T_i *has a unique measure,* μ_i, *with maximal entropy. Suppose* $h_{\mu_1}(T_1) = h_{\mu_2}(T_2)$. *If* $\phi : X_1 \to X_2$ *is a bimeasurable bijection with* $\phi \circ T_1 = T_2 \circ \phi$ *then* $\mu_1 \circ \phi^{-1} = \mu_2$ *(and so* ϕ *is an isomorphism between the measure-preserving transformations* T_i *on* $(X_i, \mathscr{B}(X_i), \mu_i)$).

PROOF. By Theorem 4.11 $h_{\mu_1 \circ \phi^{-1}}(T_2) = h_{\mu_1}(T_1)$, so $h_{\mu_1 \circ \phi^{-1}}(T_2) = h_{\mu_2}(T_2)$ so $\mu_2 = \mu_1 \circ \phi^{-1}$. $\square$

In the proof we only used the fact that T_2 has a unique measure with maximal entropy. The fact that T_1 also does follows from the existence of ϕ.

In the next section we shall show that if $T : G \to G$ is an affine transformation of a compact metrisable group then $m \in M_{\max}(G, T)$ where m denotes Haar measure. In this case it is known that if $h_m(T) < \infty$ then $M_{\max}(G, T) = \{m\}$ iff m is ergodic (Berg [1] Conze [1], Walters [4]).

We shall now prove this last statement in the special case when T is the two-sided shift (which is a group automorphism).

Theorem 8.9. *Let* $Y = \{0, 1, \ldots, k - 1\}$, $X = \prod_{-\infty}^{\infty} Y$ *and let* $T : X \to X$ *be the two-sided shift. Then* T *has a unique measure with maximal entropy and this unique measure is the* $(1/k, 1/k, \ldots, 1/k)$-*product measure.*

PROOF. We know $h(T) = \log k$. Suppose $h_\mu(T) = \log k$. Let $\xi = \{A_0, \ldots, A_{k-1}\}$ be the natural generator (i.e. $A_j = \{\{x_n\}_{-\infty}^{\infty} \,|\, x_0 = j\}$). Then $\log k = h_\mu(T) \leq (1/n) H_\mu(\bigvee_{i=0}^{n-1} T^{-i}\xi) \leq (1/n) \log k^n = \log k$ by Theorem 4.10 and Corollary 4.2.1.

Therefore $H_\mu(\bigvee_{i=0}^{n-1} T^{-i}\xi) = \log k^n$ so, by Corollary 4.2.1 each member of $\bigvee_{i=0}^{n-1} T^{-i}\xi$ has measure $(1/k^n)$. Hence μ is the $(1/k, \ldots, 1/k)$-product measure. $\square$

This was generalised to the case of topological Markov chains by Parry. Recall from §6.6 that if $T : X_A \to X_A$ is a two-sided topological Markov chain and A is an irreducible matrix then there is a canonically defined Markov measure given by a probability vector $(p_0, \ldots, p_{k-1})$ and stochastic matrix (p_{ij}) as follows. If λ is the largest positive eigenvalue of A and $(u_0, \ldots, u_{k-1})$ is a strictly positive left eigenvector and $(v_0, \ldots, v_{k-1})$ is a strictly positive right eigenvector with $\sum_{i=0}^{k-1} u_i v_i = 1$ then $p_i = u_i v_i$ and $p_{ij} = a_{ij} v_j / \lambda v_i$. We call this measure the Parry measure for $T : X_A \to X_A$.

Theorem 8.10. *If* $T : X_A \to X_A$ *is a two-sided topological Markov chain, where* A *is an irreducible matrix, then the Parry measure is the unique measure with maximal entropy for* T.

PROOF. We know $h(T) = \log \lambda$ by Theorem 7.13.

Let μ denote the Parry measure. We first show $\mu \in M_{\max}(X_A, T)$ by showing $h_\mu(T) = \log \lambda$. By the formula for the entropy of a Markov measure

(Theorem 4.27) we have

$$
\begin{aligned}
h_\mu(T) &= - \sum_{i,j=0}^{k-1} u_i v_i \frac{a_{ij} v_j}{\lambda v_i} \log\left(\frac{a_{ij} v_j}{\lambda v_i}\right) \\
&= - \sum_{i,j=0}^{k-1} \frac{u_i a_{ij} v_j}{\lambda} \left[\log a_{ij} + \log v_j - \log \lambda - \log v_i\right] \\
&= 0 - \sum_{j=0}^{k-1} u_j v_j \log v_j + \log \lambda + \sum_{i=0}^{k-1} u_i v_i \log v_i \quad \text{since } a_{ij} \in \{0,1\} \\
&= \log \lambda.
\end{aligned}
$$

We now show μ is the only measure with maximal entropy. We know μ is ergodic since the matrix (a_{ij}) and hence (p_{ij}) is irreducible. By Theorem 8.2 and 8.7(v) we know that if $\{\mu\} \neq M_{\max}(X_A, T)$ there is another ergodic member m of $M_{\max}(X_A, T)$. By Theorem 6.10(iv) m and μ are mutually singular so $\exists E \in \mathcal{B}(X_A)$ with $\mu(E) = 0$ and $m(E) = 1$.

Let $\xi = \{A_0, \dots, A_{k-1}\}$ denote the natural generator i.e.

$$
A_j = \{\{x_n\}_{-\infty}^{\infty} \in X_A \mid x_0 = j\}.
$$

Since $\mathcal{A}(\bigvee_{i=n}^{n} T^{-i}\xi) \nearrow \mathcal{B}(X_A)$ we can choose $E_n \in \mathcal{A}(\bigvee_{i=-(n-1)}^{n-1} T^{-i}\xi)$ with $(m + \mu)(E_n \bigtriangleup E) \to 0$. Hence $\mu(E_n) \to 0$ and $m(E_n) \to 1$.

Let η_n denote the partition $\eta_n = \{E_n, X \backslash E_n\}$. Then

$$
\begin{aligned}
\log \lambda = h_m(T) &\leq \frac{1}{2n-1} H_m\left(\bigvee_{i=0}^{2n-2} T^{-i}\xi\right) = \frac{1}{2n-1} H_m\left(\bigvee_{i=-(n-1)}^{n-1} T^{-i}\xi\right) \\
&\leq \frac{1}{2n-1}\left[H_m(\eta_n) + H_m\left(\left(\bigvee_{i=-(n-1)}^{n-1} T^{-i}\xi\right)\Big/ \eta_n\right)\right] \\
&\leq \frac{1}{2n-1}\big[-m(E_n)\log(E_n) - (1 - m(E_n))\log(1 - m(E_n)) \\
&\qquad + m(E_n)\log \theta_n(E_n) + (1 - m(E_n))\log \theta_n(X \backslash E_n)\big]
\end{aligned}
$$

where $\theta_n(B)$ denotes the number of elements of $\bigvee_{i=-(n-1)}^{n-1} T^{-i}\xi$ that intersect $B \in \mathcal{A}(\bigvee_{i=-(n-1)}^{n-1} T^{-i}\xi)$. (Here we have used Corollary 4.2 to estimate the entropy of the partitions of the sets E_n, $X \backslash E_n$ induced by

$$
\bigvee_{i=-(n-1)}^{n-1} T^{-i}\xi.)
$$

Therefore

$$
\log \lambda \leq \frac{1}{2n-1}\left[m(E_n)\log\frac{\theta_n(E_n)}{m(E_n)} + (1 - m(E_n))\log\frac{\theta_n(X\backslash E_n)}{1 - m(E_n)}\right]. \quad (*)
$$

However if $C \in \bigvee_{i=-(n-1)}^{n-1} T^{-i}\xi$, say

$$
C = \{\{x_n\}_{-\infty}^{\infty} \mid (x_{-(n-1)} \cdots x_{n-1}) = (j_{-(n-1)}, \dots, j_{n-1})\}
$$

then

$$\mu(C) = u_{j-(n-1)} v_{j-(n-1)} \prod_{p=-(n-1)}^{n-2} \frac{a_{j_p j_{p+1}}}{\lambda v_{j_p}} v_{j_{p+1}}$$

$$= \frac{u_{j-(n-1)} v_{j_{n-1}}}{\lambda^{2n-1}} \quad \text{since } a_{j_p j_{p+1}} = 1.$$

So if

$$a = \min_{0 \le i, j \le k-1} u_i u_j, \qquad b = \max_{0 \le i, j \le k-1} u_i v_j$$

then

$$\frac{a}{\lambda^{2n-1}} \le \mu(C) \le \frac{b}{\lambda^{2n-1}} \quad \forall C \in \bigvee_{i=-(n-1)}^{n-1} T^{-i} \xi.$$

Therefore

$$\frac{a}{\lambda^{2n-1}} \theta_n(B) \le \mu(B) \le b \frac{\theta_n(B)}{\lambda^{2n-1}}, \quad \forall B \in \mathscr{A}\left(\bigvee_{i=-(n-1)}^{n-1} T^{-i} \xi \right).$$

Using this with $B = E_n$ and $B = X \backslash E_n$ in equality (∗) gives

$$\log \lambda \le \frac{1}{2n-1} \left[m(E_n) \log \left(\frac{\mu(E_n) \lambda^{2n-1}}{am(E_n)} \right) \right.$$

$$\left. + (1 - m(E_n)) \log \left(\frac{(1 - \mu(E_n)) \lambda^{2n-1}}{a(1 - m(E_n))} \right) \right]$$

and hence

$$0 \le m(E_n) \log \left(\frac{\mu(E_n)}{am(E_n)} \right)$$

$$+ (1 - m(E_n)) \log(1 - \mu(E_n)) - (1 - m(E_n)) \log(a(1 - m(E_n))).$$

When n tends to ∞ the limits of the three right-hand terms are $-\infty$, 0, 0 respectively. This contradiction shows that $M_{\max}(X_A, T) = \{\mu\}$. ☐

Remark. If $T: X \to X$ has a unique measure, μ, with maximal entropy then one would expect μ to be important because it was characterised in a natural way from the variational principle. We shall generalise the variational principle in Chapter 9 and this allows us to characterise other measures in a similar way.

§8.4 Entropy of Affine Transformations

In this section we study the relationship between topological entropy and the Haar measure entropy of an affine transformation. We also calculate these entropies for an affine transformation of an n-torus. The first result shows Haar measure is a member of $M_{\max}(G, T)$ when $T: G \to G$ is an affine transformation of a compact metric group.

Theorem 8.11. *Let G be a compact metrisable group and $T = a \cdot A$ an affine transformation of G. Let m denote (normalised) Haar measure on G. Then $h_m(T) = h_m(A) = h(A) = h(T)$. If d denotes a left-invariant metric on G then $h(T) = \lim_{\varepsilon \to 0} \limsup_{n \to \infty} - (1/n) \log m(\bigcap_{i=0}^{n-1} A^{-i} B(e; \varepsilon))$, where e denotes the identity element of G and $B(e; \varepsilon)$ is the open ball centre e and radius ε with respect to the metric d. (This limit clearly exists or is ∞.)*

PROOF. By Theorem 8.6 we have $h_m(T) \leq h(T)$. Suppose d is a left invariant metric on G. Put $D_n(x, \varepsilon, T) = \bigcap_{k=0}^{n-1} T^{-k} B(T^k x; \varepsilon)$. By induction we shall show that $T^{-k} B(T^k x; \varepsilon) = x \cdot (A^{-k} B(e; \varepsilon))$. It is true for $k = 0$ by the invariance of the metric d. Assuming it holds for k we prove it for $k + 1$:

$$
\begin{aligned}
T^{-(k+1)} B(T^{k+1} x; \varepsilon) &= T^{-1}(T^{-k} B(T^k (Tx); \varepsilon) \\
&= T^{-1}(Tx \cdot A^{-k} B(e; \varepsilon) \\
&= x \cdot (A^{-(k+1)} B(e; \varepsilon).
\end{aligned}
$$

Hence $D_n(x, \varepsilon, T) = x \cdot \bigcap_{k=0}^{n-1} A^{-k} B(e; \varepsilon) = x \cdot D_n(e, \varepsilon, A)$ and

$$m(D_n(x, \varepsilon, T)) = m(D_n(e, \varepsilon, A)).$$

Let $\varepsilon > 0$. Let $\xi = \{A_1, \ldots, A_k\}$ be a partition of G into Borel sets of diameter $< \varepsilon$. If $x \in \bigcap_{j=0}^{n-1} T^{-j} A_{i_j}$ then $\bigcap_{j=0}^{n-1} T^{-j} A_{i_j} \subseteq x \cdot D_n(e, \varepsilon, A)$, since if $y \in \bigcap_{j=0}^{n-1} T^{-j} A_{i_j}$ then $T^j(x)$, $T^j(y) \in A_{i_j}$ and hence $y \in T^{-j} B(T^j x; \varepsilon) \ \forall j$, and so $y \in D_n(x, \varepsilon, T) = x \cdot D_n(e, \varepsilon, A)$. Thus $m(\bigcap_{j=0}^{n-1} T^{-j} A_{i_j}) \leq m(D_n(e, \varepsilon, A))$ and taking logarithms we see that

$$
\begin{aligned}
\sum_{i_0, \ldots, i_{n-1} = 1}^{k} m\left(\bigcap_{j=0}^{n-1} T^{-j} A_{i_j} \right) & \log m\left(\bigcap_{j=0}^{n-1} T^{-j} A_{i_j} \right) \\
&\leq \sum_{i_0, \ldots, i_{n-1} = 1}^{k} m\left(\bigcap_{j=0}^{n-1} T^{-j} A_{i_j} \right) \log m(D_n(e, \varepsilon, A)) \\
&= \log m(D_n(e, \varepsilon, A)).
\end{aligned}
$$

Therefore

$$
\begin{aligned}
h_m(T) \geq h_m(T, \xi) &= \lim_{n \to \infty} \frac{1}{n} H_m\left(\bigvee_{j=0}^{n-1} T^{-j} \xi \right) \\
&\geq \limsup_n \left[-\frac{1}{n} \log m(D_n(e, \varepsilon, A)) \right].
\end{aligned}
$$

Since ε was arbitrary we have

$$h_m(T) \geq \lim_{\varepsilon \to 0} \limsup_n \left[-\frac{1}{n} \log m(D_n(e, \varepsilon, A)) \right].$$

(The limit clearly exists.) Consider now an (n, ε)-separated set E, with respect to T, having maximal cardinality. Then

$$\bigcup_{x \in E} D_n(x, \varepsilon/2, T) = \bigcup_{x \in E} x \cdot D_n(e, \varepsilon/2, A)$$

is a disjoint union because of the choice of E. Therefore

$$s_n(\varepsilon, X) \cdot m(D_n(e, \varepsilon/2, A)) \le 1$$

and so

$$s_n(\varepsilon, X) \le \frac{1}{m(D_n(e, \varepsilon/2, A))}.$$

Therefore $s(\varepsilon, X, T) \le \lim \sup_n [-(1/n) \log m(D_n(e, \varepsilon/2, A))]$, and letting $\varepsilon \to 0$ we see that

$$h(T) = h_d(T, X) \le \lim_{\varepsilon \to 0} \lim \sup_n \left[-\frac{1}{n} \log m(D_n(e, \varepsilon/2, A)) \right]$$

$$\le h_m(T).$$

Thus

$$h_m(T) = h(T) = \lim_{\varepsilon \to 0} \lim \sup_n \left[-\frac{1}{n} \log m(D_n(e, \varepsilon, A)) \right].$$

This expression also equals $h_m(A)$ and $h(A)$ since the right hand side is independent of a.   □

The formula

$$h(T) = \lim_{\varepsilon \to 0} \lim \sup_n \left[-\frac{1}{n} \log m(D_n(e, \varepsilon, A)) \right]$$

illustrates how $h(T)$ measures "the amount of expansion" in T.

We shall now compute the topological entropy (and hence by Theorem 8.11 the measure theoretic entropy) of affine transformations of finite-dimensional tori. Recall (§0.8) that we can view the p-torus K^p either multiplicatively as $K \times K \times \cdots \times K$ (p factors) or additively as R^p/Z^p. Each endomorphism A of K^p onto K^p is given, in additive notation, by

$$A(x + Z^p) = [A] \cdot x + Z^p \quad x \in R^p,$$

where $[A]$ is an $p \times p$ non-singular matrix with integer entries. $[A]$ determines a linear transformation $\tilde{A}$ of R^p and $\pi \tilde{A} = A\pi$ where $\pi: R^p \to K^p$ is the natural projection given by $\pi(x) = x + Z^p$.

Let $\|\cdot\|$ denote the usual Euclidean norm on R^p. We define a metric d on R^p/Z^p by

$$d(x + Z^p, y + Z^p) = \inf_{v \in Z^p} \|x - y + v\| \quad x, y \in R^p.$$

The metric d is left and right invariant and for every $x \in R^p \pi$ maps the ball of radius $\frac{1}{4}$ about x in R^p isometrically onto the ball of radius $\frac{1}{4}$ about $\pi(x)$ in R^p/Z^p.

The next theorem deals with such a situation and asserts that $h_d(A) = h_{\bar{d}}(\tilde{A})$ in this case, where $\bar{d}$ denotes the metric on R^p induced by the Euclidean norm $\|\cdot\|$. (Since $\|\tilde{A}x - \tilde{A}y\| \le \|\tilde{A}\| \cdot \|x - y\|$ we know $\tilde{A} \in UC(R^p, \bar{d})$.) This will equate the problem of calculating the entropy of A to that of calculating the entropy of $\tilde{A}$.

Theorem 8.12. *Let* (X,d), $(\tilde{X},\tilde{d})$ *be metric spaces and* $\pi\colon \tilde{X} \to X$ *a continuous surjection such that there exists* $\delta > 0$ *with*

$$\pi|_{B(\tilde{x};\delta)}\colon B(\tilde{x};\delta) \to B(\pi(\tilde{x});\delta)$$

an isometric surjection for all $\tilde{x} \in \tilde{X}$. *If* $T \in UC(X,d)$ *and* $\tilde{T} \in UC(\tilde{X},\tilde{d})$ *satisfy* $\pi\tilde{T} = T\pi$ *then*

$$h_d(T) = h_{\tilde{d}}(\tilde{T}).$$

PROOF. If $\tilde{K}$ is compact in $\tilde{X}$ and $\operatorname{diam}(\tilde{K}) < \delta$ then $\pi(\tilde{K})$ is compact in X and $\operatorname{diam}(\pi(\tilde{K})) < \delta$. Every compact subset of X of diameter $< \delta$ is of this form. Let $\varepsilon > 0$ be such that $\varepsilon < \delta$ and if $\tilde{d}(\tilde{x},\tilde{y}) < \varepsilon$ then $\tilde{d}(\tilde{T}\tilde{x},\tilde{T}\tilde{y}) < \delta$.

Suppose $\tilde{E} \subseteq \tilde{K}$ is an (n,ε)-separating set with respect to $\tilde{T}$. We first prove that $\pi(\tilde{E})$ is an (n,ε)-separating subset of $\pi(\tilde{K})$ with respect to T. To prove this, let $\tilde{x} \neq \tilde{y}$ belong to $\tilde{E}$. Then $\pi(\tilde{x}) \neq \pi(\tilde{y})$. Let i_0 be chosen so that $\tilde{d}(\tilde{T}^i\tilde{x},\tilde{T}^i\tilde{y}) \le \varepsilon$ if $i \le i_0$ and $(\tilde{T}^{i_0+1}\tilde{x},\tilde{T}^{i_0+1}\tilde{y}) > \varepsilon$. By our choice of ε, $\tilde{d}(\tilde{T}^{i_0+1}\tilde{x},\tilde{T}^{i_0+1}\tilde{y}) < \delta$ and so

$$d(T^{i_0+1}\pi(\tilde{x}), T^{i_0+1}\pi(\tilde{y})) = \tilde{d}(\tilde{T}^{i_0+1}\tilde{x},\tilde{T}^{i_0+1}\tilde{y}) > \varepsilon.$$

Thus $\pi(\tilde{E})$ is (n,ε)-separated with respect to T. Therefore

$$s_n(\varepsilon,\tilde{K},\tilde{T}) \le s_n(\varepsilon,\pi(\tilde{K}),T).$$

To prove the converse inequality, suppose E is an (n,ε)-separated subset of $\pi(\tilde{K}) \subset X$ with respect to T, where $\tilde{K}$ is compact and of diameter $< \delta$. Let $\tilde{E} = \pi^{-1}(E) \cap \tilde{K}$. Then $\tilde{E}$ is an (n,ε)-separated set with respect to $\tilde{T}$ since if $d(\tilde{T}^i\tilde{x},\tilde{T}^i\tilde{y}) \le \varepsilon$ where $\tilde{x},\tilde{y} \in \tilde{E}$ then $d(T^i\pi(\tilde{x}),T^i\pi(\tilde{y})) \le \varepsilon$. Hence

$$s_n(\varepsilon,\pi(\tilde{K}),T) \le s_n(\varepsilon,\tilde{K},\tilde{T}).$$

Therefore

$$s_n(\varepsilon,\tilde{K},\tilde{T}) = s_n(\varepsilon,\pi(\tilde{K}),T)$$

and hence

$$h_{\tilde{d}}(\tilde{T},\tilde{K}) = h_d(T,\pi(\tilde{K})).$$

By Corollary 7.5.1

$$h_{\tilde{d}}(\tilde{T}) = h_d(T). \qquad \square$$

Corollary 8.12.1. *If* $A\colon K^p \to K^p$ *is an endomorphism then* $h_d(A) = h_{\tilde{d}}(\tilde{A})$ *where* $\tilde{A}$ *is the linear map of* R^p *covering* A, $\tilde{d}$ *is the metric on* R^p *determined from the Euclidean norm and* d *is any metric on* K^p.

We shall now proceed towards calculating the entropy of a linear map of R^p.

We first show the analogue of Theorem 8.11.

Lemma 8.13. *Let* $A: R^p \to R^p$ *be linear, let* m *denote Lebesgue measure on* R^p *and let* ρ *be a metric on* R^p *determined by a norm. Then*

$$h_\rho(A) = \lim_{\varepsilon \to 0} \limsup_{n \to \infty} \left[-\frac{1}{n} \log m(D_n(0, \varepsilon, A)) \right]$$

where

$$D_n(0, \varepsilon, A) = \bigcap_{i=0}^{n-1} A^{-i} B_\rho(0; \varepsilon) \quad \text{and} \quad B_\rho(0; \varepsilon) = \{x \in R^p \, | \, \rho(x, 0) < \varepsilon\}.$$

Also, $h_\rho(A)$ *does not depend on the norm chosen.*

PROOF. Since all norms on R^p are equivalent they induce uniformly equivalent metrics on R^p so by Theorem 7.4 $h_\rho(A) = h_d(A)$, where d is the Euclidean distance. Also by comparing balls in different norms it is clear that the expression given in the theorem is also independent of the norm. We may as well suppose ρ is the Euclidean distance. The proof is similar to that of Theorem 8.11.

We have $m(A(B)) = |\det A| m(B) \ \forall B \in \mathscr{B}(R^p)$ (Parthasarathy [2], p. 176). Let K be a compact subset of R^p with $m(K) > 0$. If $F(n, \varepsilon)$ spans K then

$$K \subset \bigcup_{x \in F} D_n(x, 2\varepsilon, A) = \bigcup_{x \in F} x + D_n(0, 2\varepsilon, A),$$

where

$$D_n(x, \varepsilon, A) = \bigcap_{i=0}^{n-1} A^{-i} B_\rho(A^i(x); \varepsilon).$$

Therefore $m(K) \le r_n(\varepsilon, K) m(D_n(0, 2\varepsilon, A))$. This gives

$$r_n(\varepsilon, K) \ge \frac{m(K)}{m(D_n(0, 2\varepsilon, A))}$$

and hence $h_\rho(T) \ge r(\varepsilon, K, A) \ge \limsup_{n \to \infty} -(1/n) \log m(D_n(0, 2\varepsilon, A))$. Therefore

$$h_\rho(T) \ge \lim_{\varepsilon \to 0} \limsup_{n \to \infty} -\frac{1}{n} \log m(D_n(0, 2\varepsilon, A)).$$

Let K_q be the closed p-cube with centre $0 \in R^p$ and side length $2q$. If E is a (n, ε) separated subset of K_q then $\bigcup_{x \in E} D_n(x, \varepsilon/2, A)$ is a disjoint union and $\bigcup_{x \in E} D_n(x, \varepsilon/2, A) = \bigcup_{x \in E} (x + D_n(0, \varepsilon/2, A)) \subset K_{q+\varepsilon}$. Therefore

$$s_n(\varepsilon, K_q) \cdot m(D_n(0, \varepsilon/2, A)) \le 2^p (q + \varepsilon)^p$$

and so $s(\varepsilon, K_q, A) \le \limsup_{n \to \infty} -(1/n) \log m(D_n(0, \varepsilon/2, A))$. If K is any compact subset of R^p then $K \subset K_q$ for some q so

$$s(\varepsilon, K, A) \le s(\varepsilon, K_q, A) \le \limsup_{n \to \infty} -\frac{1}{n} m(D_n(0, \varepsilon/2, A)).$$

Therefore

$$h_\rho(A) = \sup_K \lim_{\varepsilon \to 0} s(\varepsilon, K, A) \le \lim_{\varepsilon \to 0} \limsup_{n \to \infty} -\frac{1}{n} \log m(D_n(0, \varepsilon, A)). \qquad \square$$

Theorem 8.14. *Let V be a p-dimensional vector space, let $A : V \to V$ be linear and let ρ be a metric on V induced by a norm on V. Then*

$$h_\rho(A) = \sum_{\{i : |\lambda_i| > 1\}} \log|\lambda_i|,$$

where $\lambda_1, \ldots, \lambda_p$ are the eigenvalues of A. (Some of the λ_i can be equal.)

PROOF. Write V as a direct sum of two subspaces $V = E_1 \oplus E_2$ where $AE_i \subset E_i$, $i = 1, 2$ and $A_1 = A|_{E_1}$ has all its eigenvalues with modulus greater than 1 and $A_2 = A|_{E_2}$ has all its eigenvalues of modulus less than or equal to 1. By choosing a basis in E_i we can suppose $E_i = R^{p_i}$. Let m_i be the Lebesgue measure on $R^{p_i}(i = 1, 2)$. We can consider V as $R^p = R^{p_1} \oplus R^{p_2}$ and $m_1 \times m_2$ is the Lebesgue measure on R^p. Let ρ_i be the Euclidean metric on $R^{p_i}(i = 1, 2)$. Then $d((x_1, x_2), (y_1, y_2)) = \max_{i = 1, 2} \rho_i(x_i, y_i)$ gives the metric on R^p coming from the norm $\|(x_1, x_2)\| = \max(\|x_1\|_1, \|x_2\|_2)$ where $\|\cdot\|_i$ is the Euclidean norm on R^{p_i}. By Lemma 8.13, using d to evaluate the expression given in Lemma 8.13, we have

$$h_\rho(A) = \lim_{\varepsilon \to 0} \limsup_{n \to \infty} \left[-\frac{1}{n} \log m_1(D_n(0, \varepsilon, A_1)) - \frac{1}{n} \log m_2(D_n(0, \varepsilon, A_2)) \right]$$

where $D_n(0, \varepsilon, A_i) = \bigcap_{j=0}^{n-1} A_i^{-j} B_{\rho_i}(0; \varepsilon)$. We have

$$m_1(D_n(0, \varepsilon, A_1)) \le m_1(A_1^{-(n-1)} B_{\rho_1}(0; \varepsilon)) = |\det A_1|^{-(n-1)} m_1(B_{\rho_1}(0; \varepsilon))$$

and $m_2(D_n(0, \varepsilon, A_2)) \le m_2(B_{\rho_2}(0; \varepsilon))$ so that

$$-\frac{1}{n} \log m_1(D_n(0, \varepsilon, A_1)) - \frac{1}{n} \log m_2(D_n(0, \varepsilon, A_2)) \ge \left(\frac{n-1}{n} \right) \log|\det A_1|$$

$$-\frac{1}{n} \log m_1(B_{\rho_1}(0; \varepsilon))$$

$$-\frac{1}{n} \log m_2(B_{\rho_2}(0; \varepsilon)).$$

Therefore $h(A) \ge \log|\det A_1| = \sum_{\{i | |\lambda_i| > 1\}} \log|\lambda_i|$.

We now prove the opposite inequality. By the Jordan decomposition theorem we can write V as a direct sum of subspaces, $V = V_1 \oplus \cdots \oplus V_k$ where $AV_i \subset V_i$ and the eigenvalues of $A_i = A|_{V_i}$ all have the same absolute values τ_i, $1 \le i \le k$. By choosing a basis in V_i we can suppose $V_i = R^{p_i}$ and then $V = R^p = R^{p_1} \oplus \cdots \oplus R^{p_k}$. Let m_i denote Lebesgue measure on R^{p_i} and then $m_1 \times m_2 \times \cdots \times m_k$ is Lebesgue measure on R^p. By Lemma 8.13, if ρ_i

denotes the metric on R^{p_i} induced by any norm on R^{p_i} and if ρ is the metric on R^p induced by any norm on R^p we have

$$h_\rho(A) = \lim_{\varepsilon \to 0} \limsup_{n \to \infty} \sum_{i=1}^{k} -\frac{1}{n} \log m_i(D_n(0, \varepsilon, A_i))$$

where

$$D_n(0, \varepsilon, A_i) = \bigcap_{j=0}^{n-1} A_i^{-j} B_{\rho_i}(0; \varepsilon).$$

We shall show

$$\lim_{\varepsilon \to 0} \limsup_{n \to \infty} -\frac{1}{n} \log m_i(D_n(0, \varepsilon, A_i)) \leq \max\{0, p_i \log \tau_i\}$$

and this will complete the proof.

Fix i and let $\delta > 0$. Let $\|\cdot\|_i$ denote the Euclidean norm on R^{p_i} and define a new norm on R^{p_i} by

$$|\|x\|| = \sum_{n=0}^{\infty} \frac{\|A^n x\|_i}{(\tau_i + \delta)^n}.$$

This series converges by the n-th root test since if $x \neq 0$

$$\sqrt[n]{\frac{\|A^n x\|_i}{(\tau_i + \delta)^n}} \leq \frac{\|A^n\|_i^{1/n} \|x\|_i^{1/n}}{\tau_i + \delta} \to \frac{\tau_i}{\tau_i + \delta}.$$

Since

$$\|x\|_i \leq |\|x\|| \leq \|x\|_i \sum_{n=0}^{\infty} \frac{\|A^n\|_i}{(\tau_i + \delta)^n} = c^{-1} \|x\|_i$$

the two norms on R^{p_i} are equivalent. Also $|\|Ax\|| \leq (\tau_i + \delta)|\|x\||$. If $B_2(0; \varepsilon)$ denotes the open ball of centre 0 and radius ε in the norm $|\|\cdot\||$ then

$$A^{-j} B_2(0; \varepsilon) \supset B_2\left(0; \frac{\varepsilon}{(\tau_i + \delta)^j}\right).$$

If $D_n(0, \varepsilon, A_i)$ is computed using the norm $|\|\cdot\||$ on R^{p_i} we have

$$D_n(0, \varepsilon, A_i) \supset B_2\left(0; \frac{\varepsilon}{(\tau_i + \delta)^{n-1}}\right) \supset B\left(0; \frac{c\varepsilon}{(\tau_i + \delta)^{n-1}}\right)$$

if $\tau_i + \delta \geq 1$, and $D_n(0, \varepsilon, A_i) \supset B_2(0; \varepsilon) \supset B(0; c\varepsilon)$ if $\tau_1 + \delta \leq 1$. Here $B(0; \varepsilon)$ denotes the Euclidean ball. Therefore

$$m_i(D_n(0, \varepsilon, A_i)) \geq \frac{1}{(\tau_i + \delta)^{p_i(n-1)}} m_i(B(0; c\varepsilon)) \quad \text{if } \tau_i + \delta \geq 1$$

and

$$m_i(D_n(0, \varepsilon, A_i)) \geq m_i(B(0; c\varepsilon)) \quad \text{if } \tau_i + \delta \leq 1.$$

Hence

$$\limsup_{n \to \infty} -\frac{1}{n} \log m_i(D_n(0, \varepsilon, A_i)) \leq \max(0, p_i \log(\tau_i + \delta)).$$

Since this is true for each $\delta > 0$ we have

$$\limsup_{n \to \infty} -\frac{1}{n} \log m_i(D_n(0, \varepsilon, A_i)) \leq \max(0, p_i \log \tau_i). \qquad \square$$

We can now deduce the result we set out to prove.

Theorem 8.15. *Suppose* $T: K^p \to K^p$ *is an affine transformation,* $Tx = a \cdot A(x)$, *where* $a \in K^p$ *and* A *is a surjective endomorphism of* K^p. *If* m *is Haar measure then*

$$h(T) = h_m(T) = h_m(A) = h(A) = \sum_{\{i : |\lambda_i| > 1\}} \log|\lambda_i|,$$

where $\lambda_1, \dots, \lambda_p$ *are the eigenvalues of the matrix* $[A]$ *which represents* A.

PROOF. We know by Theorem 8.11 that

$$h(T) = h_m(T) = h_m(A) = h(A)$$

and by Corollary 8.12.1 that $h(A) = h(\tilde{A})$, where $\tilde{A}$ denotes the covering linear map of A. Since $\tilde{A}$ is represented by the matrix $[A]$ in the natural basis the formula for $h(\tilde{A})$ follows from Theorem 8.14. $\qquad \square$

§8.5 The Distribution of Periodic Points

If a continuous map $T: X \to X$ has a unique measure with maximal entropy one expects this measure to have strong properties and tie in with the other dynamical behaviour of T. We discuss how, for some maps T, this measure is connected with periodic points.

If $T: X \to X$ is a continuous transformation of a compact metrisable space then $N_n(T)$ will denote the cardinality of the set $F_n(T) = \{x \in X \mid T^n(x) = x\}$. We have

Theorem 8.16. *If* $T: X \to X$ *is an expansive homeomorphism of a compact metric space then* $N_n(T) < \infty$ $\forall n \geq 1$ *and* $h(T) \geq \limsup_{n \to \infty} (1/n) \log N_n(T)$.

PROOF. Let δ be an expansive constant for T. If $T^n x = x$, $T^n y = y$ and $x \neq y$ then if $d(T^j(x), T^j(y)) \leq \delta, 0 \leq j \leq n - 1$, then $d(T^j(x), T^j(y)) \leq \delta \; \forall j \in Z$ and hence $x = y$. Therefore the set $F_n(T) = \{x \mid T^n x = x\}$ is (n, δ) separated and so $N_n(T) \leq s_n(X, \delta) < \infty$. Hence

$$\limsup_{n \to \infty} \frac{1}{n} \log N_n(T) \leq \limsup_{n \to \infty} \frac{1}{n} \log s_n(X, \delta) \leq h(T). \qquad \square$$

We are interested in the distribution of the periodic points so we make the following definition.

Definition 8.4. Let $T: X \to X$ be a continuous map of a compact metrisable space with $N_n(T) < \infty$ $\forall n \geq 1$. A measure $\mu \in M(X, T)$ *describes the distribution of the periodic points of* T if

$$\frac{1}{N_n(T)} \sum_{x \in F_n(T)} \delta_x \to \mu \quad \text{in } M(X).$$

Theorem 8.17. *Suppose* $T: X_A \to X_A$ *is a two sided topological Markov chain where A is an irreducible matrix. Then $h(T) = \lim_{n \to \infty} (1/n) \log N_n(T)$ and the unique measure with maximal entropy describes the distribution of the periodic points of* T.

PROOF. If $\{x_j\}_{-\infty}^{\infty}$ is a point of X_A then it belongs to $F_n(T)$ iff $x_j = x_{j+n}$ $\forall j \in Z$. Therefore

$$N_n(T) = \sum_{i_0, \ldots, i_{n-1} = 0}^{k-1} a_{i_0 i_1} a_{i_1 i_2} \cdots a_{i_{n-2} i_{n-1}} a_{i_{n-1} i_0}$$

$$= \text{trace of } A^n = \sum_{i=1}^{k} \lambda_i^n$$

where $\lambda_1, \ldots, \lambda_k$ are the eigenvalues of A. Therefore

$$\lim_{n \to \infty} \frac{N_n(T)}{\lambda^n} = \lim_{n \to \infty} \left(\frac{\sum_{i=1}^{k} \lambda_i^n}{\lambda^n} \right) = 1$$

where λ is the largest positive eigenvalue of A which is simple because A is irreducible. Hence

$$\lim_{n \to \infty} \frac{1}{n} \log N_n(T) = \log \lambda = h(T).$$

Let μ denote the unique measure with maximal entropy (see Theorem 8.10; we use the same notation as there). To show

$$\mu_n = \frac{1}{N_n(T)} \sum_{x \in F_N(T)} \delta_x \to \mu$$

it suffices to show $\int f \, d\mu_n \to \int f \, d\mu$ for functions of the form $f = \chi_C$ where $C = \{\{x_j\}_{-\infty}^{\infty} \mid x_r = i_r, \ldots, x_s = i_s\}$ for some r, s $(r \leq s)$ and some $i_r, \ldots, i_s$. This is because finite linear combinations of such functions are dense in $C(X_A)$ by the Stone–Weierstrass theorem. Let $f = \chi_C$ be as above. If $n > s - r$

$$\int f \, d\mu_n = \sum_{i_{s+1}, \ldots, i_{n+r-1} = 0}^{k-1} a_{i_r i_{r+1}} \cdots a_{i_{s-1} i_s} a_{i_s i_{s+1}} \cdots a_{i_{n+r-1} i_r} N_n(T)^{-1}$$

$$= \frac{a_{i_r i_{r+1}} \cdots a_{i_{s-1} i_s} (A^{n+r-s})_{i_s i_r}}{N_n(T) \lambda^{-n} \lambda^n} \to \frac{a_{i_r i_{r+1}} \cdots a_{i_{s-1} i_s} u_{i_r} v_{i_s}}{\lambda^{s-r}}$$

by the above and Theorem 0.17. Therefore $\int f \, d\mu_n \to \int f \, d\mu$.   □

A special case of this theorem says that the $(1/k, \ldots, 1/k)$ product measure describes the distribution of the periodic points of the two-sided shift.

The corresponding result holds for automorphisms of tori:

Theorem 8.18. *Suppose* $A : K^p \to K^p$ *is an automorphism of* K^n *which is expansive (i.e. $[A]$ has no eigenvalues of absolute value 1). Then* $h(A) = \lim_{n \to \infty}$ $(1/n) \log N_n(A)$ *and the Haar measure m describes the distribution of the periodic points of A. (The same conclusions hold when A is merely ergodic but the proof is more delicate.)*

We refer to Bowen [4] for the proof of the second part and to Bowen [5] and §22 of Denker, Grillenberger and Sigmund [1] for a discussion of generalisations to a wider class of homeomorphisms. We now prove that $h(A) = \lim_{n \to \infty} (1/n) \log N_n(A)$.

PROOF. In the proof we shall use the fact that if $B : K^p \to K^p$ is an endomorphism of K^p onto K^p then the kernel of B contains $|\det [B]|$ points. (This is because the image $\tilde{B}(I)$ of the unit square I of R^p under the linear map $\tilde{B}$ has Lebesgue measure $|\det [B]|$ and so when $\tilde{B}(I)$ is reduced mod Z^p each point of I is covered by $|\det [B]|$ points of $\tilde{B}(I)$).

Since A is expansive, $A^n - I : K^p \to K^p$ (using additive notation) is an endomorphism of K^p onto K^p with corresponding matrix $[A^n - I] = [A]^n - I$. Therefore $\det[A^n - I] = \prod_{i=1}^p (\lambda_i^n - 1)$ where $\lambda_1, \ldots, \lambda_p$ are the eigenvalues of $[A]$. Then $N_n(A)$ is the cardinality of the kernel of $A^n - I$ so $N_n(A) = \prod_{i=1}^p |\lambda_i^n - 1|$ and

$$\frac{1}{n} \log N_n(A) = \sum_{i=1}^p \frac{1}{n} \log|\lambda_i^n - 1|.$$

If $|\lambda_i| > 1$ then

$$\frac{1}{n} \log|\lambda_i^n - 1| = \frac{1}{n} [\log|\lambda_i|^n + \log|1 - \lambda_i^{-n}|] \to \log|\lambda_i|.$$

If $|\lambda_i| < 1$ then $(1/n) \log|\lambda_i^n - 1| \to 0$. Hence

$$\lim_{n \to \infty} \frac{1}{n} \log N_n(A) = \sum_{\{i : |\lambda_i| > 1\}} \log|\lambda_i| = h(A). \qquad \square$$

§8.6 Definition of Measure-Theoretic Entropy Using the Metrics d_n

Let (X, d) be a compact metric space and let $T : X \to X$ be a homeomorphism. In §7.2 we introduced the metrics d_n on X by $d_n(x, y) = \max_{0 \leq i \leq n-1}$ $d(T^i(x), T^i(y))$. We then defined $r_n(\varepsilon, X)$ to be the minimum number of ε-balls, in the d_n metric, whose union covers X, and we showed $h(T) = \lim_{\varepsilon \to 0}$

$\limsup_{n\to\infty} (1/n)\log r_n(\varepsilon, X) = \lim_{\varepsilon\to 0} \liminf_{n\to\infty}(1/n)\log r_n(\varepsilon, X)$. A. B. Katok has given an analogous description of measure-theoretic entropy.

Theorem 8.19. *Let* $T: X \to X$ *be a homeomorphism of the compact metric space* (X, d). *Let* $m \in M(X, T)$ *and let* m *be ergodic. For* $\varepsilon > 0$, $\delta > 0$ *let* $r_n(\varepsilon, \delta, m)$ *denote the minimum number of ε-balls in the d_n metric whose union has m-measure more than or equal to* $1 - \delta$. *Then, for each* $\delta > 0$, *we have*

$$h_m(T) = \lim_{\varepsilon\to 0} \limsup_{n\to\infty} \frac{1}{n}\log r_n(\varepsilon, \delta, m) = \lim_{\varepsilon\to 0} \liminf_{n\to\infty} \frac{1}{n}\log r_n(\varepsilon, \delta, m).$$

We refer to Katok [1] for the proof.

CHAPTER 9

Topological Pressure and Its Relationship with Invariant Measures

Let $T: X \to X$ be a continuous transformation of a compact metric space (X, d). Let $C(X, R)$ denote the Banach algebra of real-valued continuous functions of X equipped with the supremum norm. The topological pressure of T will be a map $P(T, \cdot): C(X, R) \to R \cup \{\infty\}$ which will have good properties relative to the structures on $C(X, R)$. It contains topological entropy in the sense that $P(T, 0) = h(T)$ where 0 denotes the member of $C(X, R)$ which is identically zero. A generalisation of the variational principle of §8.2 is true and this sometimes gives a natural way of choosing important members of $M(X, T)$. In this theory ideas from mathematical statistical mechanics are used and the theory has important applications to other fields. We shall mention in §10.1, one important application to differentiable dynamical systems.

The concept of pressure in this type of setting was introduced by Ruelle [1] and studied in the general case in Walters [3].

§9.1 Topological Pressure

Let (X, d) be a compact metric space, $C(X, R)$ the space of real-valued continuous functions of X and $T: X \to X$ a continuous transformation. We shall use natural logarithms. The definition of pressure can be given by using open covers or spanning sets or separated sets. Since X is compact we can generalise the definition of $h(T)$ given in the remarks following Corollary 7.5.2. For $f \in C(X, R)$ and $n \geq 1$ we denote $\sum_{i=0}^{n-1} f(T^i x)$ by $(S_n f)(x)$.

Definition 9.1. For $f \in C(X, R)$, $n \geq 1$ and $\varepsilon > 0$ put

$$Q_n(T, f, \varepsilon) = \inf \left\{ \sum_{x \in F} e^{(S_n f)(x)} \, \middle| \, F \text{ is a } (n, \varepsilon) \text{ spanning set for } X \right\}.$$

Remarks

(1) $0 < Q_n(T, f, \varepsilon) \leq \|e^{S_n f}\| r_n(\varepsilon, X) < \infty$ (see Remark (1) of §7.2).
(2) If $\varepsilon_1 < \varepsilon_2$ then $Q_n(T, f, \varepsilon_1) \geq Q_n(T, f, \varepsilon_2)$.
(3) $Q_n(T, 0, \varepsilon) = r_n(\varepsilon, X)$.
(4) In Definition 9.1 it suffices to take the infinium over those (n, ε) spanning sets which don't have proper subsets that (n, ε) span X. This is because $e^{(S_n f)(x)} > 0$.

Definition 9.2. For $f \in C(X, R)$ and $\varepsilon > 0$ put

$$Q(T, f, \varepsilon) = \limsup_{n \to \infty} \frac{1}{n} \log Q_n(T, f, \varepsilon).$$

Remarks

(5) $Q(T, f, \varepsilon) \leq \|f\| + r(\varepsilon, X, T) < \infty$ (see Remark (1) and Theorem 7.7(ii)).
(6) If $\varepsilon_1 < \varepsilon_2$ then $Q(T, f, \varepsilon_1) \geq Q(T, f, \varepsilon_2)$ (by Remark (2)).

Definition 9.3. If $f \in C(X, R)$ let $P(T, f)$ denote $\lim_{\varepsilon \to 0} Q(T, f, \varepsilon)$.

Remarks

(7) By Remark (6), $P(T, f)$ exists but could be ∞.

Definition 9.4. The map $P(T, \cdot) : C(X, R) \to R \cup \{\infty\}$ defined above is called the *topological pressure of* T.

Clearly $P(T, 0) = h(T)$. We shall obtain some equivalent ways of giving the definition.

Definition 9.5. For $f \in C(X, R)$, $n \geq 1$ and $\varepsilon > 0$ put

$$P_n(T, f, \varepsilon) = \sup \left\{ \sum_{x \in E} e^{(S_n f)(x)} \, \middle| \, E \text{ is a } (n, \varepsilon) \text{ separated subset of } X \right\}.$$

Remarks

(8) If $\varepsilon_1 < \varepsilon_2$ then $P_n(T, f, \varepsilon_1) \geq P_n(T, f, \varepsilon_2)$.
(9) $P_n(T, 0, \varepsilon) = s_n(\varepsilon, X)$.
(10) In Definition 9.5 it suffices to take the supremum over all the (n, ε) separated sets which fail to be (n, ε) separated when any point of X is added. This is because $e^{(S_n f)(x)} > 0$.
(11) We have $Q_n(T, f, \varepsilon) \leq P_n(T, f, \varepsilon)$. This follows from Remark (10) and the fact that a (n, ε) separated set which cannot be enlarged to a (n, ε) separated set must be a (n, ε) spanning set for X.

(12) If $\delta > 0$ is such that $d(x, y) < \varepsilon/2$ implies $|f(x) - f(y)| < \delta$ then $P_n(T, f, \varepsilon) \leq e^{n\delta}Q_n(T, f, \varepsilon/2)$.

PROOF. Let E be a (n, ε) separated set and F a $(n, \varepsilon/2)$ spanning set. Define $\phi : E \to F$ by choosing, for each $x \in E$, some point $\phi(x) \in F$ with $d_n(x, \phi(x)) \leq \varepsilon/2$ (using the notation $d_n(x, y) = \max_{0 \leq i \leq n-1} d(T^i(x), T^i(y))$). Then ϕ is injective so

$$\sum_{y \in F} e^{(S_n f)(y)} \geq \sum_{y \in \phi E} e^{(S_n f)(y)} \geq \left(\min_{x \in E} e^{(S_n f)(\phi x) - (S_n f)(x)} \right) \sum_{x \in E} e^{(S_n f)(x)}$$

$$\geq e^{-n\delta} \sum_{x \in E} e^{(S_n f)(x)}.$$

Therefore $Q_n(T, f, \varepsilon/2) \geq e^{-n\delta} P_n(T, f, \varepsilon)$. $\qquad\qquad\square$

Definition 9.6. For $f \in C(X, R)$ and $\varepsilon > 0$ put

$$P(T, f, \varepsilon) = \limsup_{n \to \infty} \frac{1}{n} \log P_n(T, f, \varepsilon).$$

Remarks

(13) $Q(T, f, \varepsilon) \leq P(T, f, \varepsilon)$ (by Remark (11)).
(14) If δ is such that $d(x, y) < \varepsilon/2$ implies $|f(x) - f(y)| < \delta$ then $P(T, f, \varepsilon) \leq \delta + Q(T, f, \varepsilon)$ (by Remark (12)).
(15) If $\varepsilon_1 < \varepsilon_2$ then $P(T, f, \varepsilon_1) \geq P(T, f, \varepsilon_2)$.

Theorem 9.1. *If $f \in C(X, R)$ then $P(T, f) = \lim_{\varepsilon \to 0} P(T, f, \varepsilon)$.*

PROOF. The limit exists by Remark 15. By Remark 13 we have $P(T, f) \leq \lim_{\varepsilon \to 0} P(T, f, \varepsilon)$.
By Remark 14, for any $\delta > 0$ we have $\lim_{\varepsilon \to 0} P(T, f, \varepsilon) \leq \delta + P(T, f)$ so $\lim_{\varepsilon \to 0} P(T, f, \varepsilon) \leq P(T, f)$. $\qquad\qquad\square$

To obtain definitions of pressure involving open covers we generalise Theorem 7.7. We need the following definitions.

Definition 9.7. If $f \in C(X, R)$, $n \geq 1$ and α is an open cover of X put

$$q_n(T, f, \alpha) = \inf \left\{ \sum_{B \in \beta} \inf_{x \in B} e^{(S_n f)(x)} \Big| \beta \right.$$

is a finite subcover of $\bigvee_{i=0}^{n-1} T^{-i}\alpha\}$ and

$$p_n(T, f, \alpha) = \inf \left\{ \sum_{B \in \beta} \sup_{x \in B} e^{(S_n f)(x)} \Big| \beta \right.$$

is a finite subcover of $\bigvee_{i=0}^{n-1} T^{-1}\alpha\}$.
Clearly $q_n(T, f, \alpha) \leq p_n(T, f, \alpha)$.

Theorem 9.2. *Let* $T:X \to X$ *be continuous and* $f \in C(X,R)$.

(i) *If* α *is an open cover of* X *with Lebesgue number* δ *then* $q_n(T,f,\alpha) \le Q_n(T,f,\delta/2) \le P_n(T,f,\delta/2)$.

(ii) *If* $\varepsilon > 0$ *and* γ *is an open cover with* $\mathrm{diam}(\gamma) \le \varepsilon$ *then* $Q_n(T,f,\varepsilon) \le P_n(T,f,\varepsilon) \le p_n(T,f,\gamma)$.

PROOF. We know from Remark 13 that $Q_n(T,f,\varepsilon) \le P_n(T,f,\varepsilon)$ for all $\varepsilon > 0$.

(i) If F is an $(n,\delta/2)$ spanning set then $X = \bigcup_{x \in F} \bigcap_{i=0}^{n-1} T^{-i}\bar{B}(T^i x; \delta/2)$. Since each $\bar{B}(T^i x; \delta/2)$ is a subset of a member of α we have $q_n(T,f,\alpha) \le \sum_{x \in F} e^{(S_n f)(x)}$ and hence $q_n(T,f,\alpha) \le Q_n(T,f,\delta/2)$.

(ii) Let E be a (n,ε) separated subset of X. Since no member of $\bigvee_{i=0}^{n-1} T^{-i}\gamma$ contains two elements of E we have $\sum_{x \in E} e^{(S_n f)(x)} \le p_n(T,f,\gamma)$. Therefore $P_n(T,f,\varepsilon) \le p_n(T,f,\gamma)$.   □

Remarks

(16) If α, γ are open covers of X and $\alpha < \gamma$ then $q_n(T,f,\alpha) \le q_n(T,f,\gamma)$.

(17) If $d(x,y) < \mathrm{diam}(\alpha)$ implies $|f(x) - f(y)| \le \delta$ then $p_n(T,f,\alpha) \le e^{n\delta} q_n(T,f,\alpha)$.

Lemma 9.3. *If* $f \in C(X,R)$ *and* α *is an open cover of* X *then*

$$\lim_{n \to \infty} \frac{1}{n} \log p_n(T,f,\alpha)$$

exists and equals $\inf_n (1/n) \log p_n(T,f,\alpha)$.

PROOF. By Theorem 4.9 it suffices to show $p_{n+k}(T,f,\alpha) \le p_n(T,f,\alpha) \cdot p_k(T,f,\alpha)$. If β is a finite subcover of $\bigvee_{i=0}^{n-1} T^{-i}\alpha$ and γ is a finite subcover of $\bigvee_{i=0}^{k-1} T^{-i}\alpha$ then $\beta \vee T^{-n}\gamma$ is a finite subcover of $\bigvee_{i=0}^{k+n-1} T^{-i}\alpha$, and we have

$$\sum_{D \in \beta \vee T^{-n}\gamma} \sup_{x \in D} e^{(S_{n+k} f)(x)} \le \left(\sum_{B \in \beta} \sup_{x \in B} e^{(S_n f)(x)} \right)\left(\sum_{C \in \gamma} \sup_{x \in C} e^{(S_k f)(x)} \right).$$

Therefore $p_{n+k}(T,f,\alpha) \le p_n(T,f,\alpha) \cdot p_k(T,f,\alpha)$.   □

The following gives definitions of pressure using open covers.

Theorem 9.4. *If* $T:X \to X$ *is continuous and* $f \in C(X,R)$ *then each of the following equals* $P(T,f)$.

(i) $\lim_{\delta \to 0} \left[\sup_\alpha \{ \lim_{n \to \infty} (1/n) \log p_n(T,f,\alpha) | \, \alpha \text{ is an open cover of } X \text{ with } \mathrm{diam}(\alpha) \le \delta \} \right]$.

(ii) $\lim_{k \to \infty} \left[\lim_{n \to \infty} (1/n) \log p_n(T,f,\alpha_k) \right]$ *if* $\{\alpha_k\}$ *is a sequence of open covers with* $\mathrm{diam}(\alpha_k) \to 0$.

(iii) $\lim_{\delta \to 0} \left[\sup_\alpha \{ \liminf_{n \to \infty} (1/n) \log q_n(T,f,\alpha) | \, \alpha \text{ is an open cover of } X \text{ with } \mathrm{diam}(\alpha) \le \delta \} \right]$.

(iv) $\lim_{\delta \to 0} [\sup_\alpha \lim\sup_{n \to \infty} (1/n) \log q_n(T, f, \alpha) | \alpha$ is an open cover of X with $\operatorname{diam}(\alpha) \leq \delta\}]$.

(v) $\lim_{k \to \infty} [\lim\sup_{n \to \infty} (1/n) q_n(T, f, \alpha_k)]$ if $\{\alpha_k\}$ is a sequence of open covers with $\operatorname{diam}(\alpha_k) \to 0$.

(vi) $\sup_\alpha \{\lim\sup_{n \to \infty} (1/n) \log q_n(T, f, \alpha) | \alpha$ is an open cover of $X\}$.

(vii) $\lim_{\varepsilon \to 0} \lim\inf_{n \to \infty} (1/n) Q_n(T, f, \varepsilon)$.

(viii) $\lim_{\varepsilon \to 0} \lim\inf_{n \to \infty} (1/n) \log P_n(T, f, \varepsilon)$.

PROOF

(i) If $\delta > 0$ and γ is an open cover with $\operatorname{diam}(\gamma) \leq \delta$ then $P_n(T, f, \delta) \leq p_n(T, f, \gamma)$ (Theorem 9.2(ii)). Therefore

$$P(T, f, \delta) \leq \sup\left\{ \lim_{n \to \infty} \frac{1}{n} \log p_n(T, f, \gamma) | \gamma \right.$$

is an open cover of X with $\operatorname{diam}(\gamma) \leq \delta\}$, using Lemma 9.3. Therefore $P(T, f)$ is no larger than the expression in (i).

If α is a cover and δ is a Lebesgue number for α then $q_n(T, f, \alpha) \leq P_n(T, f, \delta/2)$ by Theorem 9.2(i). Also if $\tau_\alpha = \sup\{|f(x) - f(y)| : d(x, y) \leq \operatorname{diam}(\alpha)\}$ then $p_n(T, f, \alpha) \leq e^{n\tau_\alpha} q_n(T, f, \alpha)$, by Remark 17. Hence

$$p_n(T, f, \alpha) \leq e^{n\tau_\alpha} P_n(T, f, \delta/2)$$

so

$$\lim_{n \to \infty} \frac{1}{n} \log p_n(T, f, \alpha) \leq \tau_\alpha + P(T, f),$$

and

$$\lim_{\eta \to 0} \left(\sup_\alpha \left\{ \lim_{n \to \infty} \frac{1}{n} \log p_n(T, f, \alpha) \middle| \operatorname{diam}(\alpha) \leq \eta \right\} \right) \leq P(T, f).$$

Therefore (i) is proved. The same reasoning proves (ii).

(iii) We know $q_n(T, f, \alpha) \leq p_n(T, f, \alpha)$ for all α. Also if

$$\tau_\alpha = \sup\{|f(x) - f(y)| : d(x, y) \leq \operatorname{diam}(\alpha)\} \quad \text{then } p_n(T, f, \alpha) \leq e^{n\tau_\alpha} q_n(T, f, \alpha)$$

(Remark 17). Therefore

$$e^{-n\tau_\alpha} p_n(T, f, \alpha) \leq q_n(T, f, \alpha) \leq p_n(T, f, \alpha)$$

so

$$-\tau_\alpha + \lim_{n \to \infty} \frac{1}{n} \log p_n(T, f, \alpha) \leq \lim\inf_{n \to \infty} \frac{1}{n} \log q_n(T, f, \alpha)$$

$$\leq \lim\sup_{n \to \infty} \log q_n(T, f, \alpha)$$

$$\leq \lim_{n \to \infty} \frac{1}{n} \log p_n(T, f, \alpha).$$

The formulae in (iii), (iv), and (v) follow from (i) and (ii).

(vi) Let α be an open cover of X and let 2ε be a Lebesgue number for α. By Theorem 9.2 $q_n(T,f,\alpha) \leq Q_n(T,f,\varepsilon)$ so that $\limsup_{n\to\infty}(1/n)q_n(T,f,\alpha) \leq Q(T,f,\varepsilon) \leq P(T,f)$. Therefore the expression in (v) is majorised by $P(T,f)$. The opposite inequality follows from (iv).

(vii) and (viii) Let α_ε denote the cover of X be all open balls of radius 2ε and γ_ε denote any cover by balls of radius $(\varepsilon/2)$. Then, by Theorem 9.2 and Remark 17,

$$e^{-n\tau_{4\varepsilon}}p_n(T,f,\alpha_\varepsilon) \leq q_n(T,f,\alpha_\varepsilon) \leq Q_n(T,f,\varepsilon) \leq P_n(T,f,\varepsilon) \leq p_n(T,f,\gamma_\varepsilon)$$

where $\tau_{4\varepsilon} = \sup\{|f(x) - f(y)|:d(x,y) \leq 4\varepsilon\}$. Then (vii) and (viii) follow by taking lim infs in this expression and using (ii). $\qquad\square$

Remarks

(18) For some examples $\sup\{\lim_{n\to\infty}(1/n)\log p_n(T,f,\alpha)|\alpha$ is an open cover of $X\}$ is strictly larger than $P(T,f)$.

(19) From (vi) of Theorem 9.4 we see that $P(T,f)$ does not depend on the metric on X.

As one may expect, from our knowledge of topological entropy, the definition of pressure can be simplified for expansive homeomorphisms. We shall need the following lemma.

Lemma 9.5. *If* $T:X \to X$ *is a continuous transformation of a compact metrisable space and* α *is an open cover of* X *then for* $k > 0$ *and* $f \in C(X,R)$

$$\limsup_{n\to\infty} \frac{1}{n}\log q_n(T,f,\alpha) = \limsup_{n\to\infty}\frac{1}{n}\log q_n\left(T,f,\bigvee_{i=0}^{k} T^{-i}\alpha\right)$$

and $\lim_{n\to\infty}(1/n)\log p_n(T,f,\alpha) = \lim_{n\to\infty}(1/n)\log p_n(T,f,\bigvee_{i=0}^{k} T^{-i}\alpha)$. *If* T *is a homeomorphism and* $k,m \geq 0$ *then these formulae hold with* $\bigvee_{i=0}^{k-1}T^{-i}\alpha$ *replaced by* $\bigvee_{i=-m}^{k}T^{-i}\alpha$.

PROOF. One readily gets

$$e^{-(k+1)\|f\|}q_{n+k}(T,f,\alpha) \leq q_n\left(T,f,\bigvee_{i=0}^{k}T^{-i}\alpha\right) \leq e^{(k+1)\|f\|}q_{n+k}(T,f,\alpha)$$

and

$$e^{-(k+1)\|f\|}p_{n+k}(T,f,\alpha) \leq p_n\left(T,f,\bigvee_{i=0}^{k}T^{-i}\alpha\right) \leq e^{(k+1)\|f\|}p_{n+k}(T,f,\alpha).$$

The first results follow from this. To obtain the second result it suffices to show

$$\limsup_{n\to\infty}\frac{1}{n}\log q_n(T,f,T^{-1}\beta) = \limsup_{n\to\infty}\frac{1}{n}\log q_n(T,f,\beta)$$

and

$$\lim \frac{1}{n} \log p_n(T, f, T^{-1}\beta) = \lim \frac{1}{n} \log p_n(T, f, \beta)$$

when T is a homeomorphism and β is an open cover of X. Since γ is a finite subcover of $\bigvee_{i=0}^{n-1} T^{-i}\beta$ iff $T^{-1}\gamma$ is a finite subcover $\bigvee_{i=1}^{n} T^{-i}\beta$ we have

$$\frac{q_n(T, f, \beta)}{q_n(T, f, T^{-1}\beta)} \geq \inf \left\{ \frac{\sum\limits_{C \in \gamma} \inf\limits_{x \in C} e^{S_n f(x)}}{\sum\limits_{C \in \gamma} \inf\limits_{y \in T^{-1}C} e^{S_n f(y)}} \middle| \gamma \text{ is a finite subcover of } \bigvee_{i=0}^{n-1} T^{-i}\beta \right\}$$

$$= \inf \left\{ \frac{\sum\limits_{C \in \gamma} \inf\limits_{x \in C} e^{S_n f(x)}}{\sum\limits_{C \in \gamma} \inf\limits_{x \in C} e^{S_n f(T^{-1}x)}} \middle| \gamma \text{ is a finite subcover of } \bigvee_{i=0}^{n-1} T^{-i}\beta \right\}$$

$$\geq \inf \left\{ \min_{C \in \gamma} \frac{\inf\limits_{x \in C} e^{S_n f(x)}}{\inf\limits_{x \in C} e^{S_n f(T^{-1}x)}} \middle| \gamma \text{ is a finite subcover of } \bigvee_{i=0}^{n-1} T^{-i}\beta \right\}$$

$$\geq e^{-2\|f\|}.$$

A similar proof gives

$$\frac{q_n(T, f, T^{-1}\beta)}{q_n(T, f, \beta)} \geq e^{-2\|f\|}$$

so the result follows. The result for p_n follows by a similar proof. □

The following result generalises Theorem 7.11.

Theorem 9.6. *Let* $T: X \to X$ *be an expansive homeomorphism of a compact metric space* (X, d).

(i) *If* α *is a generator for* T *then*

$$P(T, f) = \lim_{n \to \infty} \frac{1}{n} \log p_n(T, f, \alpha)$$

$$= \limsup_{n \to \infty} \frac{1}{n} \log q_n(T, f, \alpha) \quad \forall f \in C(X, R).$$

(ii) *If* δ *is an expansive constant for* T *then* $P(T, f) = P(T, f, \delta_0) = Q(T, f, \delta_0)$ *for all* $\delta_0 < \delta/4$ *and all* $f \in C(X, R)$.

PROOF

(i) Let α be a generator for T. By Theorem 5.21 we have

$$\text{diam}\left(\bigvee_{i=-k}^{k} T^{-i}\alpha \right) \to 0$$

and so by Theorem 9.4(ii)

$$P(T,f) = \lim_{k \to \infty} \lim_{n \to \infty} \frac{1}{n} \log p_n \left(T, f, \bigvee_{i=-k}^{k} T^{-i}\alpha \right).$$

An application of Lemma 9.5 gives the desired result. The formula involving q_n is proved in a similar way using Theorem 9.4(v) and Lemma 9.5.

(ii) Let $\delta_0 < \delta/4$ and choose $x_1, \ldots, x_k$ so that $X = \bigcup_{i=1}^{k} B(x_i; (\delta/2) - 2\delta_0)$. The cover $\alpha = \{B(x_i; \delta/2) \mid 1 \le i \le k\}$ has $2\delta_0$ for a Lebesgue number so by Theorem 9.2

$$\limsup_{n \to \infty} \frac{1}{n} \log q_n(T, f, \alpha) \le \limsup_{n \to \infty} \frac{1}{n} \log Q_n(T, f, \delta_0)$$

$$\le \limsup_{n \to \infty} \frac{1}{n} \log P_n(T, f, \delta_0) \le P(T, f).$$

The result follows by (i) since α is a generator. □

We can use this to calculate $P(T,f)$ when T is the two-sided shift on $X = \prod_{-\infty}^{\infty} Y$, $Y = \{0, 1, \ldots, k-1\}$, and f depends only on the 0-th coordinate i.e. $f(\{x_n\}) = a_{x_0}$ where $a_0, a_1, \ldots, a_{k-1} \in R$. If α denotes the natural generator then $q_n(T, f, \alpha) = p_n(T, f, \alpha) = (e^{a_0} + \cdots + e^{a_{k-1}})^n$ so $P(T,f) = \log(e^{a_0} + \cdots + e^{a_{k-1}})$.

§9.2 Properties of Pressure

We now study the properties of $P(T, \cdot) : C(X, R) \to R \cup \{\infty\}$. In particular we see that either $P(T, \cdot)$ never takes the value ∞ or is identically ∞.

Theorem 9.7. *Let $T : X \to X$ be a continuous transformation of a compact metrisable space X. If $f, g \in C(X, R)$, $\varepsilon > 0$ and $c \in R$ then the following are true.*

(i) $P(T, 0) = h(T)$.

(ii) $f \le g$ implies $P(T, f) \le P(T, g)$. In particular $h(T) + \inf f \le P(T, f) \le h(T) + \sup f$.

(iii) $P(T, \cdot)$ is either finite valued or constantly ∞.

(iv) $|P(T, f, \varepsilon) - P(T, g, \varepsilon)| \le \|f - g\|$, and so if $P(T, \cdot) < \infty$, $|P(T, f) - P(T, g)| \le \|f - g\|$.

(v) $P(T, \cdot, \varepsilon)$ is convex, and so if $P(T, \cdot) < \infty$ then $P(T, \cdot)$ is convex.

(vi) $P(T, f + c) = P(T, f) + c$.

(vii) $P(T, f + g \circ T - g) = P(T, f)$.

(viii) $P(T, f + g) \le P(T, f) + P(T, g)$.

(ix) $P(T, cf) \le cP(T, f)$ if $c \ge 1$ and $P(T, cf) \ge cP(T, f)$ if $c \le 1$.

(x) $|P(T, f)| \le P(T, |f|)$.

PROOF. Several times in the proofs we shall use the simple inequality

$$\frac{\sup a_j}{\sup b_j} \le \sup\left(\frac{a_j}{b_j}\right)$$

when (a_j), (b_j) are collections of positive real numbers.

(i) and (ii) are clear from the definition of pressure.

(iii) From (ii) and (i) we get

$$h(T) + \inf f \le P(T,f) \le h(T) + \sup f$$

so $P(T,f) = \infty$ iff $h(T) = \infty$.

(iv) By the inequality mentioned at the beginning of the proof we have

$$\frac{P_n(T,f,\varepsilon)}{P_n(T,g,\varepsilon)} \le \sup\left\{ \frac{\displaystyle\sum_{x \in E} e^{(S_n f)(x)}}{\displaystyle\sum_{x \in E} e^{(S_n g)(x)}} \,\middle|\, E \text{ is a } (n,\varepsilon) \text{ separated set} \right\}$$

$$\le \sup\left\{ \max_{x \in E} \frac{e^{(S_n f)(x)}}{e^{(S_n g)(x)}} \,\middle|\, E \text{ is a } (n,\varepsilon) \text{ separated set} \right\}$$

$$\le e^{n\|f - g\|}.$$

This proves (iv).

(v) By Hölder's inequality, if $p \in [0,1]$ and E is a finite subset of X, we have

$$\sum_{x \in E} e^{p(S_n f)(x) + (1-p)(S_n g)(x)} \le \left(\sum_{x \in E} e^{(S_n f)(x)} \right)^p \left(\sum_{x \in E} e^{(S_n g)(x)} \right)^{1-p}.$$

Therefore $P_n(T, pf + (1-p)g, \varepsilon) \le P_n(T,f,\varepsilon)^p \cdot P_n(T,g,\varepsilon)^{1-p}$ and (v) follows.

(vi) is clear from the definition of pressure.

(vii) We have

$$P_n(T, f + g \circ T - g, \varepsilon)$$
$$= \sup\left\{ \sum_{x \in E} e^{(S_n f)(x) + g(T^n x) - g(x)} \,\middle|\, E \text{ is a } (n,\varepsilon) \text{ separated set} \right\}$$

so that

$$e^{-2\|g\|} P_n(T,f,\varepsilon) \le P_n(T, f + g \circ T - g, \varepsilon) \le e^{2\|g\|} P_n(T,f,\varepsilon).$$

The result follows from this.

(viii) This follows because $P_n(T, f + g, \varepsilon) \le P_n(T,f,\varepsilon) \cdot P_n(T,g,\varepsilon)$.

(ix) If $a_1, \ldots, a_k$ are positive numbers with $\sum_{i=1}^k a_i = 1$ then $\sum_{i=1}^k a_i^c \le 1$ if $c \ge 1$, and $\sum_{i=1}^k a_i^c \ge 1$ if $c \le 1$. Therefore, if E is a finite subset of X we have

$$\sum_{x \in E} e^{c(S_n f)(x)} \le \left(\sum_{x \in E} e^{(S_n f)(x)} \right)^c \quad \text{if } c \ge 1$$

and

$$\sum_{x \in E} e^{c(S_n f)(x)} \ge \left(\sum_{x \in E} e^{(S_n f)(x)} \right)^c \quad \text{if } c \le 1.$$

Therefore $P_n(T, cf, \varepsilon) \le (P_n(T, f, \varepsilon))^c$ if $c \ge 1$ and $P_n(T, cf, \varepsilon) \ge (P_n(T, f, \varepsilon))^c$ if $c \le 1$.

(x) Since $-|f| < f < |f|$ we have, by (ii), $P(T, -|f|) \le P(T, f) \le P(T, |f|)$. From (ix) we have $-P(T, |f|) \le P(T, -|f|)$ so $|P(T, f)| \le P(T, |f|)$.   □

We now investigate how $P(T, \cdot)$ depends on T. Some other properties, that we could prove now, are given after we prove the variational principle in §9.3.

Theorem 9.8. *Let* $T: X \to X$ *be a continuous transformation of a compact metrisable space and let* $f \in C(X, R)$. *The following are true.*

(i) *If* $k > 0$ $P(T^k, S_k f) = kP(T, f)$. *(Here* $(S_k f)(x) = \sum_{i=0}^{k-1} f(T^i x)$.)

(ii) *If* T *is a homeomorphism* $P(T^{-1}, f) = P(T, f)$.

(iii) *If* Y *is a closed subset of* X *with* $TY \subset Y$ *then* $P(T|_Y, f|_Y) \le P(T, f)$.

(iv) *If* $T_i: X_i \to X_i$ $(i = 1, 2)$ *is a continuous map of a compact metrisable space* (X_i, d_i) *and if* $\phi: X_1 \to X_2$ *is a surjective continuous map with* $\phi T_1 = T_2 \phi$ *then* $P(T_2, f) \le P(T_1, f \circ \phi)$ $\forall f \in C(X_2, R)$. *If* ϕ *is a homeomorphism then* $P(T_2, f) = P(T_1, f \circ \phi)$ $\forall f \in C(X_2, R)$.

(v) *If* $T_i: X_i \to X_i$ $(i = 1, 2)$ *is a continuous map of a compact metrisable space* (X_i, d_i) *and if* $f_i \in C(X_i, R)$ *then* $P(T_1 \times T_2, f_1 \times f_2) = P(T_1, f_1) + P(T_2, f_2)$ *where* $f_1 \times f_2 \in C(X_1 \times X_2, R)$ *is defined by* $(f_1 \times f_2)(x_1, x_2) = f_1(x_1) + f_2(x_2)$.

PROOF

(i) If F is (nk, ε) spanning for T then F is (n, ε) spanning for T^k. Hence $Q_n(T^k, S_k f, \varepsilon) \le Q_{nk}(T, f, \varepsilon)$ so that $P(T^k, S_k f) \le kP(T, f)$. To show the opposite inequality we let $\varepsilon > 0$ and choose $\delta > 0$ so that $d(x, y) < \delta$ implies $\max_{1 \le i \le k-1} d(T^i x, T^i y) < \varepsilon$. If F is (n, δ) spanning for T^k then F is (nk, ε) spanning for T. Hence $Q_n(T^k, S_k f, \delta) \ge Q_{nk}(T, f, \varepsilon)$ and so $P(T^k, S_k f) \ge Q(T^k, S_k f, \delta) \ge kQ(T, f, \varepsilon)$. Letting ε go to zero gives $P(T^k, S_k f) \ge kP(T, f)$.

(ii) A set E is (n, ε) separated for T iff $T^{n-1} E$ is (n, ε) separated for T^{-1}. Also

$$\sum_{x \in E} e^{(S_n f)(x)} = \sum_{y \in T^{n-1} E} e^{f(y) + f(T^{-1} y) + \cdots + f(T^{-(n-1)} y)}$$

so $P_n(T, f, \varepsilon) = P_n(T^{-1}, f, \varepsilon)$ and $P(T, f) = P(T^{-1}, f)$.

(iii) This is clear since a (n, ε) separated subset of Y is a (n, ε) separated subset of X.

(iv) Let $\varepsilon > 0$ and choose $\delta > 0$ so that $d_1(x, y) < \delta$ implies $d_2(\phi(x), \phi(y)) < \varepsilon$. If F is (n, δ) spanning for T_1 then ϕF is (n, ε) spanning for T_2 and

$$\sum_{x \in F} e^{f(\phi x) + f(\phi T_1 x) + \cdots + f(\phi T_1^{n-1} x)} \ge \sum_{y \in \phi F} e^{f(y) + f(T_2 y) + \cdots + f(T_2^{n-1} y)} \ge Q_n(T_2, f, \varepsilon).$$

Therefore $P(T_1, f \circ \phi) \geq Q(T_1, f \circ \phi, \delta) \geq Q(T_2, f, \varepsilon)$ so $P(T_1, f \circ \phi) \geq P(T_2, f)$.

If ϕ is a homeomorphism then we can apply the above with T_1, T_2, ϕ, f replaced by T_2, T_1, ϕ^{-1}, $f \circ \phi$ respectively to give $P(T_2, f) \geq P(T_1, f \circ \phi)$.

(v) Consider the metric on $X_1 \times X_2$ given by $d((x_1, x_2), (y_1, y_2)) = \max\{d_1(x_1, y_1), d_2(x_2, y_2)\}$. If F_i is a (n, ε) spanning set for X_i then $F_1 \times F_2$ is a (n, ε) spanning set for $X_1 \times X_2$ with respect to $T_1 \times T_2$. Also

$$\sum_{(x_1, x_2) \in F_1 \times F_2} \exp\left(\sum_{i=0}^{n-1} (f_1 \times f_2)(T_1 \times T_2)^i(x_1, x_2)\right)$$

$$= \left(\sum_{x_1 \in F_1} \exp\left(\sum_{i=0}^{n-1} f_1(T_1^i(x_1))\right)\right)\left(\sum_{x_2 \in F_2} \exp\left(\sum_{i=0}^{n-1} f_2(T_2^i x_2)\right)\right)$$

so that

$$Q_n(T_1 \times T_2, f_1 \times f_2, \varepsilon) \leq Q_n(T_1, f_1, \varepsilon) \cdot Q_n(T_2, f_2, \varepsilon).$$

Therefore $P(T_1 \times T_2, f_1 \times f_2) \leq P(T_1, f_1) + P(T_2, f_2)$.

If E_i is a (n, ε) separated set for X_i then $E_1 \times E_2$ is a (n, ε) separated set for $X_1 \times X_2$, so that

$$P_n(T_1 \times T_2, f_1 \times f_2, \varepsilon) \geq P_n(T_1, f_1, \varepsilon) \cdot P_n(T_2, f_2, \varepsilon).$$

Since

$$\limsup_{n \to \infty} \frac{1}{n} \log P_n(T_1 \times T_2, f_1 \times f_2, \varepsilon)$$

$$\geq \liminf_{n \to \infty} \frac{1}{n} \log P_n(T_1, f_1, \varepsilon) + \limsup_{n \to \infty} \frac{1}{n} \log P_n(T_2, f_2, \varepsilon),$$

Theorem 9.4(viii) gives $P(T_1 \times T_2, f_1 \times f_2) \geq P(T_1, f_1) + P(T_2, f_2)$. $\quad\square$

§9.3 The Variational Principle

We now extend the variational principle of §8.2. This variational principle was proved for some transformations by D. Ruelle [1] and then for all transformations by P. Walters [3]. The proof we give here is similar to that of Theorem 8.6 and is due to M. Misiurewicz [2]. Other proofs have been given by M. Denker and R. Bowen.

We shall use the remarks given in §8.2 before the proof of Theorem 8.6. We shall also use the following simple result.

Lemma 9.9. Let $a_1, \ldots, a_k$ be given real numbers. If $p_i \geq 0$ and $\sum_{i=1}^{k} p_i = 1$ then

$$\sum_{i=1}^{k} p_i(a_i - \log p_i) \leq \log\left(\sum_{i=1}^{k} e^{a_i}\right)$$

and equality holds iff

$$p_i = \frac{e^{a_i}}{\sum\limits_{j=1}^{k} e^{a_j}}.$$

PROOF. Let $M = \sum_{j=1}^{k} e^{a_j}$. In Theorem 4.2 put

$$\alpha_i = \frac{e^{a_i}}{M} \quad \text{and} \quad x_i = \frac{p_i M}{e^{a_i}}.$$

Then

$$0 = \phi(1) \le \sum_{i=1}^{k} \frac{e^{a_i}}{M} \frac{p_i M}{e^{a_i}} \log\left(\frac{p_i M}{e^{a_i}}\right)$$

$$= \sum_{i=1}^{k} p_i[\log p_i + \log M - a_i].$$

Therefore $\sum p_i(a_i - \log p_i) \le \log M$, and equality holds iff $(p_i M/e^{a_i})$ is independent of i i.e. $p_i = (e^{a_i}/M)$. ∎

Theorem 9.10. *Let $T:X \to X$ be a continuous map of a compact metric space and let $f \in C(X, R)$. Then*

$$P(T, f) = \sup\left\{ h_\mu(T) + \int f \, d\mu \,\middle|\, \mu \in M(X, T) \right\}.$$

PROOF

(1) Let $\mu \in M(X, T)$. We shall show $h_\mu(T) + \int f \, d\mu \le P(T, f)$. Let $\xi = \{A_1, \ldots, A_k\}$ be a partition of $(X, \mathscr{B}(X))$. Let $a > 0$ be given. Choose $\varepsilon > 0$ so that $\varepsilon k \log k < a$. Since μ is regular there are compact sets $B_j \subset A_j$ with $\mu(A_j \backslash B_j) < \varepsilon$, $1 \le j \le k$. Let η be the partition $\eta = \{B_0, B_1, \ldots, B_k\}$ where $B_0 = X \backslash \bigcup_{j=1}^{k} B_j$. Then as in the proof of Theorem 8.6 $H_\mu(\xi/\eta) < \varepsilon k \log k < a$. Let

$$b = \min_{1 \le i \ne j \le k} d(B_i, B_j) > 0.$$

Pick $\delta > 0$ so that $\delta < b/2$ and so that $d(x, y) < \delta$ implies $|f(x) - f(y)| < \varepsilon$. Fix n and let E be an (n, δ) separated set with respect to T, which fails to be (n, δ) separated when any point is added. Then E is also (n, δ) spanning. If $C \in \bigvee_{i=0}^{n-1} T^{-i}\eta$ let $\alpha(C)$ denote $\sup\{(S_n f)(x) | x \in C\}$. Then

$$H_\mu\left(\bigvee_{i=0}^{n-1} T^{-i}\eta\right) + \int S_n f \, d\mu \le \sum_{C \in \bigvee_{i=0}^{n-1} T^{-i}\eta} \mu(C)[-\log \mu(C) + \alpha(C)]$$

$$\le \log \sum_{C \in \bigvee_{i=0}^{n-1} T^{-i}\eta} e^{\alpha(C)} \quad \text{by Lemma 9.9.}$$

For each $C \in \bigvee_{i=0}^{n-1} T^{-i}\eta$ choose some $x \in \bar{C}$ so that $(S_n f)(x) = \alpha(C)$. Since E is (n, δ) spanning choose $y(C) \in E$ with $d(T^i x, T^i y(C)) \le \delta$, $0 \le j \le n - 1$.

Then $\alpha(C) \leq (S_n f)(y(C)) + n\varepsilon$. Also each ball of radius δ meets the closures of at most two members of η so if $y \in E$ then $\{C \in \bigvee_{i=0}^{n-1} T^{-i}\eta \mid y(C) = y\}$ has cardinality at most 2^n. Therefore

$$\sum_{C \in \bigvee_{i=0}^{n-1} T^{-i}\eta} e^{\alpha(C) - n\varepsilon} \leq \sum_{C \in \bigvee_{i=0}^{n-1} T^{-i}\eta} e^{(S_n f)(y(C))} \leq 2^n \sum_{y \in E} e^{(S_n f)(y)}$$

and so

$$\log \left(\sum_{C \in \bigvee_{i=0}^{n-1} T^{-i}\eta} e^{\alpha(C)} \right) - n\varepsilon \leq n \log 2 + \log \sum_{y \in E} e^{(S_n f)(y)}$$

Hence

$$\frac{1}{n} H_\mu \left(\bigvee_{i=0}^{n-1} T^{-i}\eta \right) + \int f \, d\mu = \frac{1}{n} H_\mu \left(\bigvee_{i=0}^{n-1} T^{-i}\eta \right) + \frac{1}{n} \int S_n f \, d\mu$$

$$\leq \varepsilon + \log 2 + \frac{1}{n} \log \sum_{y \in E} e^{(S_n f)(y)}$$

$$\leq \varepsilon + \log 2 + \frac{1}{n} \log P_n(T, f, \delta)$$

and therefore

$$h_\mu(T, \eta) + \int f \, d\mu \leq \varepsilon + \log 2 + P(T, f, \delta)$$

$$\leq \varepsilon + \log 2 + P(T, f).$$

Now $h_\mu(T, \xi) \leq h_\mu(T, \eta) + H_\mu(\xi/\eta)$ (Theorem 4.12(iv)) so that $h_\mu(T, \xi) + \int f \, d\mu \leq 2a + \log 2 + P(T, f)$ and hence $h_\mu(T) + \int f \, d\mu \leq 2a + \log 2 + P(T, f)$. This holds for all continuous maps T and all $f \in C(X, R)$ so we can apply it to T^n and $S_n f = \sum_{i=0}^{n-1} f \circ T^i$ to get $n[h_\mu(T) + \int f \, d\mu] \leq 2a + \log 2 + nP(T, f)$ (by Theorem 9.8(i)). Since this holds for all n we have

$$h_\mu(T) + \int f \, d\mu \leq P(T, f).$$

(2) Let $\varepsilon > 0$. We shall find $\mu \in M(X, T)$ with $h_\mu(T) + \int f \, d\mu \geq P(T, f, \varepsilon)$, and this clearly implies $\sup\{h_\mu(T) + \int f \, d\mu \mid \mu \in M(X, T)\} \geq P(T, f)$.

Let E_n be an (n, ε) separated set with

$$\log \sum_{y \in E_n} e^{(S_n f)(y)} \geq \log P_n(T, f, \varepsilon) - 1.$$

Let $\sigma_n \in M(X)$ be the atomic measure concentrated on E_n by the formula

$$\sigma_n = \frac{\sum_{y \in E_n} e^{(S_n f)(y)} \delta_y}{\sum_{z \in E_n} e^{(S_n f)(z)}}.$$

Let $\mu_n \in M(X)$ be defined by

$$\mu_n = \frac{1}{n} \sum_{i=0}^{n-1} \sigma_n \circ T^{-i}.$$

Since $M(X)$ is compact we can choose a subsequence $\{n_j\}$ of the natural numbers such that

$$\lim_{j \to \infty} \frac{1}{n_j} \log P_{n_j}(T, f, \varepsilon) = P(T, f, \varepsilon)$$

and $\{\mu_{n_j}\}$ converges in $M(X)$ to some $\mu \in M(X)$. By Theorem 6.9 we know $\mu \in M(X, T)$. We shall show $h_\mu(T) + \int f \, d\mu \geq P(T, f, \varepsilon)$.

By Lemma 8.5 choose a partition $\xi = \{A_1, \ldots, A_k\}$ of $(X, \mathscr{B}(X))$ so that $\mathrm{diam}(A_i) < \varepsilon$ and $\mu(\partial A_i) = 0$ for $1 \leq i \leq k$. Since each element of $\bigvee_{j=0}^{n-1} T^{-j}\xi$ contains at most one element of E_n we have

$$H_{\sigma_n}\left(\bigvee_{j=0}^{n-1} T^{-i}\xi \right) + \int S_n f \, d\sigma_n$$

$$= \sum_{y \in E_n} \sigma_n(\{y\})((S_n f)(y) - \log \sigma_n(\{y\}))$$

$$= \log \sum_{y \in E_n} e^{(S_n f)(y)} \quad \text{by definition of } \sigma_n \text{ and Lemma 9.9.}$$

Fix natural members q, n with $1 \leq q < n$ and as in Remark 2 of §8.2 define $a(j)$, for $0 \leq j \leq q - 1$, by $a(j) = [(n - j)/q]$. Fix $0 \leq j \leq q - 1$. From Remark 2(i) of §8.2 we have

$$\bigvee_{i=0}^{n-1} T^{-i}\xi = \bigvee_{r=0}^{a(j)-1} T^{-(rq+j)} \bigvee_{i=0}^{q-1} T^{-i}\xi \vee \bigvee_{l \in S} T^{-l}\xi$$

and S has cardinality at most $2q$. Therefore

$$\log \sum_{y \in E_n} e^{(S_n f)(y)} = H_{\sigma_n}\left(\bigvee_{j=0}^{n-1} T^{-j}\xi \right) + \int S_n f \, d\sigma_n$$

$$\leq \sum_{r=0}^{a(j)-1} H_{\sigma_n}\left(T^{-(rq+j)} \bigvee_{i=0}^{q-1} T^{-i}\xi \right) + H_{\sigma_n}\left(\bigvee_{k \in S} T^{-k}\xi \right) + \int S_n f \, d\sigma_n$$

$$\leq \sum_{r=0}^{a(j)-1} H_{\sigma_n \circ T^{-(rq+j)}}\left(\bigvee_{i=0}^{q-1} T^{-i}\xi \right) + 2q \log k + \int S_n f \, d\sigma_n.$$

Summing this over j from 0 to $q - 1$ and using Remark 2(iii) of §8.2 gives

$$q \log \sum_{y \in E_n} e^{(S_n f)(y)} \leq \sum_{p=0}^{n-1} H_{\sigma \circ T^{-p}}\left(\bigvee_{i=0}^{q-1} T^{-i}\xi \right) + 2q^2 \log k + q \int S_n f \, d\sigma_n.$$

Now divide by n and use Remark (1) of §8.2 to get

$$\frac{q}{n} \log \sum_{y \in E_n} e^{(S_n f)(y)} \leq H_{\mu_n}\left(\bigvee_{i=0}^{q-1} T^{-i}\xi \right) + \frac{2q^2}{n} \log k + q \int f \, d\mu_n. \qquad (*)$$

Because $\mu(\partial A_i) = 0$ all i we have by Remark 3 of §8.2 that

$$\lim_{j \to \infty} H_{\mu_{n_j}}\left(\bigvee_{i=0}^{q-1} T^{-i}\xi \right) = H_\mu\left(\bigvee_{i=0}^{q-1} T^{-i}\xi \right).$$

So replacing n by n_j in (∗) we get

$$qP(T,f,\varepsilon) \leq H_\mu\left(\bigvee_{i=0}^{q-1} T^{-i}\xi\right) + q \int f \, d\mu.$$

Dividing by q and letting $q \to \infty$ gives

$$P(T,f,\varepsilon) \leq h_\mu(T,\xi) + \int f \, d\mu \leq h_\mu(T) + \int f \, d\mu. \qquad \square$$

Corollary 9.10.1. *Let* $T:X \to X$ *be a continuous map of a compact metrisable space and let* $f \in C(X,R)$. *Then*

(i) $P(T,f) = \sup\left\{h_\mu(T) + \int f \, d\mu \,\middle|\, \mu \in E(X,T)\right\}.$

(ii) $P(T,f) = P(T|_{\Omega(T)}, f|_{\Omega(T)}).$

(iii) $P(T,f) = P(T|_{\bigcap_{n=0}^{\infty} T^n X}, f|_{\bigcap_{n=0}^{\infty} T^n X}).$

PROOF. The proofs are simple generalisations of the proofs of the corresponding statements in Corollary 8.6. $\square$

The variational principle helps to calculate the pressure of some examples. It follows readily from the variational principle (or it can be easily proved from the definition of pressure) that if T is the identity map of X we have $P(T,f) = \sup f$. (The ergodic invariant measures for T are the Dirac delta measures, δ_x, $x \in X$, so the formula follows from Corollary 9.10.1(i).) The following calculates $P(T,f)$ when T is an ergodic rotation of a compact metrisable group.

Corollary 9.10.2. *If* $T:X \to X$ *is uniquely ergodic and* $M(X,T) = \{m\}$ *then* $P(T,f) = h_m(T) + \int f \, dm.$

So for a rotation $Tz = az$ of a compact metric group G with $\{a^n\}$ dense in G we have $P(T,f) = \int f \, dm$ where m is Haar measure on G.

§9.4 Pressure Determines $M(X, T)$

We shall show how $P(T, \cdot)$ determines the members of $M(X,T)$ when $T:X \to X$ is a continuous map of a compact metrisable space. Recall that a finite signed measure on X is a map $\mu:\mathscr{B}(X) \to R$ which is countably additive.

Theorem 9.11. *Let* $T:X \to X$ *be a continuous map of a compact metrisable space with* $h(T) < \infty$. *Let* $\mu:\mathscr{B}(X) \to R$ *be a finite signed measure. Then* $\mu \in M(X,T)$ *iff* $\int f \, d\mu \leq P(T,f) \quad \forall f \in C(X,R).$

PROOF. If $\mu \in M(X, T)$ then $\int f \, d\mu \leq P(T, f)$ by the variational principle.

Now suppose μ is a finite signed measure and $\int f \, d\mu \leq P(T, f) \, \forall f \in C(X, R)$. We first show μ takes only non-negative values. Suppose $f \geq 0$. If $\varepsilon > 0$ and $n > 0$ we have

$$\int n(f + \varepsilon) \, d\mu = -\int -n(f + \varepsilon) \, d\mu \geq -P(T, -n(f + \varepsilon))$$

$$\geq -[h(T) + \sup(-n(f + \varepsilon))] \text{ by Theorem 9.7(ii)}$$

$$= -h(T) + n \inf(f + \varepsilon)$$

$$> 0 \text{ for large } n.$$

Therefore $\int (f + \varepsilon) \, d\mu > 0$ so that $\int f \, d\mu \geq 0$. Hence μ is a measure.

We now show $\mu(X) = 1$. If $n \in Z$ then $\int n \, d\mu \leq P(T, n) = h(T) + n$, so that $\mu(X) \leq 1 + h(T)/n$ if $n > 0$ and hence $\mu(X) \leq 1$, and $\mu(X) \geq 1 + h(T)/n$ if $n < 0$ and hence $\mu(X) \geq 1$. Therefore $\mu(X) = 1$.

Lastly we show $\mu \in M(X, T)$. If $n \in Z$ and $f \in C(X, R)$, $n \int (f \circ T - f) \, d\mu \leq P(T, n(f \circ T - f)) = h(T)$ by Theorem 9.7(vii). If $n \geq 1$ then dividing by n and letting n go to ∞ gives $\int (f \circ T - f) \, d\mu \leq 0$, and if $n \leq -1$ then dividing by n and letting n go to $-\infty$ gives $\int (f \circ T - f) \, d\mu \geq 0$. Therefore $\int f \circ T \, d\mu = \int f \, d\mu \, \forall f \in C(X, R)$ so $\mu \in M(X, T)$. $\qquad \square$

Theorem 9.11 says that when $h(T) < \infty$ the pressure of T determines the set $M(X, T)$. We now investigate when the pressure of T determines the measure theoretic entropies of T. In the proof we use the fact that if K_1, K_2 are disjoint closed convex subsets of a locally convex linear topological space V and if K_1 is compact there exists a continuous real-valued linear functional F on V such that $F(x) < F(y) \, \forall x \in K_1, y \in K_2$ (Dunford and Schwartz [1], p. 417). In our application V will be $C(X, R)^* \times R$ where $C(X, R)^*$ is the dual space of $C(X, R)$ and is equipped with the weak*-topology.

Theorem 9.12. *Let $T : X \to X$ be a continuous map of a compact metrisable space with $h(T) < \infty$ and let $\mu_0 \in M(X, T)$. Then $h_{\mu_0}(T) = \inf\{P(T, f) - \int f \, d\mu_0 \, | \, f \in C(X, R)\}$ iff the entropy map of T is upper semi-continuous at μ_0.*

PROOF. Suppose $h_{\mu_0}(T) = \inf\{P(T, f) - \int f \, d\mu_0 \, | \, f \in C(X, R)\}$. Let $\varepsilon > 0$ be given and choose $g \in C(X, R)$ such that $P(T, g) - \int g \, d\mu_0 < h_{\mu_0}(T) + \varepsilon/2$. Let $V_{\mu_0}(g; \varepsilon/2) = \{\mu \in M(X, T) | |\int g \, d\mu - \int g \, d\mu_0| < \varepsilon/2\}$. If $\mu \in V_{\mu_0}(g; \varepsilon/2)$ then

$$h_\mu(T) \leq P(T, g) - \int g \, d\mu, \quad \text{by the variational principle,}$$

$$< P(T, g) - \int g \, d\mu_0 + \varepsilon/2$$

$$< h_{\mu_0}(T) + \varepsilon.$$

Therefore the entropy map is upper semi-continuous at μ_0.

Now suppose the entropy map is upper semi-continuous at μ_0. By the variational principle we have $h_{\mu_0}(T) \leq \inf\{P(T, f) - \int f \, d\mu_0 \, | \, f \in C(X, R)\}$.

We now prove the opposite inequality. Let $b > h_{\mu_0}(T)$ and let $C = \{(\mu, t) \in M(X, T) \times R \mid 0 \le t \le h_\mu(T)\}$. By Theorem 8.1 C is a convex set. If we consider C as a subset of $C(X, R)^* \times R$, where the weak*-topology is used on $C(X, R)^*$, then $(\mu_0, b) \notin \bar{C}$ by the upper semi-continuity of the entropy map at μ_0. Applying the result quoted above to the disjoint convex sets $\bar{C}$ and (μ_0, b) there is a continuous linear functional $F: C(X, R)^* \times R \to R$ such that $F((\mu, t)) < F((\mu_0, b)) \; \forall (\mu, t) \in \bar{C}$. Since we are using the weak*-topology on $C(X, R)^*$ we know that F has the form $F(\mu, t) = \int f \, d\mu + td$ for some $f \in C(X, R)$ and some $d \in R$. Therefore $\int f \, d\mu + dt < \int f \, d\mu_0 + db \; \forall (\mu, t) \in \bar{C}$, so $\int f \, d\mu + dh_\mu(T) < \int f \, d\mu_0 + db \; \forall \mu \in M(X, T)$. If we put $\mu = \mu_0$ then $dh_{\mu_0}(T) < db$ so $d > 0$. Hence

$$h_\mu(T) + \int \frac{f}{d} \, d\mu < b + \int \frac{f}{d} \, d\mu_0 \; \forall \mu \in M(X, T)$$

so, by the variational principle,

$$P(T, f/d) \le b + \int \frac{f}{d} \, d\mu_0.$$

Rearranging gives

$$b \ge P(T, f/d) - \int f/d \, d\mu_0 \ge \inf\left\{P(T, g) - \int g \, d\mu_0 \,\middle|\, g \in C(X, R)\right\}.$$

Therefore $h_{\mu_0}(T) \ge \inf\{P(T, g) - \int g \, d\mu_0 \mid g \in C(X, R)\}$. □

Remarks

(1) The same proof shows that if $T: X \to X$ is a continuous map with $h(T) < \infty$ and $\mu_0 \in M(X, T)$ then $\inf\{P(T, f) - \int f \, d\mu_0 \mid f \in C(X, R)\} = \sup\{\limsup_{n \to \infty} h_{\mu_n}(T) \mid \{\mu_n\}$ is a sequence in $M(X, T)$ with $\mu_n \to \mu_0\}$.

(2) Theorems 9.11 and 9.12 show that if the entropy map of T is upper semi-continuous at every point of $M(X, T)$ and $h(T) < \infty$ then $P(T, \cdot)$ determines the set $M(X, T)$ and the entropy $h_\mu(T)$ for all $\mu \in M(X, T)$. Combining this with the variational principle we see that when the entropy map of T is upper semi-continuous on $M(X, T)$ (for example, when T is an expansive homeomorphism (Theorem 8.2.)) and $h(T) < \infty$ then knowledge of $P(T, \cdot): C(X, R) \to R$ is equivalent to the knowledge of $M(X, T)$ and $h_\mu(T)$ for all $\mu \in M(X, T)$. Hence the pressure contains a lot of information.

(3) It is not difficult to show that if $h(T) < \infty$ then T is uniquely ergodic iff $P(T, \cdot): C(X, R) \to R$ is (Fréchet) differentiable at each point of $C(X, R)$.

§9.5 Equilibrium States

The variational principle gives a natural way of selecting members of $M(X, T)$. The concept extends the idea of measure with maximal entropy.

Definition 9.8. Let $T:X \to X$ be a continuous map of a compact metrisable space X and let $f \in C(X,R)$. A member μ of $M(X,T)$ is called an *equilibrium state for* f if $P(T,f) = h_\mu(T) + \int f \, d\mu$. Let $M_f(X,T)$ denote the collection of all equilibrium states for f.

Remarks

(1) A measure with maximal entropy is precisely an equilibrium state for 0. Hence $M_{max}(X,T)$ is the same as $M_0(X,T)$.

(2) As we know from §8.3 the set $M_f(X,T)$ can be empty.

(3) If $h(T) = \infty$ then $M_f(X,T) = \{\mu \in M(X,T) | h_\mu(T) = \infty\}$ $\forall f \in C(X,R)$, so $M_f(X,T) \neq \emptyset$ by Theorem 8.7(iv).

We have the following generalisation of Theorem 8.7.

Theorem 9.13. *Let* $T:X \to X$ *be a continuous map of a compact metrisable space and let* $f \in C(X,R)$. *Then*

(i) $M_f(X,T)$ *is convex.*

(ii) *If* $h(T) < \infty$ *the extreme points of* $M_f(X,T)$ *are precisely the ergodic members of* $M_f(X,T)$.

(iii) *If* $h(T) < \infty$ *and* $M_f(X,T) \neq \emptyset$ *then* $M_f(X,T)$ *contains an ergodic measure.*

(iv) *If the entropy map is upper semi-continuous then* $M_f(X,T)$ *is compact and non-empty.*

(v) *If* $f, g \in C(X,R)$ *and if there exists* $c \in R$ *such that* $f - g - c$ *belongs to the closure of the set* $\{h \circ T - h | h \in C(X,R)\}$ *in* $C(X,R)$, *then* $M_f(X,T) = M_g(X,T)$.

PROOF. The first four parts are proved in the same way as the corresponding parts of Theorem 8.7. To prove (v) we notice that $\forall \mu \in M(X,T) \int f \, d\mu = \int g \, d\mu + c$. Therefore $h_\mu(T) + \int f \, d\mu = h_\mu(T) + \int g \, d\mu + c$ and $P(T,f) = P(T,g) + c$. Hence $M_f(X,T) = M_g(X,T)$. □

Part (iv) implies that when T is an expansive homeomorphism every $f \in C(X,R)$ has an equilibrium state. This also follows from Theorem 9.6 and the proof of the variational principle which also give a way of obtaining an equilibrium state as a limit of atomic measures on separated sets.

The notion of equilibrium state is tied in with the notion of tangent functional to the convex function $P(T,\cdot):C(X,R) \to R$.

Definition 9.9. Let $T:X \to X$ be a continuous map of a compact metrisable space with $h(T) < \infty$ and let $f \in C(X,R)$. A *tangent functional* to $P(T,\cdot)$ at f is a finite signed measure $\mu:\mathscr{B}(X) \to R$ such that $P(T,f+g) - P(T,f) \geq \int g \, d\mu$ $\forall g \in C(X,R)$. Let $t_f(X,T)$ denote the collection of all tangent functionals to $P(T,\cdot)$ at f.

Remarks

(1) It follows from the Riesz representation theorem (Theorem 6.3) that the dual space $C(X, R)^*$ of $C(X, R)$ can be identified with the collection of all finite signed measures on $(X, \mathscr{B}(X))$. This is because each $L \in C(X, R)^*$ is of the form $L(f) = \int f \, d\mu \; \forall f \in C(X, R)$. (Kingman and Taylor [1] p. 253). Hence we can think of the tangent functionals to $P(T, \cdot)$ at f as those members L of $C(X, R)^*$ satisfying $L(g) \leq P(T, f + g) - P(T, f) \; \forall g \in C(X, R)$.

(2) For each $f \in C(X, R)$ we have $t_f(X, T) \neq \varnothing$. This follows from Remark (1) and the Hahn-Banach theorem, since we can extend the identity map on R to an element of $C(X, R)^*$ dominated by the convex function $g \rightarrow P(T, f + g) - P(T, f)$.

Theorem 9.14. *Let $T : X \rightarrow X$ be a continuous map of a compact metrisable space X with $h(T) < \infty$ and let $f \in C(X, R)$. Then $M_f(X, T) \subset t_f(X, T) \subset M(X, T)$.*

PROOF. Let $\mu \in M_f(X, T)$. If $g \in C(X, R)$,

$$P(T, f + g) - P(T, f) \geq h_\mu(T) + \int f \, d\mu + \int g \, d\mu - h_\mu(T) - \int f d\mu = \int g \, d\mu,$$

by the variational principle. Therefore $M_f(X, T) \subset t_f(X, T)$.

We now show $t_f(X, T) \subset M(X, T)$. Let $\mu \in t_f(X, T)$. We first show μ takes only non-negative values. Suppose $g \geq 0$. If $\varepsilon > 0$ we have

$$\int (g + \varepsilon) \, d\mu = -\int -(g + \varepsilon) \, d\mu$$

$$\geq -P(T, f - (g + \varepsilon)) + P(T, f)$$

$$\geq -[P(T, f) - \inf((g + \varepsilon))] + P(T, f) \quad \text{by Theorem 9.7(ii)}$$

$$= (\inf g + \varepsilon) > 0.$$

Therefore $\int g \, d\mu \geq 0$ so μ takes non-negative values. We next show $\mu(X) = 1$. If $n \in Z$ then $\int n d\mu \leq P(T, f + n) - P(T, f) = n$ so if $n \geq 1$ then $\mu(X) \leq 1$ and if $n \leq -1$ then $\mu(X) \geq 1$. Finally we show $\mu \in M(X, T)$. If $n \in Z$ and $g \in C(X, R)$,

$$n \int (g \circ T - g) \, d\mu \leq P(T, f + n(g \circ T - g)) - P(T, f)$$

$$= 0 \quad \text{by Theorem 9.13(v).}$$

If $n > 0$ this gives $\int g \circ T \, d\mu \leq \int g \, d\mu$ and if $n < 0$ this gives $\int g \circ T \, d\mu \geq \int g \, d\mu$. Therefore $\int g \circ T \, d\mu = \int g \, d\mu$ so $\mu \in M(X, T)$. □

With an extra assumption we get the equality of $t_f(X, T)$ and $M_f(X, T)$.

Theorem 9.15. *Let $T : X \rightarrow X$ be a continuous map of a compact metrisable space with $h(T) < \infty$ and let $f \in C(X, R)$. If the entropy map of T is upper semi-continuous at the members of $t_f(X, T)$ then $t_f(X, T) = M_f(X, T)$.*

PROOF. It remains to show $t_f(X, T) \subset M_f(X, T)$. Let $\mu \in t_f(X, T)$. Then $P(T, f + g) - \int (f + g) \, d\mu \geq P(T, f) - \int f \, d\mu \; \forall g \in C(X, R)$ so $P(T, h) - \int h \, d\mu \geq P(T, f) - \int f \, d\mu \; \forall h \in C(X, R)$. Theorem 9.12 then implies $h_\mu(T) \geq P(T, f) - \int f \, d\mu$ so that $P(T, f) = h_\mu(T) + \int f \, d\mu$ by the variational principle.   □

Remark. Without the upper semi-continuity assumption one can show $t_f(X, T) = \bigcap_{n=1}^{\infty} \{\mu \in M(X, T) \, | \, h_\mu(T) + \int f \, d\mu > P(T, f) - 1/n\}$.

We know that the two-sided shift homeomorphism is expansive and so the following deduction from Theorem 9.15 holds for the shift.

Corollary 9.15.1. *Let $T : X \to X$ be a continuous map of a compact metrisable space and suppose the entropy map of T is upper semi-continuous at each point of $M(X, T)$. Then there is a dense subset of $C(X, R)$ such that each member f of this subset has a unique equilibrium state (i.e. $M_f(X, T)$ has just one member).*

PROOF. We use the theorem that a convex function on a separable Banach space has a unique tangent functional at a dense set of points (Dunford and Schwartz [1], p. 450). This combined with Theorem 9.15 gives the result. □

If $\mu \in M(X, T)$ is the only equilibrium state for some $f \in C(X, R)$ then this gives a natural way of characterising μ: the only measure with $h_\mu(T) + \int f \, d\mu = P(T, f)$. It turns out in many cases that such a measure μ has very strong ergodic properties. In many cases T is a Bernoulli automorphism of the probability space $(X, \mathcal{B}(X), \mu)$. When T is a specific homeomorphism (such as a shift or an Axiom A diffeomorphism) results are known which give conditions on $f \in C(X, R)$ to ensure f has a unique equilibrium state (see Bowen [2]). These results are important in the study of diffeomorphisms (see §10.1). For example let $T : X \to X$ ($X = \prod_{-\infty}^{\infty} Y$, $Y = \{0, 1, \ldots, k - 1\}$) be the shift homeomorphism. Let us denote points of X by $x = \{x_n\}_{-\infty}^{\infty}$ and $y = \{y_n\}_{-\infty}^{\infty}$. For $f \in C(X, R)$ and $n \geq 1$ let

$$\text{var}_n(f) = \sup\{|f(x) - f(y)| \, | \, x, y \in X; \, x_i = y_i \text{ when } -(n - 1) \leq i \leq n - 1\}.$$

Since f is continuous we know $\text{var}_n(f) \to 0$. Let us suppose f has the stronger requirement that $\text{var}_n(f)$ goes to 0 at an exponential rate i.e. $\exists C > 0, \alpha \in (0, 1)$ with $\text{var}_n(f) < C\alpha^n \; \forall n \geq 1$. (If $f(x)$ only depends on a finite number of co-ordinates of x then f satisfies this assumption.) It can be shown that f has a unique equilibrium state μ_f and that the measure-preserving transformation T on the probability space $(X, \mathcal{B}(X), \mu_f)$ is a Bernoulli automorphism. Also if $f, g \in C(X, R)$ both satisfy the exponential condition then $\mu_f = \mu_g$ iff $\exists c \in R$ and $h \in C(X, R)$ with $f - g = c + h \circ T - h$. This tells us many different measures are characterised as unique equilibrium states. This gives a dense subset of $C(X, R)$ each member of which has a unique equilibrium state. We shall prove a special case of this which generalises Theorem 8.9.

Theorem 9.16. *Let* $T: X \to X$ *be the two-sided shift homeomorphism of the space* $X = \prod_{-\infty}^{\infty} Y$, $Y = \{0, 1, \ldots, k - 1\}$. *Let* $a_0, a_1, \ldots, a_{k-1} \in R$ *and define* $f \in C(X, R)$ *by* $f(x) = a_{x_0}$ *where* $x = \{x_n\}_{-\infty}^{\infty}$. *Then* f *has a unique equilibrium state which is the product measure defined by the measure on* Y *which gives the point* i *measure*

$$\frac{e^{a_i}}{\sum\limits_{j=0}^{k-1} e^{a_j}}.$$

PROOF. Let $\xi = \{A_0, \ldots, A_{k-1}\}$ denote the natural generator i.e. $A_i = \{\{x_n\}_{-\infty}^{\infty} \mid x_0 = i\}$. We know $h_\mu(T) = h_\mu(T, \xi) \le H_\mu(\xi) \, \forall \mu \in M(X, T)$ (Theorem 4.17). We know from the end of §9.1 that $P(T, f) = \log(\sum_{j=0}^{k-1} e^{a_j})$. Also $M_f(X, T) \ne \emptyset$ by Theorem 9.13(iv). Let $\mu \in M_f(X, T)$. Put $p_i = \mu(A_i)$, $0 \le i \le k - 1$. Then

$$\log\left(\sum_{j=0}^{k-1} e^{a_j}\right) = h_\mu(T) + \sum_{j=0}^{k-1} a_j p_j$$

$$\le H_\mu(\xi) + \sum_{j=0}^{k-1} a_j p_j = \sum_{j=0}^{k-1} p_j(a_j - \log p_j)$$

$$\le \log\left(\sum_{j=0}^{k-1} e^{a_j}\right) \text{ by Lemma 9.9.}$$

By Lemma 9.9 we must have

$$p_i = \frac{e^{a_i}}{\sum\limits_{j=0}^{k-1} e^{a_j}}.$$

Also $h_\mu(T) = H_\mu(\xi)$ so since

$$h_\mu(T) = h(T, \xi) \le \frac{1}{n} H_\mu\left(\bigvee_{i=0}^{n-1} T^{-i}\xi\right) \le H_\mu(\xi) \quad \text{(Theorem 4.10)}$$

we have $H_\mu(\bigvee_{i=0}^{n-1} T^{-i}\xi) = nH_\mu(\xi)$. Theorem 4.4(ii) implies μ is a product measure, and therefore μ is the product measure which gives A_i measure

$$\frac{e^{a_i}}{\sum\limits_{j=0}^{k-1} e^{a_j}}. \qquad \square$$

Remarks

Let $T:X \to X$ be the two-sided shift homeomorphism as in Theorem 9.16.

(1) There exist $f \in C(X, R)$ which have more than one equilibrium state (for a nice description of this see Hofbauer [1]).

(2) $\bigcup_{f \in C(X,R)} M_f(X, T)$ is a dense subset of $M(X, T)$ for the norm topology on $C(X, R)^*$ (and hence in the weak*-topology) (Israel [1], p. 117; Ruelle [2], p. 52).

(3) If $\mu_1, \ldots, \mu_n \in E(X, T)$ there exists $f \in C(X, R)$ with $\{\mu_1, \ldots, \mu_n\} \subset M_f(X, T)$ (Israel [1], p. 117; Ruelle [2], p. 52). Therefore every ergodic measure for the shift is an equilibrium state. This statement is not true for an arbitrary homeomorphism T.

(4) See Ruelle [2] for the connection of topological pressure and equilibrium states with the corresponding notions in physics. Ya. G. Sinai was the first to use equilibrium states to study diffeomorphisms (see §10.1).

CHAPTER 10
Applications and Other Topics

In this chapter we briefly describe some applications of the concepts introduced in the earlier chapters and mention some topics we have not discussed.

§10.1 The Qualitative Behaviour of Diffeomorphisms

In the subject of differentiable dynamics one tries to understand the behaviour of T^n for large n when $T: M \to M$ is a diffeomorphism of a compact differentiable manifold M. (see Smale [1]). Therefore one would like to know about the orbits $\{T^n(x) | n \in Z\}$ of a large set of points $x \in M$. There is a natural notion of set of measure zero in M: — a Borel subset A of M has smooth measure zero if the intersection $A \cap U$ with every coordinate chart U has zero Lebesgue measure (i.e. if $\phi: U \to R^p$ is the coordinate map then $\phi(A \cap U)$ has Lebesgue measure zero in R^p). So it would be natural to try to understand the orbits of a set of points whose compliment is a set of smooth measure zero. It turns out that this problem is closely connected to the study of equilibrium states of $T: M \to M$.

For a certain class of diffeomorphisms $T: M \to M$ the following result holds (see Bowen [2]). Let $T: M \to M$ be an Axiom A diffeomorphism of a compact manifold M. There is a finite collection $\{\mu_1, \dots, \mu_r\}$ of members of $M(M, T)$ for which the following statements hold.

(i) The set

$$B_j = \left\{ x \in M \left| \frac{1}{n} \sum_{i=0}^{n-1} \delta_{T^i x} \to \mu_j \right. \right\}$$

has positive smooth measure for each j, $1 \le j \le r$, and $M \backslash \bigcup_{j=1}^r B_j$ has smooth measure zero.

(ii) There is a natural function $\phi \in C(M, R)$ such that $\{\mu_1, \ldots, \mu_r\}$ are exactly the ergodic equilibrium states for ϕ.

The condition $(1/n) \sum_{i=0}^{n-1} \delta_{T^i x} \to \mu_j$ means that for every $f \in C(M, R)$ $(1/n) \sum_{i=0}^{n-1} f(T^i x) \to \int f \, d\mu_j$ so that the average value of each "observable" f on the orbit of x is calculable by the measure μ_j. So this result connects the study of equilibrium states of T to the understanding of the asymptotic behaviour of the orbits of most points in M. D. Ruelle has suggested a programme of extending this, and related results to non Axiom A diffeomorphisms (see Ruelle [3], [4], [5]) and even to the case of infinite dimensional manifolds in an attempt to give a description of hydrodynamic turbulence. At the basis of these extensions are two ergodic theorems which we describe in the next section.

§10.2 The Subadditive Ergodic Theorem and the Multiplicative Ergodic Theorem

In order to motivate the two ergodic theorems let us consider a differentiable transformation $T: M \to M$ of a compact C^∞ manifold M. For each point $x \in M$ let $\tau_x M$ denote the vector space of all tangent vectors to M at x. The tangent bundle $\tau M = \bigcup_{x \in M} \tau_x M$ can be given a natural manifold structure and is a C^∞ vector bundle over M. If $T: M \to M$ is C^r differentiable ($r \geq 1$) then its tangent map $\tau T: \tau M \to \tau M$ is a C^{r-1} differentiable map and for each $x \in M$ the transformation $\tau_x T = \tau T|_{\tau_x M}: \tau_x M \to \tau_{Tx} M$ is linear. We think of τT as the linearisation of T, and some (non-linear) problems about T on M can be lifted to linear problems about τT on τM. We sometimes write $\tau(T)$ if it is ambiguous to write τT. The tangent bundle τM always possesses a Riemannian metric i.e. there is an inner product $(,)_x$ on $\tau_x M$ for each x and these inner products depend smoothly on x. A Riemannian metric on M determines a norm $\|\cdot\|_x$ on $\tau_x M$ and all Riemannian metrics on M are equivalent in the sense that if $\|\|\cdot\|\|_x$ denotes the norm on $\tau_x M$ coming from another Riemannian metric then there are positive constants a, b such that $a\|\|v\|\|_x \leq \|v\|_x \leq b\|\|v\|\|_x \, \forall v \in \tau_x M \, \forall x \in M$. The results we state will not depend on which Riemannian metric on M is chosen.

We are trying to understand the behaviour of T^n for large n. One of the most important aspects of this is to understand the expansion and contraction that T creates. The linearised form of this problem is to study contraction and expansion of $\tau(T^n)$. We could state the problem as follows: if a Riemannian metric is chosen on M try to understand, for each $x \in M$ and each $v \in \tau_x M$, how $\|\tau_x(T^n)v\|_{T^n x}$ varies with n.

To begin with let us consider the problem of how $\|\tau_x(T^n)\|$ varies with n, where $\|\tau_x(T^n)\|$ denotes the norm of the linear map $\tau_x(T^n): \tau_x M \to \tau_{T^n x} M$ calculated using the norms $\|\cdot\|_x$ on $\tau_x M$ and $\|\cdot\|_{T^n x}$ on $\tau_{T^n x} M$. By the chain rule we have

$$\tau_x(T^n) = \tau_{T^{n-1} x}(T) \circ \cdots \circ \tau_{Tx}(T) \circ \tau_x(T)$$

so that

$$\|\tau_x(T^n)\| \le \prod_{i=0}^{n-1} \|\tau_{T^i x}(T)\|.$$

If this inequality were an equality then we could put $f(x) = \log\|\tau_x(T)\|$ and then $(1/n)\log\|\tau_x(T^n)\|$ would equal $(1/n)\sum_{i=0}^{n-1} f(T^i x)$ so we could use the ergodic theorem (Theorem 1.14) if we knew $\|\tau_x(T)\|$ was bounded below (so that f would be integrable). So we have two reasons why we cannot use Theorem 1.14: firstly we cannot express our quantity as $\sum_{i=0}^{n-1} f(T^i x)$, and secondly the functions $\log\|\tau_x(T)\|$ may not be integrable and may even take the value $-\infty$. However if we put $f_n(x) = \log\|\tau_x(T^n)\|$ then $f_{n+k}(x) \le f_n(x) + f_k(T^n x)$ by the chain rule and certainly $\max(0, f_n(x))$ is integrable for any probability measure on $(M, \mathscr{B}(m))$ because f_n is bounded above. This problem is one motivation for the following theorem of J. F. C. Kingman (Kingman [1], [2]).

If $f:M \to R$ is a function we put $f^+(x) = \max(0, f(x))$.

Theorem 10.1 (Subadditive Ergodic Theorem). *Let $(X, \mathscr{B}, m)$ be a probability space and let $T:X \to X$ be measure-preserving. Let $\{f_n\}_1^\infty$ be a sequence of measurable functions $f_n:X \to R \cup \{-\infty\}$ satisfying the conditions:*

(a) $f_1^+ \in L^1(m)$
(b) *for each $k, n \ge 1$ $f_{n+k} \le f_n + f_k \circ T^n$ a.e.*

Then there exists a measurable function $f:X \to R \cup \{-\infty\}$ such that

$$f^+ \in L^1(m), \quad f \circ T = f \text{ a.e.,} \quad \lim_{n \to \infty} \frac{1}{n} f_n = f \text{ a.e.,} \quad and$$

$$\lim_{n \to \infty} \frac{1}{n} \int f_n \, dm = \inf_n \frac{1}{n} \int f_n \, dm = \int f \, dm.$$

Remarks

(1) It follows from (a) and (b) that $f_n^+ \in L^1(m)$. So either $f_n \in L^1(m)$ or $\int f_n \, dm = -\infty$. The same statement holds for f, so that some of the integrals in the last statement of the theorem could be $-\infty$.

(2) From (b) we have $\int f_{n+k} \, dm \le \int f_n \, dm + \int f_k \, dm$ so

$$\lim_{n \to \infty} \frac{1}{n} \int f_n \, dm = \inf_n \frac{1}{n} \int f_n \, dm \quad \text{by Theorem 4.9.}$$

(3) The subadditive ergodic theorem generalises the Birkhoff ergodic theorem (Theorem 1.14) because if $g \in L^1(m)$ is given we can take $f_n(x) = \sum_{i=0}^{n-1} g(T^i x)$.

(4) Another proof of a special case of this theorem has been given by Derriennic [1] and Theorem 10.1 can easily be reduced to this special case (see Appendix A of Ruelle [4]).

The following is an immediate corollary to Theorem 10.1, where we use $L(R^k, R^k)$ to denote the space of all linear operators on R^k and $\|\cdot\|$ to denote any norm on it. By choosing a basis in R^k we could of course consider $L(R^k, R^k)$ as the space of all $k \times k$ real matrices.

Corollary 10.1.1. *Let T be a measure-preserving transformation of the probability space $(X, \mathscr{B}, m)$. Let $A: X \to L(R^k, R^k)$ be a measurable function such that $(\log\|A(x)\|)^+ \in L^1(m)$. There exists a measurable function $\chi: X \to R \cup \{-\infty\}$ such that $\chi^+ \in L^1(m)$, $\chi \circ T = \chi$ a.e.,*

$$\lim_{n \to \infty} \frac{1}{n} \log\|A(T^{n-1}x) \circ \cdots \circ A(Tx) \circ A(x)\| = \chi(x) \text{ a.e.}$$

and

$$\lim_{n \to \infty} \frac{1}{n} \int \log\|A(T^{n-1}x) \circ \cdots \circ A(Tx) \circ A(x)\| \, dm$$

$$= \inf_n \frac{1}{n} \int \log\|A(T^{n-1}x) \circ \cdots \circ A(Tx) \circ A(x)\| \, dm$$

$$= \int \chi(x) \, dm.$$

This result was first proved by Furstenberg and Kesten.

We also get the following corollary of Theorem 10.1 for a differentiable map T of a smooth manifold M. Recall that $M(M, T)$ denotes the collection of all T-invariant probability measures on the σ-algebra, $\mathscr{B}(M)$, of Borel subsets of M.

Corollary 10.1.2. *Let $T: M \to M$ be a C^1-differentiable map of the compact manifold M and take any Riemannian metric on M. There exists $B \in \mathscr{B}(M)$ with $TB \subset B$ and $m(B) = 1 \,\forall m \in M(m, T)$, and a measurable function $\chi: M \to R \cup \{-\infty\}$ such that*

$$\lim_{n \to \infty} \frac{1}{n} \log\|\tau_x(T^n)\| = \chi(x) \,\forall x \in B.$$

We have $\chi(x) \le \sup\{\|\tau_y T\| : y \in M\}$, $\chi(Tx) = \chi(x) \,\forall x \in B$ and

$$\lim_{n \to \infty} \frac{1}{n} \int \log\|\tau_x(T^n)\| \, dm = \inf_n \frac{1}{n} \int \log\|\tau_x(T^n)\| \, dm$$

$$= \int \chi(x) \, dm \,\forall m \in M(M, T).$$

The set B and the function χ do not depend on the Riemannian metric chosen.

Let us now turn to the problem of understanding how $\log\|\tau_x(T^n)v\|$ varies with n when $v \in \tau_x M$. This motivates the following result of V. I. Oseledets (Oseledets [1]).

Theorem 10.2 (Multiplicative Ergodic Theorem). *Let T be measure-preserving transformation of the probability space* $(X, \mathcal{B}, m)$. *Let* $A : X \to L(R^k, R^k)$ *be measurable and suppose* $(\log \|A(x)\|)^+ \in L^1(m)$. *There exists* $B \in \mathcal{B}$ *with* $TB \subset B$ *and* $m(B) = 1$ *with the following properties.*

(a) *There is a measurable function* $s : B \to Z^+$ *with* $s \circ T = s$.

(b) *If* $x \in B$ *there are real numbers* $\lambda^{(1)}(x) < \lambda^{(2)}(x) < \cdots < \lambda^{(s(x))}(x)$, *where* $\lambda^{(1)}(x)$ *could be* $-\infty$.

(c) *If* $x \in B$ *there are linear subspaces* $\{0\} = V^{(0)}(x) \subset V^{(1)}(x) \subset \cdots \subset V^{(s(x))}(x) = R^k$, *of* R^k.

(d) *if* $x \in B$ *and* $1 \le i \le s(x)$ *then*

$$\lim_{n \to \infty} \frac{1}{n} \log \|A(T^{n-1}x) \circ \cdots \circ A(Tx) \circ A(x)(v)\|$$

$$= \lambda^{(i)}(x) \quad \text{for all } v \in V^{(i)}(x) \backslash V^{(i-1)}(x).$$

(e) *The function* $\lambda^{(i)}$ *is defined and measurable on* $\{x \mid s(x) \ge i\}$ *and* $\lambda^{(i)}(Tx) = \lambda^{(i)}(x)$ *on this set.*

(f) $A(x)(V^{(i)}(x)) \subset V^{(i)}(Tx)$ *if* $s(x) \ge i$.

Remarks

(1) For $x \in B$, Theorem 10.2 gives the behaviour of $\log \|A(T^{n-1}x) \circ \cdots \circ A(Tx) \circ A(x)v\|$ for all $v \in R^k$ and the behaviour is determined by the $s(x)$ numbers $\lambda^{(1)}(x), \ldots, \lambda^{(s(x))}(x)$.

(2) If we take $(X, \mathcal{B}, m)$ to be the trivial probability space consisting of one point and T is the identity transformation then the result reduces to saying that for a linear transformation A of R^k these are subspaces $\{0\} = V^{(0)} \subset V^{(1)} \subset V^{(2)} \subset \cdots \subset V^{(s)} = R^k$ such that $\|A^n v\|^{1/n} \to e^{\lambda^{(i)}}$ if $v \in V^{(i)} \backslash V^{(i-1)}$, where $e^{\lambda^{(1)}} < e^{\lambda^{(2)}} < \cdots < e^{\lambda^{(s)}}$ are the district numbers which are absolute values of eigenvalues of A.

(3) When $k = 1$, Theorem 10.2 can be stated: if $g : X \to R$ is measurable and $g^+ \in L^1(m)$ then $\lim_{n \to \infty} (1/n) \sum_{i=0}^{n-1} g(T^i x) = g_*(x)$ exists a.e. but could take on the value $-\infty$.

(4) If T is ergodic then $s(x)$ is constant a.e. and each $\lambda^{(i)}(x)$ is constant a.e.

(5) The numbers $\lambda^{(1)}(x), \ldots, \lambda^{(s(x))}(x)$ are called the (Liapunov) characteristic exponents of the system (T, A) at x, and $V^{(1)}(x) \subset V^{(2)}(x) \subset \cdots \subset V^{(s(x))}(x) = R^k$ is called the associated filtration. The number $m^{(i)}(x) = \dim V^{(i)}(x) - \dim V^{(i-1)}(x)$ is called the multiplicity of $\lambda^{(i)}(x)$.

(6) It follows that the function $\chi(x)$ occurring in corollary 10.1.1 is $\lambda^{(s(x))}(x)$.

(7) Because of its use by Margulis in the study of algebraic groups Raghunathan (Raghunathan [1]) gave another proof of Theorem 10.2 that is valid for local fields. A version of this proof is given in Ruelle [4].

When each $A(x)$ is an invertible linear transformation we have the following result. We use $GL(R^k)$ to denote the space of invertible linear transformation of R^k.

Theorem 10.3. *Let T be an invertible measure-preserving transformation of the probability space $(X, \mathcal{B}, m)$. Let $A: X \to GL(R^k)$ be a measurable function with $(\log \|A(x)\|)^+ \in L^1(m)$ and $(\log \|(A(x))^{-1}\|)^+ \in L^1(m)$. There exists $C \in \mathcal{B}$ with $TC = C$ and $m(C) = 1$ such that for each $x \in C$ there is a direct sum decomposition of R^k into linear subspaces $R^k = W^{(1)}(x) \oplus W^{(2)}(x) \oplus \cdots \oplus W^{(s(x))}(x)$ with*

$$\lim_{n \to \infty} \frac{1}{n} \log \|A(T^{n-1}x) \circ \cdots \circ A(Tx) \circ A(x)(v)\| = \lambda^{(i)}(x)$$

and

$$\lim_{n \to \infty} \frac{1}{n} \log \|(A(T^{-1}x) \circ \cdots \circ A(T^{-n}x))^{-1}(v)\| = -\lambda^{(i)}(x) \quad \text{if } 0 \neq v \in W^{(i)}(x).$$

The function $\lambda^{(1)}(x)$ is never $-\infty$, and $A(x)W^{(i)}(x) = W^{(i)}(Tx)$ if $i \leq s(x)$.

Remarks

(1) When $(X, \mathcal{B}, m)$ consists of one point and $A \in GL(R^k)$ this theorem reduces to the decomposition $R^k = W^{(1)} \oplus W^{(2)} \oplus \cdots \oplus W^{(s)}$ into subspaces such that $AW^{(i)} = W^{(i)}$ and all eigenvalues of $A | W^{(i)}$ have the same absolute value $e^{\lambda^{(i)}}$.

(2) See Ruelle [4] for the proof.

There is the following version of the multiplicative ergodic theorem for differentiable maps.

Theorem 10.4. *Let M be a compact C^∞ manifold and let $T: M \to M$ be a C^1 differentiable map. Choose a Riemannian metric on M. Then there exists $B \in \mathcal{B}(M)$ with $TB \subset B$ and $m(B) = 1$ $\forall m \in M(M, T)$ with the following properties.*

(a) *There is a measurable function $s: B \to Z^+$ with $s \circ T = s$.*

(b) *If $x \in B$ there are real numbers $\lambda^{(1)}(x) < \lambda^{(2)}(x) < \cdots < \lambda^{(s(x))}(x)$, where $\lambda^{(1)}(x)$ could be $-\infty$.*

(c) *If $x \in B$ there are linear subspaces, $\{0\} = V^{(0)}(x) \subset V^{(1)}(x) \subset \cdots \subset V^{(s(x))}(x) = R^k$, of R^k*

(d) *If $x \in B$ and $1 \leq i \leq s(x)$ then $\lim_{n \to \infty} (1/n) \log \|\tau_x(T^n)v\| = \lambda^{(i)}(x)$ for all $v \in V^{(i)}(x) \backslash V^{(i-1)}(x)$.*

(e) *$\lambda^{(i)}(x)$ is defined and measurable on $\{x \in B | s(x) \geq i\}$ and $\lambda^{(i)}(Tx) = \lambda^{(i)}(x)$.*

(f) *$\tau_x(T)V^{(i)}(x) \subset V^{(i)}(Tx)$ if $i \leq s(x)$.*
The objects B, s, $\lambda^{(i)}$, $V^{(i)}$ do not depend on the choice of Riemannian metric.

If T is a diffeomorphism then $\lambda^{(1)}(x)$ is never $-\infty$ and we also have $\tau_x(T)V^{(i)}(x) = V^{(i)}(Tx)$.

Remarks

(1) This theorem says there is a "large" subset of M ("large" in the sense that $m(B) = 1 \; \forall m \in M(M, T)$) such that the behaviour of $\|\tau_x(T^n)v\|$ is known for all $x \in B$ and all $v \in \tau_x M$. One can produce examples to show that one cannot obtain this result for all $x \in M$. The functions $\lambda^{(i)}$ are called the Liapunov exponents of the diffeomorphism T. As one might expect they are connected with the entropies of T (see Ruelle [3]).

(2) This result can be generalised to the case of a vector bundle map covering a continuous map of a compact metric space (see Ruelle [4], Appendix D).

(3) There is also a version of Theorem 10.4 for diffeomorphisms that says the set of points C where the conclusion of Theorem 10.3 hold is large in the sense that $m(C) = 1 \; \forall m \in M(M, T)$.

(4) If $\lambda^{(i)}(x) < 0$ then for every $0 \neq v \in V^{(i)}(x)$ and every $\varepsilon > 0$ we have $\|\tau_x(T^n)v\| < e^{n(\lambda^{(i)}(x) + \varepsilon)}$ for all large n. Therefore if $\lambda^{(i)}(x) < 0$ then all elements of $V^{(i)}(x)$ converge exponentially to zero with rate at most $e^{\lambda^{(i)}(x)}$ under the iterates of the tangent map to T. Similarly if $\lambda^{(j)}(x) > 0$ then all elements of $\tau_x M \backslash V^{(j-1)}(x)$ have lengths which become infinite exponentially with rate at least $e^{\lambda^{(j)}(x)}$ under the iterates of $\tau(T)$. If $\lambda^{(p)}(x) = 0$ then we know the elements of $V^{(p)}(x) \backslash V^{(p-1)}(x)$ do not converge exponentially to 0 or ∞ under the iterates of $\tau(T)$.

Suppose T is a diffeomorphism and consider the decomposition $\tau_x M = W^{(1)}(x) \oplus \cdots \oplus W^{(s(x))}(x)$ over the "large" set C (see Remark (3)). Let $q(x)$ be such that $\lambda^{(q(x))}(x) < 0 \leq \lambda^{(q(x)+1)}(x)$ and $p(x)$ such that $\lambda^{(p(x)-1)}(x) \leq 0 < \lambda^{(p(x))}(x)$. Either $p(x) = q(x) + 1$ or $p(x) = q(x) + 2$. Put $U^c(x) = \{0\}$ in the first case and $U^c(x) = W^{(q(x)+1)}(x)$ in the second case, and

$$U^s(x) = \bigoplus_{i=1}^{q(x)} W^{(i)}(x), \quad U^u(x) = \bigoplus_{j=p(x)}^{s(x)} W^{(j)}(x).$$

(Note that $U^s(x) = V^{(q(x))}(x)$ in the previous notation.) Then $\tau_x M = U^s(x) \oplus U^c(x) \oplus U^u(x)$ and for each $v \neq 0, v \in U^s(x), \|\tau_x(T^n)v\|$ goes to 0 with exponential rate at most $\lambda^{(q(x))}(x)$; for each $v \neq 0, v \in U^u(x), \|\tau_x(T^n)v\|$ goes to ∞ with exponential rate at least $\lambda^{(p(x))}(x)$; and for each $v \neq 0, v \in U^c(x), \|\tau_x(T^n)v\|$ does not converge to 0 or ∞ with any exponential rate. It is the contraction caused by the "stable" subspaces $U^s(x)$ and the expansion determined by the "unstable" subspaces $U^u(x)$ that make the interesting behaviour of $T: M \to M$.

Let us now consider the interpretation of the above results for two examples.

Consider first the north–south diffeomorphism $T: K \to K$ of the unit circle. We know $M(K, T)$ consists of all convex combinations of δ_N, δ_S (where N is the north pole of the circle and S is the south pole) so for a set $B \in \mathcal{B}(K)$ to have $m(B) = 1 \; \forall m \in M(K, T)$ means that N and S belong to B. The set $\{N, S\}$, although only containing two members, is large from the point of view of the dynamics of T because we know that all other points move away from N towards S under iteration of T. The tangent bundle τK can be considered as the product $K \times R$ and we can write $\tau(T): K \times R \to K \times R$ by $\tau(T)(z, v) = (T(z), T'(z) \cdot v)$ where $T': K \to R$ is a function. Therefore $\tau(T^n)(x, v) = (T^n(z), T'(T^{n-1}z) \cdots T'(z)v)$ so if $\|\cdot\|$ denotes the trivial Riemannian metric on K ($\|(z, v)\| = |v|$) then

$$\frac{1}{n} \log \|\tau(T^n)(z, v)\| = \frac{1}{n} \sum_{i=0}^{n-1} \log(|T'(T^i z)| \cdot |v|) \to \log(|T'(s)|) \quad \text{if } z \neq N,$$

because $T^n(z) \to S$ and hence $\log|T'(T^n z)| \to \log|T'(S)|$. So in fact we could take $B = K$ in Theorem 10.4 and $\lambda^{(1)}(N) = \log|T'(N)| > 0$, $V^{(1)}(N) = \tau_N K$, and if $z \neq N$, $\lambda^{(1)}(z) = \log|T'(S)| < 0$, $V^{(1)}(z) = \tau_z K$. This shows the function $\lambda^{(1)}$ need not be continuous.

Now let $A: K^p \to K^p$ be an automorphism of the p-dimensional torus. The tangent bundle τK^p can be represented as the product $K^p \times R^p$ and then the tangent map $\tau A: K^p \times R^p \to K^p \times R^p$ is given by $\tau(A)(x, v) = (A(x), \tilde{A}v)$ where $\tilde{A}: R^p \to R^p$ is the linear transformation that covers A. Take the Riemannian metric where $\|(x, v)\|$ is the Euclidean length of $v \in R^p$. Then $\|\tau(A^n)(x, v)\| = \|(A^n(x), \tilde{A}^n(v))\| = \|\tilde{A}^n(v)\| \; \forall n \geq 1$. Therefore if $\lambda^{(1)} < \lambda^{(2)} < \cdots < \lambda^{(s)}$ are the numbers such that $e^{\lambda^{(1)}}, \ldots, e^{\lambda^{(s)}}$ are the distinct absolute values of eigenvalues of $\tilde{A}$ and $V^{(1)} \subset \cdots \subset V^{(s)} = R^p$ is the corresponding filtration of R^p then, for all x, $s(x) = s$, $\lambda^{(i)}(x) = \lambda^{(i)}$ and $V^{(i)}(x) = \{x\} \times V^{(i)}$.

We know that for each diffeomorphism $T: M \to M$ we have a nice theory of expansion and contraction for the linearised situation $\tau(T): \tau M \to \tau M$. To tackle problems about the action of T on M it is desirable to have a nonlinear version of this theory on M. Let us consider contraction. In the linearised situation this is determined by the subspaces $V^{(q(x))}(x)$ of $\tau_x M$ for $x \in B$ where $q(x)$ is the largest natural number with $\lambda^{(q(x))}(x) < 0$ (put $q(x) = 0$ if no such natural number exists). One would like to "integrate" the family $\{V^{(q(x))}(x): x \in B\}$ of subspaces by finding a family of smooth submanifolds of M which are disjoint and such that for each $x \in B$ the space $V^{(q(x))}(x)$ is the tangent space to the submanifold containing x. This can be done and we refer the reader to Ruelle [4] or Pesin ([1], [2]) for the statement of such a theorem. An infinite dimensional version of this theory appears in Ruelle [5]. See Ruelle [3] for further discussion and references. Katok [1] connects Liapunov exponents with existence of periodic points.

§10.3 Quasi-invariant Measures

In some situations one needs to study a measurable transformation $T: X \to X$ of a probability space $(X, \mathcal{B}, m)$ where T is not measure-preserving but does preserve sets of zero measure (i.e. whenever $m(B) = 0$ then $m(T^{-1}B) = 0$). This occurs when T is a differentiable map of the interval $[0, 1]$ (or of a compact manifold) with a continuous derivative. We then say m is a quasi-invariant measure and that T is a non-singular transformation of $(X, \mathscr{B}, m)$. This raises the problems of studying such transformations and, in particular, of deciding if there is another measure μ on $(X, \mathscr{B})$ which is equivalent to m (i.e. μ and m have the same sets of measure zero) and which is an invariant measure for T. There are examples of non-singular transformations having no equivalent invariant probability measure, and examples with no equivalent invariant σ-finite measure. A description of results on these problems is in Friedman's book (Friedman [1]).

Let us consider invertible non-singular transformations from now on (i.e. T is bijective and bimeasurable and $m(B) = 0$ iff $m(T^{-1}B) = 0$). Notice that the definition of ergodicity makes sense for non-singular transformations. The class of all ergodic invertible measure-preserving transformations can be partitioned into four classes as follows. Class I consists of these tranformations $T: X \to X$ where $m(\bigcup_{n=-\infty}^{\infty} T^n x) = 1$ for some $x \in X$ (i.e. m is concentrated on one orbit). If $T:(X, \mathscr{B}, m) \to (X, \mathscr{B}, m)$ is not in class I then it belongs to class II_1 if there is a probability measure μ on $(X, \mathscr{B})$ which is equivalent to m and invariant for T. It belongs to class II_∞ if there is an infinite measure equivalent to m and invariant for T. Finally T belongs to class III if there is no measure which is equivalent to m and invariant for T. This terminology is used because there is a natural way of associating to each T a von Neumann algebra which is a factor, and then the above decomposition corresponds to the Murray-von Neumann classification of factors.

A natural notion of isomorphism of non-singular transformations is orbit equivalence. If $T_i: X_i \to X_i$ is an invertible non-singular transformation $i = 1, 2$, we say T_1 is orbit equivalent to T_2 if there is an invertible non-singular transformation $\phi: X_1 \to X_2$ such that for almost all $x \in X_1$ ϕ maps the set $\{T_1^n x \mid n \in Z\}$ onto the set $\{T_2^n \phi(x) \mid n \in Z\}$.

H. Dye proved that any two transformations of class II_1 are orbit equivalent and any two transformations of class II_∞ are orbit equivalent. W. Krieger introduced the idea of the ratio set of T and this allowed class III to be further divided into classes III_λ for $\lambda \in [0, 1]$. He showed that if $\lambda \in (0, 1]$ then any two members of III_λ are orbit equivalent. For the class III_0 the situation is different. These results are described in Sutherland [1].

The above results apply to the non-singular actions of other countable groups on $(X, \mathscr{B}, m)$. If G is a topological group then an action of G on $(X, \mathscr{B}, m)$ is a measurable map $\phi: G \times X \to X$ such that $\phi(e, x) = x \; \forall x \in X$ (where e is the identity element of G) and $\phi(g_1, \phi(g_2, x)) = \phi(g_1 g_2, x) \; \forall g_1, g_2 \in G$ $\forall x \in X$. Every action of the integers Z is determined by a bijection $T: X \to X$ by the formula $Z \times X \to X:(n, x) \to T^n(x)$. An action of G is non-singular if for each $g \in G$ the transformation $x \to \phi(g, x)$ is a non-singular transformation of $(X, \mathscr{B}, m)$.

The results described above hold for actions of a class of countable groups that includes all countable abelian groups. Similar questions can be considered for actions of Lie groups on $(X, \mathscr{B}, m)$ and then one can consider other types of orbit equivalence; for example we can require that the map ϕ, in the definition of orbit equivalence, be infinitely differentiable on almost all orbits. Quite a lot is known about orbit equivalence of actions of R (Ornstein [2]).

One can also consider certain questions for countable equivalence relations (see Feldman and Moore [1]) and also for foliations (see Bowen and Marcus [1], Plante [1]). In particular the ideas of quasi-invariant measure and invariant measure for a foliation has proved useful in the study of differentiable dynamical systems (see references above) and geometry of manifolds (see Thurston's work on isotopy classes of diffeomorphisms of surfaces).

§10.4 Other Types of Isomorphism

When dealing with measure-preserving transformations which are homeomorphisms of compact metric spaces it makes sense to try to use a more restrictive kind of isomorphism than the usual one. One would say that T_1 is isomorphic to T_2 in the new sense if $\phi T_1 = T_2 \phi$ for some invertible measure-preserving transformation ϕ which also ties in with the topological structure in some way. To require that ϕ be a homeomorphism would be too restrictive. Let us suppose that T_1 is the two-sided shift on $X_1 = \prod_{-\infty}^{\infty} \{0, 1, \ldots, k-1\}$ and T_2 is the two-sided shift on $X_2 = \prod_{-\infty}^{\infty} \{0, 1, \ldots, l-1\}$ and that m_1, m_2 are shift invariant measures on X_1, X_2 respectively. Suppose $\phi : X_1 \to X_2$ is an invertible measure-preserving transformation with $\phi T_1(x) = T_2 \phi(x)$ a.e.. We could consider ϕ to be computable if for almost all $x \in X_1$ we could compute $(\phi(x))_0$, the zero-th coordinate of $\phi(x)$, from just a finite number of coordinates of x. This leads to the notion of finitary isomorphism studied by M. Keane and M. Smorodinsky (Keane and Smorodinsky [1], [2]).

They showed that two Bernoulli shifts with equal entropy are finitary isomorphic, strengthening the theorem of Ornstein (Theorem 4.28) in this case. They also extended this result to irreducible aperiodic Markov shifts.

A topological version of Ornstein's theorem has been obtained by R. L. Adler and B. Marcus [1]. They showed that for the class of irreducible aperiodic topological Markov chains topological entropy is a complete invariant for the equivalence relation of "almost topological conjugacy".

§10.5 Transformations of Intervals

The study of transformations of intervals can arise in many ways from studying other problems. Consideration of billiard problems (see Sinai [2]) leads to the consideration of maps of intervals, so does the study of the well-known Lorenz differential equation (see Williams [1]), and certain problems in complex function theory can be reduced in a similar way (see Bowen [6]). The family of maps $T_a(x) = ax(1 - x)$ mod 1 of $[0, 1)$, where $a > 0$, have received a lot of attention. One of the main problems is to decide which maps preserve a measure which is absolutely continuous relative to Lebesgue measure and then to determine the ergodic properties relative to this measure. There are many recent results in this direction.

§10.6 Further Reading

We indicate here where some topics not covered in this book can be found.

A good source for the history of ergodic theory and its close connections with probability theory, harmonic analysis and group representations is Mackey's survey article, Mackey [1].

The modern connections of ergodic theory with statistical mechanics is described in the books of Ruelle (Ruelle [2]) and Israel (Israel [1]). The connections of this theory with the study of diffeomorphisms is given in Ruelle [2] and Bowen [2], [3]. See Arnold and Avez [1] for some earlier theory.

We have not described the theory of flows of measure-preserving transformations (i.e. measure-preserving actions of R). These are studied in Hopf's book (Hopf [1]) and Sinai's book (Sinai [2]). Also presented in Sinai's book is an introduction to the theory of geodesic flows. Geodesic flows are, perhaps, the most important examples of flows. Their importance and smooth structure is described in Abraham and Marsden [1]. Another important class of flows, billiard flows, are also studied in Sinai [2].

The important topic of approximation of measure-preserving transformations by periodic transformations is described in Katok and Stepin [1].

H. Furstenberg has used ergodic theory and topological dynamics to give proofs of some important theorems in number theory. In particular he has given a proof of Szemerédi's theorem (Furstenberg [2]). Several other results are proved in Furstenberg and Weiss [1].

A detailed description of many theorems in ergodic theory known before 1967 appears in Vershik and Yuzviskii [1], and a description of much of the theory discovered between 1967 and 1974 appears in Katok, Sinai and Stepin [1]. Also many references are given there. The notes of Jacobs (Jacobs [2]) give an excellent account of ergodic theory up to 1962. A detailed account of entropy theory is given in Rohlin [3]. The proofs of some of the theorems on entropy that we stated without proof are given in Parry [2].

References

R. Abraham and J. E. Marsden
  [1] *Foundations of Mechanics*, 2nd ed. Benjamin, Reading, MA, 1978.
L. M. Abramov
  [1] Metric automorphisms with quasi-discrete spectrum. *Amer. Math. Soc. Translations* **2**, 39, 37–56 (1964).
R. L. Adler, A. G. Konheim, and M. H. McAndrew
  [1] Topological entropy, *Trans. Amer. Math. Soc.* **114**, 309–319 (1965).
R. L. Adler and B. Marcus
  [1] Topological entropy and equivalence of dynamical systems. *Memoirs of Amer. Math. Soc.* No. 219, 1979.
V. I. Arnold and A. Avez
  [1] *Ergodic problems of Classical Mechanics*. Benjamin, New York, 1968.
L. Auslander, L. W. Green and F. J. Hahn
  [1] *Flows on homogeneous spaces*. Princeton Univ. Press, *Ann. Math Studies* No. 53, 1963.
K. R. Berg
  [1] Convolution of invariant measures, maximal entropy. *Math. Systems Theory* **3**, 146–150 (1969).
P. Billingsley
  [1] *Ergodic Theory and Information*, Wiley, New York, 1965.
G. D. Birkoff
  [1] *Dynamical systems. Amer. Math. Soc. Colloqium Publications* Vol IX, 1927.
J. R. Blum and D. L. Hanson
  [1] On the isomorphism problem for Bernoulli schemes. *Bull. Amer. Math. Soc.* **69**, 221–223 (1963).
  [2] On the mean ergodic theorem for subsequences. *Bull Amer. Math. Soc.* **66**, 308–311 (1960).
R. E. Bowen
  [1] Entropy for group endomorphisms and homogeneous spaces. *Trans. Amer. Math. Soc.* **153**, 401–414 (1971).
  [2] *Equilibrium States and the Ergodic Theory of Anosov Diffeomorphism*. Springer Lecture Notes in Math. **470**, 1975.

240

[3] *On Axiom A diffeomorphisms. Amer. Math. Soc. Regional Conf. Proc.*, No. 35, 1978.

[4] Periodic points and measures for Axiom A diffeomorphism. *Trans. Amer. Math. Soc.* **154**, 377–397 (1971).

[5] Some systems with unique equilibrium states. *Math. Systems Theory* **8**, 193–202 (1974).

[6] Hausdorff dimension of quasi-circles. *IHES Pub.* **50**, 259–273 (1979).

R. E. Bowen and B. Marcus

[1] Unique ergodicity of horocycle foliations. *Israel J. Math.* **26**, 43–67 (1977).

R. E. Bowen and P. Walters

[1] Expansive one-parameter flows. *J. Diff. Equations* **12**, 180–193 (1972).

J. R. Brown

[1] *Ergodic Theory and Topological Dynamics.* Academic Press, New York, 1976.

R. V. Chacon

[1] Transformations having continuous spectrum. *J. Math and Mech.* **16**, 399–416 (1966).

H. Chu

[1] Some results on affine transformations of compact groups. *Invent. Math.* **28**, 161–183 (1975)

J. Clark

[1] A Kolmogorov shift with no roots. PhD. Thesis, Stanford Univ., Stanford, C. A. 1971.

J-P. Conze

[1] Extensions de systemes dynamiques par des endomorphismes de groupes compacts. *Ann. Inst. H. Poincaré* **268**, 33–66 (1972).

M. Denker, C. Grillenberger and K. Sigmund

[1] *Ergodic Theory on Compact Spaces.* Springer Lecture Notes in Math. **527**, 1976.

Y. Derriennic

[1] Sur le théorème ergodique sous-additif *C.R.A.S. Paris* **281A**, 985–988, (1975).

E. I. Dinaburg

[1] The relation between topological entropy and metric entropy. *Soviet Math.* **11**, 13–16 (1970)

N. Dunford and J. T. Schwartz

[1] *Linear Operator Part I.* Interscience, New York, 1958.

R. Ellis

[1] *Topological Dynamics.* Benjamin, New York, 1969.

J. Feldman and C. C. Moore

[1] Ergodic equivalence relations, cohomology and von Neumann algebras I,II. *Trans. Amer. Math. Soc.* **234**, 289–324, 325–359 (1977).

W. Feller

[1] *An Introduction to Probability Theory and Its Applications Vol. 1*, 2nd ed. Wiley, New York, 1957.

N. A. Friedman

[1] *Introduction to Ergodic Theory.* Van Nostrand, 1970.

N. A. Friedman and D. S. Ornstein

[1] On isomorphism of weak Bernoulli transformations. *Advances in Math.* **5**, 365–394 (1970)

H. Furstenberg

[1] Strict ergodicity and transformations of the torus. *Amer. J. Math* **83**, 573–601 (1961)

[2] *Recurrence in Ergodic Theory and Combinatorial Number Theory.* Princeton Univ Press. 1981

H. Furstenberg and B. Weiss

[1] Topological dynamics and combinatorial number theory. *J. d'Analyse Math* **34**, 61–85 (1978).

F. R. Gantmacher
  [1] *Applications of the Theory of Matrices.* Interscience, New York, 1959.
C. Grillenberger
  [1] Ensembles minimaux sans mesure d'entropie maximale. *Monatsh. Math.* **32**, 275–285 (1976).
F. J. Hahn and Y. Katznelson
  [1] On the entropy of uniquely ergodic transformations. *Trans. Amer. Math. Soc.* **126**, 335–360 (1967).
F. J. Hahn and W. Parry
  [1] Minimal dynamical systems with quasi-discrete spectrum. *J. London Math. Soc.* **40**, 309–323 (1965).
  [2] Some characteristic properties of dynamical systems with quasi-discrete spectrum. *Math. Systems Theory* **2**, 179–190 (1968).
P. Halmos
  [1] *Lectures on Ergodic Theory.* Chelsea, New York, 1953.
  [2] *Introduction to Hilbert Space and the Theory of Spectral Multiplicity.* Chelsea, New York, 1957.
E. Hewitt and K. A. Ross
  [1] *Abstract of Harmonic Analysis Vol 1.* Springer-Verlag, 1963.
A. H. M. Hoare and W. Parry
  [1] Affine transformations with quasi-discrete spectrum I and II. *J. London Math. Soc.* **41**, 88–96 (1966) and **41**, 529–530 (1966).
F. Hofbauer
  [1] Examples of the nonuniqueness of the equilibrium state. *Trans. Amer. Math. Soc.* **228**, 223–241 (1977).
E. Hopf
  [1] *Ergodentheorie*, Chelsea, New York, 1937.
P. Hulse
  [1] On the sequence entropy of transformations with quasi-discrete spectrum. *J. London Math. Soc.* **20**, 128–136 (1979).
R. B. Israel
  [1] *Convexity in the Theory of Lattice Gases.* Princeton Series in Physics, Princeton Univ. Press., 1979.
K. Jacobs
  [1] *Neue methode und ergebnisse der ergodentheorie*, Springer-Verlag, Berlin, 1960.
  [2] *Lecture Notes on Ergodic Theory* (2 vols.) Aarhus Univ., 1963.
R. I. Jewett
  [1] The prevalence of uniquely ergodic systems. *J. Math. and Mechanics* **19**, 717–729 (1970).
S. Kakutani
  [1] Examples of ergodic measure-preserving transformations which are weakly mixing but not strongly mixing. *Springer Lecture Notes in Math.* **318**, 143–149, (1973)
S. Kalikow
  [1] The $T - T^{-1}$ transformation is not loosely Bernoulli. *Annals Math.*, to appear.
I. Kaplansky
  [1] Groups with representations of bounded degree. *Can. J. Math.* **1**, 105–112 (1949).
A. B. Katok
  [1] Lyapunov exponents, entropy and periodic points for diffeomorphisms. *IHES Pub* **51**, 137–173 (1980).
A. B. Katok, Ya. G. Sinai and A. M. Stepin
  [1] Theory of dynamical systems and general transformation groups with invariant measure. *J. Soviet Math* **7**, 974–1041 (1977).

A. B. Katok and A. M. Stepin
[1] Approximations in ergodic theory. *Russian Math. Surveys* **22**, 77–102 (1967).

Y. Katznelson
[1] Ergodic automorphisms of T^n are Bernoulli shifts. *Israel J. Math.* **10**, 186–195 (1971).

M. Keane and M. Smorodinsky
[1] A class of finitary codes. *Israel J. Math.* **26**, 352–371 (1977).
[2] Bernoulli schemes of the same entropy are finitary isomorphic. *Ann. Math.* **109**, 397–406 (1979).

J. L. Kelly
[1] *General Topology*, Van Nostrand, 1955.

H. B. Keynes and J. B. Robertson
[1] Generators for topological entropy and expansiveness *Math. Systems Theory* **3**, 51–59, 1969.

A. I. Khinchine
[1] *Mathematical Foundations of Information Theory*. Dover Publications, New York, 1957.

J. F. C. Kingman
[1] The ergodic theory of subadditive stochastic processes. *J. Royal Stat. Soc.* **B30**, 499–510 (1968).
[2] Subadditive processes, in *Springer Lecture Notes in Math.* **539** (1976).

J. F. C. Kingman and S. J. Taylor
[1] *Introduction to measure and probability. Cambridge Univ. Press*, Cambridge, England, 1966.

A. N. Kolmogorov and V. M. Tihomirov
[1] ε-entropy and ε-capacity of sets in function spaces. *Amer. Math. Soc. Transl.* **2**, 17, 277–364 (1961)

W. Krieger
[1] On entropy and generators of measure-preserving transformations. *Trans. Amer. Math. Soc.* **149**, 453–464 (1970) Erratum: **168**, 519 (1972).
[2] On unique ergodicity. In *Proc. Sixth Berkeley Symposium*. Univ. of Calif. Press, 327–346 (1972).
[3] On the Anaki-Woods asymptotic ratio set and non-singular transformations of a measure space. *Springer Lecture Notes in Math.* **160**, 158–177 (1970).

E. Krug and D. Newton
[1] On sequence entropy of automorphisms of a Lebesgue space. *Z. Wahrscheinlichkeitstheorie verw. Geb.* **24**, 211–214, (1972).

A. G. Kushnirenko
[1] On metric invariants of entropy type. *Russian Math. Surveys* **22**, 53–61 (1967).

D. A. Lind
[1] The structure of skew products with ergodic group actions. *Israel J. Math.* **28**, 205–248 (1977).

G. W. Mackey
[1] Harmonic analysis as the exploitation of symmetry—A historical survey. *Bull. Amer. Math. Soc.* **3**, 543–697 (1980).

R. Măné
[1] Expansive homeomorphisms and topological dimension. *Trans. Amer. Math. Soc.* **252**, 313–319 (1979)

J-C. Marcuard
[1] Produits semi-directs et application aux nilvariétés et aux tores en théorie ergodique. *Bull. Math. Soc. France* **103**, 267–287 (1975).

G. Maruyama
[1] The harmonic analysis of stationary stochastic processes. *Mem. Fac. Sci. Kyushu Univ. Ser A.* **4**, 45–106 (1949).

L. D. Meshalkin
  [1] A case of isomorphism of Bernoulli schemes. *Dokl. Akad. Nauk SSSR* **128**, 41–44 (1959).
G. Miles and R. K. Thomas
  [1] The breakdown of automorphisms of compact topological groups. *Advances in Math. Supplementary Studies Vol. 2.* 207–218 (1978)
  [2] On the polynomial uniformity of translations of the n-torus. ibid, 219–230 (1978).
  [3] Generalized torus automorphisms are Bernoullian. ibid, 231–249 (1978).
M. Misiurewicz
  [1] Diffeomorphisms without any measure with maximal entropy. *Bull. Acad. Pol. Sci.* **21**, 903–910 (1973).
  [2] A short proof of the variational principle for a Z_+^N action on a compact space. *Asterique* **40**, 147–187 (1976).
J. Moser, E. Phillips and S. Varaaman
  [1] *Ergodic theory, a seminar.* Courant Institute Notes, New York, 1975.
D. S. Ornstein
  [1] *Ergodic theory, randomness and dynamical systems.* Yale Math. Monographs No. 5, Yale Univ., 1974
  [2] A survey of some recent results in ergodic theory. *Studies in Probability Theory, MAA Studies in Math. Vol.* **18**, 229–262, 1978.
D. S. Ornstein and P. Shields
  [1] An uncountable family of K-automorphisms. *Advances in Math.* **10**, 63–88 (1973).
V. I. Oseledets
  [1] A multiplicative ergodic theorm. Liapunov characteristic numbers for dynamical systems. *Trans. Moscow Math. Soc* **19**, 197–221 (1968).
J. C. Oxtoby
  [1] Ergodic sets. *Bull. Amer. Math. Soc.* **58**, 116–136 (1952).
W. Parry
  [1] Intrinsic Markov chains. *Trans. Amer. Math. Soc.* **112**, 55–65 (1964).
  [2] *Entropy and Generators in Ergodic Theory.* Benjamin, New York, 1969.
W. Parry and P. Walters
  [1] Endomorphism of a Lebesgue space. *Bull. Amer. Math. Soc.* **78**, 272–276 (1972).
K. R. Parthasarathy
  [1] *Probability measures on metric spaces.* Academic Press, New York, 1967.
  [2] *Introduction to Probability and Measure.* MacMillan, London 1977.
Ya. B. Pesin
  [1] Invariant manifold families which correspond to nonvanishing characteristic exponents. *Math. USSR Izvestija* **10**, 1261–1305 (1976).
  [2] Lyapunov characteristic exponents and smooth ergodic theory. *Russian Math. Surveys* **32**, 55–114 (1977).
R. Phelps
  [1] *Lectures on Choquet's Theorem.* Van Nostrand, Princeton, N.J., 1966.
J. Plante
  [1] Foliations with measure-preserving holonomy. *Ann. Math.* **102**, 327–361 (1975).
S. Polit
  [1] Weakly isomorphic transformations need not be isomorphic. Thesis, Stanford Univ., Stanford, CA., 1975.
M. S. Raghunathan
  [1] A proof of Oseledec's multiplicative ergodic theorem. *Israel J. Math.* **32**, 356–362 (1979).
W. L. Reddy
  [1] Lifting expansive homeomorphisms to symbolic flows. *Math. Systems Theory* **2**, 91–92 (1968).

A. W. Roberts and D. E. Varberg
[1] *Convex Functions.* Academic Press, New York, 1973.
V. A. Rohlin
[1] On the fundamental ideas of measure theory. *Amer. Math. Soc. Transl. Ser* **1**, 10, 1–54 (1962).
[2] Selected topics in the metric theory of dynamical systems. *Uspekhi Mat. Nauk.* **4**, 57–128 (1949); also *Amer. Math. Soc. Transl. Ser* **2**, 49, 171–240 (1966).
[3] Lectures on ergodic theory. *Russian Math. Surveys* **22**, 1–52 (1967).
[4] Metric properties of endomorphisms of compact commutative groups. *Amer. Math Soc. Transl. Ser 2* **64**, 244–252 (1967).
[5] On the entropy of metric automorphisms. *Dokl. Acad. Nauk. SSSR* **124**, 980–983 (1959).
H. L. Royden
[1] *Real Analysis* (2nd Ed.) MacMillan, New York, 1968.
D. Rudolph
[1] Two non-isomorphic *K*-automorphisms with isomorphic squares *J. Math* **23**, 274–287 (1976).
[2] An example of a measure-preserving map with minimal self-joinings, and applications. *J. d' Analyse Math* **35**, 97–122 (1979).
D. Ruelle
[1] Statistical Mechanics on a compact set with Z' action satisfying expansiveness and specification. *Trans. Amer. Math. Soc.* **185**, 237–251 (1973).
[2] *Thermodynamic Formalism.* Addison Wesley, Reading, MA, 1978.
[3] Sensitive dependence on initial condition and turbulent behavior of dynamical systems. *Ann. New York Acad. Sci.* **316**, 408–416 (1978).
[4] Ergodic theory of differentiable dynamical systems. *Publ. Math. IHES* **50**, 275–305 (1979).
[5] Characteristic exponents and invariant manifolds in Hilbert space. *IHES preprint.*
A. Saleski
[1] Sequence entropy and mixing. *J. Math. Anal. Appl.* **60**, 58–66 (1977).
P. Shields
[1] *The Theory of Bernoulli Shifts.* Univ. Chicago Press, Chicago, IL, 1973.
Ya. G. Sinai
[1] *Dynamical Systems* Vol. 1, Aarhus Univ. 1970.
[2] *Introduction to Ergodic Theory.* Princeton Univ. Press, 1976.
S. Smale
[1] Differentiable dynamical systems. *Bull. Amer. Math. Soc.* **73**, 747–817 (1967).
M. Smorodinsky
[1] *Ergodic theory; Entropy.* **Springer Lecture Notes 214**, 1971.
C. E. Sutherland
[1] *Notes on orbit equivalence; Krieger's theorem.* **Univ. of Oslo Lecture Notes,** No. 23, 1976.
R. K. Thomas
[1] Metric properties of transformations of *G*-spaces. *Trans. Amer. Math. Soc.* **60**, 103–117, (1971).
H. Totoki
[1] *Ergodic Theory.* **Aarhus Univ. Lecture Notes,** 1969.
A. M. Versik and S. A. Yuzvinski
[1] Dynamical Systems with invariant measure. *Progress in Math.* **8** (Math Analysis) 151–215, Plenum Press, New York (1970).
P. Walters
[1] Some results on the classification of non-invertible measure-preserving transformations. *Springer Lecture Notes in Math.* Vol. 318, 266–276 (1973).

[2] *Ergodic Theory—Introductory lectures.* **Springer Lecture Notes in Math.** Vol. 458 1975.

[3] A variational principle for the pressure of continuous transformations. *Amer. J. Math.* **17**, 937–971 (1976).

[4] Some transformations having a unique measure with maximal entropy. *Proc. London Math. Soc.* **28**, 500–516 (1974).

B. Weiss

[1] Intrinsically ergodic systems. *Bull. Amer. Math. Soc.* **76**, 1266–1269 (1970).

R. F. Williams

[1] The structure of Lorenz attractors. *IHES* Pub. **50**, 321–347 (1979).

S. A. Yuzvinskii

[1] Metric properties of endomorphisms of locally compact groups. *Russian Math. Surveys* **22**, 47–49 (1967).

[2] Metric properties of endomorphisms of a compact group. *Amer. Math. Soc. Transl.* **2**, 66, 63–98 (1968).

Index

Graduate Texts in Mathematics

9 780387 951522